Water Transmission
and Distribution

PRINCIPLES AND PRACTICES OF WATER SUPPLY OPERATIONS SERIES

Water Sources, Fourth Edition
Water Treatment, Fourth Edition
Water Transmission and Distribution, Fourth Edition
Water Quality, Fourth Edition
Basic Science Concepts and Applications, Fourth Edition

Water Transmission and Distribution

Fourth Edition

**American Water Works
Association**

Project Manager/Senior Technical Editor: Melissa Valentine
Editor: Bill Cobban
Production Editor/Cover Design: Cheryl Armstrong
Production Services: Graham High, TIPS Technical Publishing, Inc.

Disclaimer

Many of the photographs and illustrative drawings that appear in this book have been furnished through the courtesy of various product distributors and manufacturers. Any mention of trade names, commercial products, or services does not constitute endorsement or recommendation for use by the American Water Works Association or the US Environmental Protection Agency. In no event will AWWA be liable for direct, indirect, special, incidental, or consequential damages arising out of the use of information presented in this book. In particular, AWWA will not be responsible for any costs, including, but not limited to, those incurred as a result of lost revenue. In no event shall AWWA's liability exceed the amount paid for the purchase of this book.

Library of Congress Cataloging-in-Publication Data
    Mays, Larry W.
      Water transmission and distribution / by Larry Mays.—4th ed.
        p. cm.—(Principles and practices of water supply operations series)
      Rev. ed. of: Water transmission and distribution. 2003.
      Includes bibliographical references and index.
      ISBN-13: 978-1-58321-781-8 (alk. paper)
      ISBN-10: 1-58321-781-9 (alk. paper)
      1. Water—Distribution. I. American Water Works Association. II. Water transmission and distribution. III. Title.
      TD481.W384 2010
  628.1'44—dc22   2010016788

**American Water Works
Association**

6666 West Quincy Avenue
Denver, CO 80235-3098
303.794.7711
www.awwa.org

Table of Contents

Foreword

Water Transmission and Distribution is part three in a five-part series titled *Principles and Practices of Water Supply Operations.* It contains information on water distribution system design, materials, equipment, pumps, motors and engines, instrumentation and control, and public relations.

The other books in the series are

Water Sources

Water Treatment

Water Quality

Basic Science Concepts and Applications (a reference handbook)

References are made to the other books in the series where appropriate in the text.

The reference handbook is a companion to all four books. It contains basic reviews of mathematics, hydraulics, chemistry, and electricity needed for the problems and computations required in water supply operations. The handbook also uses examples to explain and demonstrate many specific problems. There is also a companion student workbook.

Acknowledgments

This fourth edition of *Water Transmission and Distribution* has been revised to include new technology and current water supply regulations. Primary emphasis for the revision was to update the references and figures and to include current operational practices. The material has also been reorganized for better coordination with the other books in the series. The author of the revision is Larry W. Mays.

Special thanks are extended to the following individuals who provided technical review of all or portions of the fourth-edition manuscript:

Jerry Anderson, CH2MHill

Ahmad Habibian, Black & Veatch

Melissa Valentine, project manager and senior technical editor, AWWA

Publication of the first edition was made possible through a grant from the US Environmental Protection Agency, Office of Drinking Water, under Grant No. T900632-01. The second edition was authored by Lyle Herman, Harry Von Huben, and Todd A. Shimoda. The third edition was authored by William C. Lauer.

Introduction

Water distribution systems are made up of the pipes, valves, and pumps through which treated water is moved from the treatment plant to homes, offices, industries, and other consumers. The distribution system also includes facilities to store water, meters to measure water use, and fire hydrants for both fire fighting and other uses.

The two most important requirements of the distribution system are that it supplies each customer with a sufficient volume of water at adequate pressure and that it delivers safe water that satisfies the quality expectations of customers.

THE OPERATOR'S ROLE IN DISTRIBUTION

Water distribution system operators have duties that vary widely. In general, operators may be expected to

- operate the system to maintain water quality by managing water age in the system, maintaining positive pressure at all times, and controlling water direction and velocity;
- maintain pipes, valves, pumps, and other facilities to ensure a continued flow of potable water;
- monitor and operate valves and pumps to vary the amount of water supplied as the demand varies;
- install connections to supply water to new customers;
- maintain main line and customer water meters;
- read customer water meters;
- sample water to ensure that its quality is maintained;
- operate a cross-connection control program to ensure that nonpotable liquid does not flow into the potable system;
- maintain system maps and records;
- keep informed on new technology and investigate the use of better equipment or methods of operation that could improve the efficiency or safety of distribution system operations; and
- recommend to superiors any repairs, replacements, or improvements that should be made to the distribution system.

To perform these duties successfully, a distribution system operator should have good judgment; strong mechanical skills; and an understanding of the fundamentals of mathematics, hydraulics, and electricity. The operator should also have training in the operation, maintenance, repair, and replacement of distribution system equipment.

The chapters in this volume discuss various types of distribution system equipment and facilities that are available. The types of materials and equipment used by each water system are usually governed by local conditions, past practices, and economics. Well-informed operators should know about all of the common types of equipment and operating methods that are available so that the system may be operated with maximum efficiency and safety.

The list of readings provided at the end of each chapter gives a number of good references that were available at the time this book was published. New publications become available every year, and many other good publications on specific subjects are available from trade associations and other sources. Operators desiring more information on any subject are urged to contact organizations listed in appendix B for a copy of their latest publications list. Many publications are free, and most others are available at nominal cost.

CHAPTER 1

Introduction to Water Distribution Systems

Water distribution systems are complex systems consisting of many components.

SYSTEM PURPOSE AND PLANNING

Drinking water distribution systems are provided for two primary purposes: (1) consumption and (2) fire protection.

There are many considerations involved in planning for a water distribution system, particularly in terms of which type of system to use, the type of system layout, the sizing of mains, and material selection.

Types of Water Systems

Water supply systems generally can be divided into three categories based on the source of water that they use. The water source, in turn, impacts the design, construction, and operation of the water distribution system. The types of systems, classified by source (Figure 1-1), are

- systems with surface water source(s),
- systems with groundwater source(s), and
- systems with purchased water source(s).

Some of the principal characteristics of each of these systems as they affect the distribution system are described below.

Surface Water Systems

In many areas, surface water is readily available for public water system use. The following are some of the special features of surface water systems:

- When a water utility operates its own surface water treatment plant, it may place the distribution system under the same supervision. This arrangement allows personnel and equipment to be shared.
- Surface water must always be treated before it can be distributed to the public, requiring at least "conventional treatment" to remove particles and other contaminants. Surface water must also be disinfected to inactivate harmful microorganisms. Water entering the distribution system therefore has extremely low turbidity and must carry a minimum disinfectant residual.

1

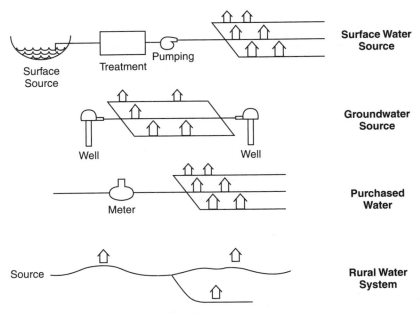

FIGURE 1-1 Types of water systems classified by source

- The treatment plant is sometimes located on one side of the distribution system, particularly when there are two or more water treatment plants. Large transmission mains are required to carry sufficient quantities of water to the far ends of the system.
- Surface water systems generally attract industries that have uses for large quantities of processed water, such as cooling, cleaning, or incorporation into a product.
- As long as rates are reasonable and use is not restricted, customers tend to use a lot of water. The amount of water used on high-use days may be several times the amount used on an average day.
- By adjusting the treatment process, the water quality can be changed so that it is non-corrosive and nonscale forming.

Groundwater Systems

Although groundwater is generally available in most of North America, the amount available for withdrawal at any particular location is usually limited. Some of the general features of distribution systems served by groundwater are as follows:

- There are very few large cities that can rely solely on groundwater for their source of supply. Small communities that start out using wells often must change to another source as the population grows and the rate of water use exceeds the aquifer capacity.

- The quality of some groundwater is good enough to use without treatment. However, the water chemistry may be corrosive or scale-forming or it may have other adverse effects on distribution piping.

- When groundwater has excessive hardness, iron, or other qualities unacceptable to the public, treatment must be provided.

- Some water systems that initially used groundwater have had to change sources or add treatment after discovering that the aquifer they were using was contaminated.

- When groundwater is generally available at any location under a community, wells are usually spaced around the distribution system. This greatly reduces the need for large transmission mains.

- Occasionally, a large quantity of groundwater is available in one area, in which case a "well field" is installed. This situation makes it possible to collect water from all wells and treat it at one location. Usually this reduces operating costs and allows closer control over water quality.

- If water enters the distribution system at only one point, the distribution system must be furnished with transmission mains similar to those in a surface water system.

Purchased Water Systems

A number of water utilities purchase treated water from another utility. Three of the principal reasons, beyond the economics, these systems have switched to a purchased water supply are (1) their well supplies became inadequate, (2) their water sources were found to be contaminated, or (3) regulatory compliance makes the supply and treatment too difficult to continue. Prime examples are the areas surrounding Chicago and Detroit, where a few large treatment plants furnish hundreds of individual water systems with water from the Great Lakes. The following are a few of the characteristics of purchased water systems:

- The operator's job is limited primarily to operating the distribution system, with little or no treatment required. In some cases, additional disinfection is required and the water must be repumped to boost pressure.

- The quality of water from bulk suppliers is generally good because the water has undergone full treatment.

- Purchasing systems must maintain particularly tight water accountability on their distribution system because all water passing the bulk meter must be paid for, including water from leaks and other wasted water.

- Purchasing systems must frequently provide a greater-than-average water storage capacity because of the possibility that they will temporarily lose their single source. Additional storage may also be required if the water-purchasing agreement involves drawing water at only a limited rate or purchasing water at a cheaper rate during the night.

System Planning Issues

Local conditions have a substantial bearing on system design. They will dictate such things as how deep piping must be buried, the fire flow required, and what types of materials are best suited in existing conditions. A few of these considerations are as follows:

- Water availability
- Source reliability
- Water quality
- Location
- State and federal requirements

Policy Considerations

Certain general planning considerations related to new system development or system expansion require local policy decisions. Although these decisions are usually made by the utility or city governing board, the operator should be aware of such concerns as

- future growth,
- costs,
- financing methods,
- ordinances,
- zoning, and
- regulatory issues.

Drinking Water Supply and Distribution Systems

The overall conventional water supply system includes all the system components to develop drinking water and distribute it to the customers. Water supply and distribution systems are discussed in many publications including Mays (2000, 2004, and 2005). Listed next are some of the common elements associated with water supply systems in the U.S. (Clark et al. 2004):

- A water source that may be a surface impoundment such as a lake, reservoir, river, or groundwater from an aquifer
- Surface supplies that generally have conventional treatment facilities including filtration, which removes particulates and potentially pathogenic microorganisms, followed by disinfection
- Transmission systems, which include tunnels, reservoirs and/or pumping facilities, and storage facilities
- A distribution system carrying finished water through a system of water mains and pipes to consumers

Figure 1-2 illustrates a conventional water supply system including the functional components: raw water sources or source development (groundwater and/or surface sources), raw water pumping and transmission, raw water storage, water treatment, high-service pumping, and water distribution.

Water distribution is composed of three major components: distribution piping, distribution storage, and distribution pumping stations. Components represent the largest functional elements in an urban water distribution system, and each of these three major components is composed of one or more subcomponents as illustrated in Figure 1-3.

Subcomponents represent the basic building blocks for components. Pumping stations can be divided into structural, electrical, pumping, and piping subcomponents. Distribution storage can be divided into tank, pipe, and valve subcomponents. Distribution piping can be divided into pipe and valve subcomponents. The sub-subcomponents are composed of one or more subcomponents integrated into a common operational element. One example is the pumping unit subcomponent, which is composed of pipes, valves, pumps, drivers, power transmission, and control sub-subcomponents.

The three major components of the water distribution system are discussed in Chapter 2, Pipe Systems and Piping Materials; Chapter 3, Water Storage; and Chapter 4, Pump Stations and Pumps. Chapter 5 describes not only hydraulics of the components but also the hydraulics of the overall water distribution system. The subcomponents and sub-subcomponents of the water distribution are described in Chapters 6–11.

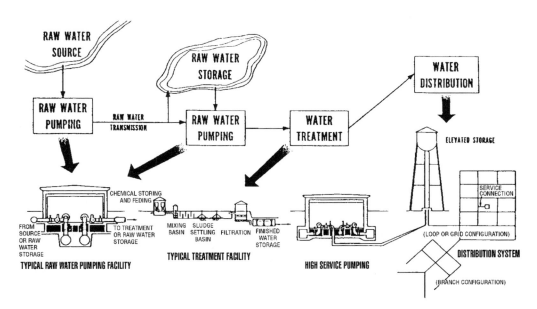

FIGURE 1-2 Conventional layout of water sources, pumping, transmission, water treatment, water distribution (Cullinane, 1989)

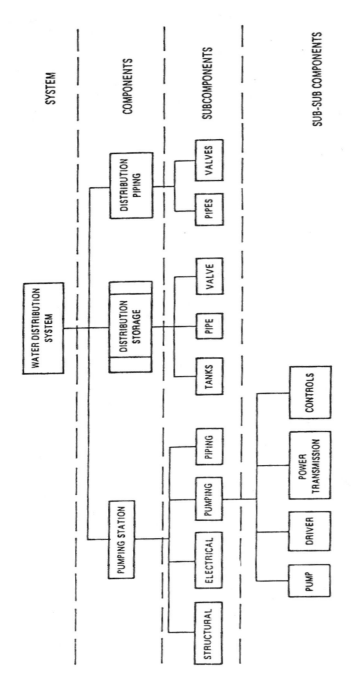

FIGURE 1-3 Hierarchical relationship of components, subcomponents, and sub-subcomponents for a water distribution system (Cullinane, 1989)

BIBLIOGRAPHY

Clark, R.M., Grayman, W.M., Buchberger, S.G., Lee, Y., and Hartman, D.J. 2004. Drinking Water Distribution Systems: An Overview, in *Water Supply Systems Security* (L.W. Mays, ed.). New York: McGraw-Hill.

Cullinane, M.J. Jr. 1989. Methodologies for the Evaluation of Water Distribution System Reliability/Availability, Ph.D. Dissertation, University of Texas at Austin.

Mays, L.W., ed. 2000. *Water Distribution Systems Handbook*. New York: McGraw-Hill.

Mays, L.W., ed . 2004. *Water Supply Systems Security*. New York: McGraw-Hill.

Mays, L.W. 2005. *Water Resources Engineering*.New York: John Wiley and Sons.

SELECTED SUPPLEMENTARY READINGS

Cesario, L. 1995. *Modeling, Analysis, and Design of Water Distribution Systems*. Denver, Colo.: American Water Works Association.

Clark, R.M., and W.M. Grayman. 1998. *Modeling Water Quality in Drinking Water Distribution Systems*. Denver, Colo.: American Water Works Association.

Design and Construction of Small Water Systems—A Guide for Managers. 1999. Denver, Colo.: American Water Works Association.

Jones, G.M. (Editor-in-Chief), Sanks, R.L., G.Tchobanoglous, and B.E. Bosserman II. 2008. *Pumping Station Design*. 3rd ed. Woburn, Mass.: Butterworth-Heinemann.

Manual M24, Dual Water Systems. Denver, Colo.: American Water Works Association (latest edition).

Manual M31, Distribution System Requirements for Fire Protection. Denver, Colo.: American Water Works Association (latest edition).

Manual M32, Distribution Network Analysis for Water Utilities. Denver, Colo.: American Water Works Association (latest edition).

Mayer, P.W., W.B. DeOreo, E.M. Opiz, J.C. Kiefer, W.Y. Davis, B. Dziegielewski, and J.O. Nelson. 1999. *Residential End Uses of Water*. Denver, Colo.: American Water Works Association Research Foundation and American Water Works Association.

Mays, L.W. 2002. *Urban Water Supply Handbook*. New York: McGraw-Hill.

Sherman, D. 2001. Your Utility Can Survive an Earthquake with Disaster Planning. *Opflow* 27(5).

Water Distribution Operator Training Handbook. 1999. Denver, Colo.: American Water Works Association.

Wood, D.J., and L.E. Ormsbee. 1989. Supply Identification for Water Distribution Systems. *Jour. AWWA* 81(7):74.

CHAPTER 2

Pipe Systems and Piping Materials

SYSTEM LAYOUT

Configuration

A distribution system layout is usually designed in one of three configurations:

1. Arterial-loop system
2. Grid system
3. Tree system

Most distribution systems are actually a combination of grid and tree systems.

Arterial-loop system

As illustrated in Figure 2-1, an arterial-loop system attempts to surround the distribution area with larger-diameter mains. The large mains then contribute water supply within the grid from several different directions.

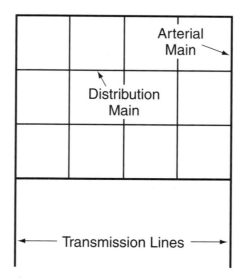

FIGURE 2-1 Arterial-loop system

Grid system

As illustrated in Figure 2-2, a grid system depends primarily on the fact that all mains are interconnected, so water drawn at any point can actually flow from several different directions. The distribution mains in the general grid system are usually 6 or 8 in. (150 or 200 mm) in diameter. They are then reinforced with larger arterial mains, and the general area is fed by still larger transmission mains.

Tree system

As illustrated in Figure 2-3, a tree system brings water into an area with a transmission main, which then branches off into smaller mains. The smaller mains generally end up as dead ends. This is not considered a good distribution system design and is generally not recommended. However, in many cases, site-specific conditions result in the selection of this type of system.

Dead ends

In general, dead-end water mains should be avoided if at all possible. Problems caused by dead ends include the following:

* A fire or large-water consumer located on a dead-end main can draw only through the single line, so flow is restricted by the single main length and pipe size. The advantage of a grid system is that water can flow to the point of demand from several different directions when a high flow rate is required.

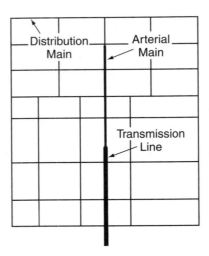

FIGURE 2-2 Grid system

- Mains are sized to provide adequate fire flow. Because the domestic use of a few houses requires much less water than fire flow, the water in the mains moves very slowly. Therefore, water often becomes stagnant or degrades in dead ends. Many water systems find it necessary to set up special programs to flush these dead-end mains periodically to maintain acceptable water quality. Mains that are looped generally flow continually in one direction or another, so they do not experience this problem.

- If repair work must be performed on a dead-end main, all customers beyond the repair lose service until the main is repaired, disinfected, and placed back in service. With a looped system, only the customers between the two closest valves would lose service.

The hydraulics of dead ends are discussed in chapter 5.

Mapping

As an initial step in system planning, a basic map should be developed to show both the existing system and the areas that may have to be served in the future. Some of the essential information that should be included is as follows:

- The existing system (including main size, water pressure, valves, and hydrants)
- Existing and planned streets
- Areas outside the system designated for future expansion
- Ground-level elevations, contour lines, and topographic features
- Existing underground utility services including sanitary sewers; storm sewers; water, gas, and steam lines; and underground electric, telephone, and television cables
- Population densities (present and projected)
- Normal water consumption (present and projected)
- Proposed additions or changes to the system

Figure 2-4 illustrates a section of a typical water system map of this type. Water system operators should realize that information about buried mains, valves, and services must be immediately recorded after installation—not simply committed to memory. It can be extremely expensive to recreate records of "lost" underground facilities. Computerized geographical information systems (GIS) are commonly used to create maps and record location information. These systems can provide needed information to system operators. More details on distribution system maps and records are provided in chapter 16.

Valving

Shutoff valves should be provided so that areas within the system can be isolated for repair or maintenance. In order to minimize service interruptions, valves should be located at regular intervals and at all branches from the arterial mains. Where mains intersect in a grid,

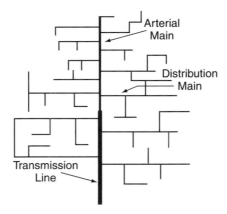

FIGURE 2-3 Tree system

at least two (preferably three) of the branch lines should be provided with valves. The distribution system should be planned so that most of the flow is maintained even if any section of the system is taken out of service.

Air-and-vacuum relief valves are required at high points, and blowoff valves are required at low points. Cross-connections between the domestic supply and other water sources should not be used. Backflow prevention devices are required by applicable regulations to prevent contamination from nonpotable sources. Chapter 6 provides additional information on valves.

SIZING MAINS

The size of a water main determines its carrying capacity. Main sizes must be selected to provide the flow (capacity) to meet peak domestic, commercial, and industrial demands in the area to be served. They must also provide for fire flow at the necessary pressure.

Quantity Requirements

The quantity of water that must be carried by a water main usually depends on two things: consumption (domestic, industrial, commercial, institutional, etc.) and fire flow requirements.

Domestic use

Requirements for domestic use can be determined either from past records or from general usage figures for the area. If rates are reasonable and water use is not restricted, the use of water for irrigation may cause domestic use to be much higher in summer months (depending on climate) than it is in winter. The determination of use must also consider projected growth factors to ensure that the system's design capacity will meet future demand. Water use considerations are discussed in greater detail in *Water Sources*, another book in this series.

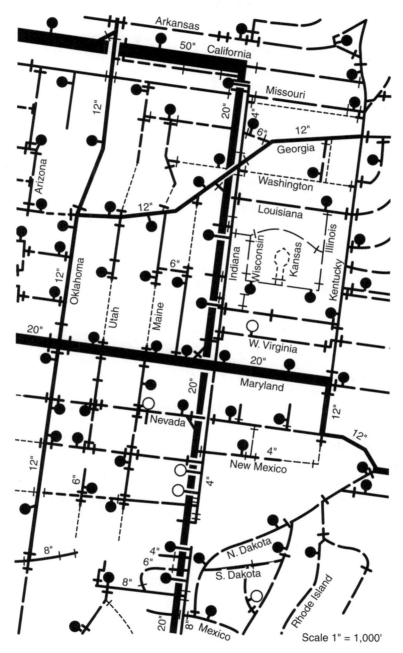

FIGURE 2-4 Portion of a typical distribution system map

Fire flow requirements

The determining factor in sizing mains, storage facilities, and pumping facilities for communities with a population of less than 50,000 is usually the need for fire protection. Fire flow requirements for each community are set by the Insurance Services Office (ISO), which is an organization representing fire insurance underwriters. This group determines the minimum flow that the system must be able to maintain for a specified period of time in order for the community to achieve a specified fire protection rating. Fire insurance rates are then based, in part, on this classification.

Many rural water systems are designed to serve only domestic water needs. Fire protection is not provided, and fire flow requirements are not considered in the design of these systems.

Fire insurance underwriters recommend that no main in the distribution system be less than 6 in. (150 mm) in diameter. They also suggest that minimum pipe sizes be governed, in part, by the type of area to be served. For example,

- High-value districts (such as sports stadiums, shopping centers, and libraries) should have minimum pipe sizes of 8 and 12 in. (200 and 300 mm).
- Residential areas should have minimum sizes of 6 and 8 in. (150 and 200 mm).
- Mains smaller than 6 in. (150 mm) should be used only when they complete a desirable grid.

These recommendations may be difficult to follow in practice but they should be a goal when system improvements are being made. The nearest ISO representative should be consulted if more specific information on fire flow requirements or a fire insurance rating is needed. Additional information on fire flow requirements can also be found in American Water Works Association (AWWA) Manual M31, *Distribution System Requirements for Fire Protection*.

Pressure Requirements

In areas requiring high fire flow capacity, the minimum static pressure required at all fire hydrants is generally 35 psi (240 kPa) or higher. The required pressure during fire flow conditions should not drop below 20 psi (140 kPa). Some insurance companies require that both a minimum water pressure and a minimum flow requirement be maintained to the buildings they insure.

Water systems located in an area with widely varying elevations usually find it necessary to divide the distribution system into two or more pressure zones. Residential water pressure of between 50 and 75 psi (345 and 517 kPa) is usually considered most desirable. Customers do not generally like higher pressure because water comes out of a quickly opened faucet with too much force.

In addition, higher pressure contributes to more main and service leaks and also hastens the failure of hot-water heaters. Excessive pressure may also damage automatic dishwashers and washing machines. Homes having very high water pressure may need to install household pressure-reducing valves.

Velocity Requirements

Flow velocity is another factor to consider when determining pipe capacity and required pipe size. Velocities should normally be limited to about 5 ft/sec (1.5 m/sec) in order to minimize friction loss as water flows through the pipe. For example, water passing through 100 ft (30.5 m) of cast-iron pipe in relatively good condition will have a head loss of about 12 ft (4 m) when flowing at a velocity of 5 ft/sec (1.5 m/sec). If the velocity is increased to 9 ft/sec (2.7 m/sec), the head loss will be tripled to about 36 ft (11 m).

It should be a goal to keep velocities low under normal operating conditions, but velocities may exceed the guidelines under fire flow conditions. Another goal concerning velocities is that water quantity and quality goals need to be balanced. Oversized pipes may cause water quality problems.

Network Analysis

Pipe carrying capacity depends on a combination of factors, including pipe size, pressure, flow velocity, and head loss resulting from friction. The amount of friction loss depends on the pipe roughness, flow velocity, and pipe diameter.

The required pipe size can be calculated when the other requirements and characteristics are known. A common method for calculating pipe size is the Hazen–Williams formula presented in Chapter 5. Charts and computer programs that use this formula (or are based on it) have been developed for various sizes and types of pipe to help in selecting proper pipe size. Additional information on this subject is included in *Basic Science Concepts and Applications*, another book in this series.

When distribution system expansion is to be extensive, it is usually prudent, if not necessary, to analyze the entire system. Consideration needs to be given to both existing and projected water demands. This analysis requires a plot of pressures and flows at points throughout the system. Most comprehensive system evaluations are performed using computerized modeling techniques discussed further in Chapter 5.

Quality Requirements

Maintaining water quality throughout the distribution system is a primary goal of system operators. The design of the piping network greatly impacts the operating and maintenance procedures that must be employed to achieve this goal. The type of materials, the design of storage facilities, the size and location of mains, and the quality of water entering the distribution system are critical factors that influence water quality. Chapter 14 describes the practices that operators can employ to maintain acceptable water quality. The distribution system design should include equal emphasis on quantity and quality factors.

MATERIAL SELECTION

An understanding of the factors that affect the choice of pipe materials is needed. Incorrect selections may result in premature failure or a deterioration of water quality.

For example, if components constructed of dissimilar metals are directly connected, a corrosion cell may be formed if the components are immersed in a conducting fluid such as water or damp soil. An insulating spacer (nonconducting material between the two metal surfaces) placed between the connection can eliminate this problem.

Selecting unprotected iron or steel pipe for use in a corrosive soil may result in pipe failure. Properly protecting the ferrous pipe or choosing a resistant material may reduce this problem.

These are only two of the many possible consequences of improper material selection. Fortunately, a number of standards apply to many distribution system components. All components and materials must conform to the appropriate standard. There are also certification programs designed to verify that a product or material meets the requirements of the standard. Utilities that require certified products or materials provide added assurance that the system components perform as expected and the drinking water quality is protected.

General factors that need to be considered in the selection of pipe include (Ysusi 2000):

- Service conditions
  - Pressure (including surges and transients)
  - Soil loads, bearing capacity of soil, potential settlement
  - Corrosion potential of soil
  - Potential corrosive nature of some waters
- Availability
  - Local availability and experienced installation personnel
  - Sizes and thicknesses (pressure ratings and classes)
  - Compatibility with available fittings
- Properties of the pipe
  - Strength (static and fatigue, especially for water hammer)
  - Ductility
  - Corrosion resistance
  - Fluid friction resistance (more important in transmission pipelines)
- Economics
  - Cost (installed cost, including freight to jobsite and installation)
  - Required life
  - Cost of maintenance and repairs

ANSI/AWWA Standards

AWWA has developed consensus standards since 1908. Currently, AWWA has more than 130 standards for products and procedures used in the water industry. The standards are developed under procedures accredited by the American National Standards Institute (ANSI) and are thus registered by this international standards body.

AWWA standards define the minimum requirements for many aspects of drinking water systems. Compliance with these standards is not mandated by AWWA, but the standards have been made mandatory by utilities and regulatory agencies. AWWA doesn't test any product in developing or approving a standard, and no products are "AWWA approved." AWWA standards are available in these major categories:

- Wells
- Treatment
  - Filtration
  - Softening
  - Disinfection chemicals
  - Coagulation
  - Scale and corrosion control
  - Taste and odor control
  - Fluorides
- Pipe and accessories
  - Ductile-iron pipe and fittings
  - Steel pipe
  - Concrete pipe
  - Valves and hydrants
- Pipe installation
- Disinfection of facilities
- Meters
- Service lines
- Plastic pipe
- Water storage
- Pumps
- Plant equipment
- Utility management

ANSI/NSF Standard 61 and Certification

The ANSI/NSF Standard 61 covers materials that come in contact with potable water and focuses on the issue of whether contaminants leach or migrate from the product/material into the drinking water, resulting in unacceptable levels of contaminants. Certain materials that may have the potential for microbiological growth (some coatings, gaskets, lubricants, etc.) are also evaluated. The standard covers valves, pipes, protective materials, joining and sealing materials, mechanical devices, process media, and related products. Most materials or components that may come in contact with drinking water are covered in this standard.

There are several certification programs (NSF International, Underwriters Laboratories, Inc. [UL], Canadian Standards Association, etc.) based on this standard that include product testing, evaluation of materials, and factory inspections. Testing is conducted in accordance with the requirements of the standard. Contaminant concentrations in laboratory testing are converted to tap levels and assessed against the maximum allowable level in the standard. If the material or product satisfies the requirements, it is certified by the testing company as acceptable for use in potable water systems. To remain certified, production facilities may be inspected at any time and periodic retesting is required (usually on an annual basis).

Many utilities and most state and provincial regulatory agencies require certification to ANSI/NSF Standard 61 for all system components that come in contact with drinking water. Utilities should be aware of these requirements and check products to make sure they comply.

Selecting Pipeline Materials

Some questions that should be considered when selecting pipe material include:

- What pipe materials are available?
- Does one material perform better than another?
- Are all materials acceptable?
- What pipeline material does the distribution system have now?
- Why was the present pipe material originally selected?
- Are the materials for the existing pipe and the proposed new pipe compatible?

The following paragraphs discuss general requirements for distribution pipe.

Pipe characteristics

Some characteristics of a pipe that need to be considered are

- strength,
- pressure rating,
- durability,

- corrosion resistance,
- smoothness of the inner surface,
- ease of tapping and repair, and
- water quality maintenance.

Strength. Water distribution pipe must have adequate strength to handle external load and internal pressure. Pipe strength is expressed in terms of tensile strength and flexural strength.

External load is the pressure exerted on a pipe after it has been buried in a trench. This pressure is a result of the backfill (the material that is filled back into the trench after the pipe has been laid), the traffic load (the weight or impact of the traffic passing over the pipe), and longitudinal thrust loads.

The pipe must be able to resist a reasonable amount of external damage from impact during installation. It must also be capable of resisting crushing or undue deflection due to external load. External load is expressed in pounds per linear foot (lb/lin ft) or pounds per square inch (psi). (In metric units, external load is expressed in kilograms per linear meter [kg/m (linear)] or kilopascals [kPa].)

Internal pressure is the hydrostatic pressure within the pipe. Normal water pressure depends on local conditions and requirements but is usually in the 40–100 psi (280–690 kPa) range.

Surge, also known as *water hammer,* is a sudden repeated increase and decrease in pressure that continues until dissipated by friction losses. It occurs from a sudden change in water velocity often caused by the opening or closing of valves or hydrants too quickly, or the sudden starting or stopping of pumps due to loss of power. Water hammer is a transient pressure wave that travels very rapidly through the pipe and may cause pressure several times normal pressure. It can cause extensive damage, such as ruptured pipe and damaged fittings.

Tensile strength is a measure of the resistance a material has to longitudinal, or lengthwise, pull before that material fails.

Flexural strength is a measure of the ability of a material to bend or flex without breaking.

Pipe shear breakage or beam breakage may occur when a force exerted on a pipe causes stresses that exceed the material's tensile or flexural strength. A shear break occurs when the earth shifts. Beam breakage, which resembles a shear break, may occur when a pipe is unevenly supported along its length. A pipe that is resting on a rock, for instance, may break if it is weak in tensile and flexural strength. Figure 2-5 shows these effects.

Pressure rating. Pipe should be carefully chosen or designed to ensure that its pressure rating is adequate for handling the pressures in a specific system. Pressure ratings can be calculated using various formulas and tables found in current AWWA standards. Distribution system pipe should have a pressure rating of 2.5 to 4 times the normal operating pressure.

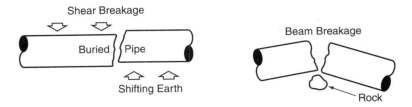

FIGURE 2-5 Shear and beam breakage

When a section of pipe is being replaced, the new piece must have a pressure rating equal to or greater than that of the piece being replaced.

Specific minimum requirements or standards for all types of pipe have been established and published by AWWA to ensure adequate and consistent quality of water mains. Other agencies that have established standards for pipe include federal and state governments, UL, NSF International, ASTM International, and the manufacturers themselves. These standards, which cover the method of design, manufacture, and installation in detail, should be used for selecting pipe.

Durability. Durability is the degree to which a pipe will provide satisfactory and economical service under the conditions of use. It implies long life, toughness, and the ability to maintain tight joints with little or no maintenance.

Corrosion resistance. Consideration must be given to a pipe's resistance to both internal and external corrosion. If a pipe is made of material that might be corroded by the water being carried, the pipe is usually lined. Some pipe materials will be vulnerable to corrosive soil unless special coatings or wrappings are applied to the pipe exterior and/or cathodic protection is provided.

Smoothness of the inner surface.   Smooth pipe walls ensure maximum flow capacity for water pipe. The C value of a pipe is a measure of the pipe wall roughness that retards flow because of friction. Higher C values correspond to smoother the pipe. Figure 2-6 illustrates the difference between smooth pipe and pipe with an interior surface roughened by tuberculation (the buildup of corrosion products on the pipe walls).

Ease of tapping and repair. The pipe selected should be easy to repair and tap for service connections. It should hold the service connection firmly without cracking, breaking, or leaking. The tapping connection should be easily replaceable or at least repairable.

Water quality maintenance. The pipe must be able to maintain the quality of the water distributed by the system. It should not add taste, odor, chemicals, or other undesirable qualities to the water.

Smooth Pipe
Courtesy of J-M Manufacturing Co., Inc.

Tuberculated Pipe
Courtesy of Girard Industrie

FIGURE 2-6 Effect of tuberculation

Economics

Several considerations in addition to price must be made in determining the most economical choice of pipe for an installation. Some questions that should be considered include

* Are the sizes and types of pipe being considered readily available?
* Are the necessary tees, elbows, and other accessories readily available?
* Is there a difference in the cost of installing different types of pipe?

Installation cost. Installation is a major part of the cost of a project, whether the utility does the work with its own forces or contracts for it to be done. The projected cost, as well as the additional unplanned costs or savings, will vary significantly for the same job based on factors such as amount and quality of advance planning and engineering, reliability of delivery times, and weather conditions. Some installation features that must be considered include

* variability of pressure requirements over pipeline route,
* pipe section length,
* weight of the pipe for handling,
* coatings and linings that have to be added during installation,
* pipe strength,
* ease of assembling joints, and
* maximum deflection allowed by the joints.

Some of the pipe installation conditions that might have to be considered are

- unusual soil conditions,
- uneven terrain,
- high groundwater and dewatering requirements,
- high bedrock,
- river or highway crossings,
- proximity to sewer lines,
- proximity to other utility services,
- cathodic protection,
- thrust design,
- air-release valves,
- soil types and need for imported materials for pipe base, pipe zone, and backfill,
- site restoration and backfill requirements, and
- tolerance for trench settlement.

Piping systems

Four general types of piping systems (Figure 2-7) are used by water utilities. Each use has certain characteristics that dictate, to some degree, the system that will be most economical and best suited for the installation. For example,

- Transmission lines carry large quantities of water from a source of supply to a treatment plant, or from a treatment plant or pumping station to a distribution system. They generally run in a rather straight line from point to point, have few side connections, and are not tapped for customer services.
- In-plant piping systems are the pipes located in pump stations and treatment plants. The piping is generally exposed and has many valves, outlets, and bends that must be secured against movement.
- Distribution mains are the pipelines that carry water from transmission lines and distribute it throughout a community. They have many side connections and are frequently tapped for customer connections.
- Service lines, or "services," are small-diameter pipes that run from the distribution mains to the customer's premises. They are discussed in chapter 15.

TYPES OF PIPE MATERIALS

Very early water systems often had water mains made of wood. These mains were usually logs with a hole burned or bored down the center, and the logs were joined with metal sleeves. Older city water systems still occasionally dig up pieces of log pipe during distribution system repairs.

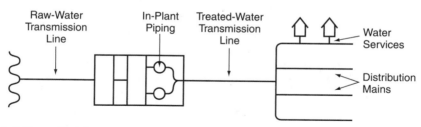

FIGURE 2-7 Types of piping systems

The logs were not usually reinforced, so the pipe did not withstand much pressure. Log mains were therefore primarily used to furnish water to local fountains or reservoirs where residents could draw water, rather than to distribute water to homes as we see today. A section of log pipe is shown in Figure 2-8.

From the late 1800s until about World War II, cast-iron, wrought steel, riveted steel, and wood stave pipe were common. The principal types of pipe in use today for water system transmission, distribution, and plant piping fall into the following general categories and the appropriate AWWA standards:

- ductile iron(DIP)   AWWA C151
- steel   AWWA C200
    - Cement–mortar-lined steel pipes   AWWA C200, C205, C207, & C208
- polyvinyl chloride (PVC)   AWWA C900 & C905
- reinforced concrete pressure pipe (RCPP)
    - steel cylinder   AWWA C300
    - prestressed, steel cylinder   AWWA C301
    - noncylinder   AWWA C302
    - pretensioned, steel cylinder   AWWA C303
      (also called concrete cylinder pipe (CCP)
- high-density polyethylene (HDPE)   AWWA C906

Asbestos–cement pipe (ACP) was available in the United States starting around 1930 but is no longer used in the water industry because of the attention focused on the hazards of asbestos. ACP is made by mixing Portland cement and asbestos fiber under pressure and heating it to produce a hard, strong, machinable product. It is estimated that over 300,000 miles of ACP is in service in the United States (Ysusi 2000). The US Environmental Protection Agency (USEPA) banned most uses of asbestos in 1989. This is not so much due to the danger of specific products as it is to the overall exposure of people involved in the mining, production, installation, and ultimate removal and disposal of asbestos products.

FIGURE 2-8 Section of an old log pipe

Gray cast-iron pipe (CIP) is another type of pipe that is no longer used in the United States. The oldest installation of cast-iron pipe (CIP) on record is in Versailles, France, and was constructed in 1664. Gray CIP has been used for water distribution systems in the United States and Canada for many years. There are more miles of this pipe in use today than of any other type. Many water systems have cast-iron mains that are more than 100 years old and still function well in daily use. In the 1920s, manufacturers began to make gray CIP using the centrifugal process (Figure 2-9). In this method, molten iron is distributed into a rotating horizontal mold, which forms the outside of the pipe. Lead joints were commonly used to connect gray CIP until the 1920s.

Standards for the design, manufacture, and installation of pressure pipe have been developed by the American Water Works Association (AWWA), American National Standards Institute (ANSI), and ASTM International. These specifications have been established by experimentation, testing, and experience in practice. Pipe manufacturers also publish product literature that is useful in pipe selection and installation. Further details on standards and approval of materials are given in appendix A. The names and addresses of associations representing manufacturers of the various types of pipes are listed in appendix B.

Table 2-1 summarizes the advantages and disadvantages of pipeline materials, and Table 2-2 summarizes the types of joints available and their applications. Other considerations that may have a bearing on the type of pipe selected for use in a system are

- state and local regulations,
- local soil conditions,
- local weather conditions,
- likelihood of earthquake activity,

FIGURE 2-9 Pipe being centrifugally cast by the de Lavaud process
Courtesy of American Cast Iron Pipe Company

- whether the system serves a community with fire protection or rural customers without fire protection,
- the type of pipe already in use in the system,
- whether the pipe is to be exposed to the weather or sunlight, and
- the possibility of the pipe being exposed to fire.

Ductile-Iron Pipe

Ductile-iron pipe (DIP) resembles CIP and has many of the same characteristics. It is produced in the same manner, but it differs in that the graphite is distributed in the metal in spheroidal or nodular form rather than in flake form (Figure 2-10). Adding an inoculant, usually magnesium, to the molten iron allows this distribution to be achieved. DIP is much stronger and tougher than CIP.

Although unlined ductile iron has a certain resistance to corrosion, aggressive waters can cause the pipe to lose carrying capacity through corrosion and tuberculation. The development of a process for lining pipe with a thin coating of cement mortar has virtually eliminated these problems. The cement–mortar lining adheres closely to the pipe wall, as illustrated in Figure 2-11. The lined pipe may be cut or tapped without damage to the lining.

Bituminous external coatings and polyethylene wraps are commonly used to reduce external corrosion.

TABLE 2-1 Comparison of transmission and distribution pipeline materials

Material	Common Sizes—Diameter		Normal Maximum Working Pressure		Advantages	Disadvantages
	in.	(mm)	psi	(kPa)		
Ductile iron	3–64	(76–1,625)	350	(2,413)	Durable, strong, high flexural strength, good corrosion resistance, lighter weight than cast iron, greater carrying capacity for same external diameter, easily tapped	Subject to general corrosion if installed unprotected in a corrosive environment
Concrete (reinforced)	12–168	(305–4,267)	250	(1,724)	Durable with low maintenance, good corrosion resistance, good flow characteristics, O-ring joints are easy to install, high external load capacity, minimal bedding and backfill requirements	Requires heavy lifting equipment for installation, may require special external protection in high-chloride soils
Concrete (prestressed)	16–144	(406–3,658)	350	(2,413)	Same as for reinforced concrete	Same as for reinforced concrete
Steel	4–120	(100–3,048)	High		Lightweight, easy to install, high tensile strength, low cost, good hydraulically when lined, adapted to locations where some movement may occur	Subject to general corrosion if installed unprotected in a corrosive environment; poor corrosion resistance unless properly lined, coated, and wrapped
Polyvinyl chloride	4–36	(100–914)	200	(1,379)	Lightweight, easy to install, excellent resistance to corrosion, good flow characteristics, high tensile strength and impact strength	Difficult to locate underground so tracer tape can be used, requires special care during tapping, susceptible to damage during handling, requires special care in bedding
High-density Polyethylene	4–63	(100–6000)	250	(1750)	Lightweight, very durable, very smooth, liners and wrapping not required, can use ductile-iron fittings	Relatively new product, thermal butt-fusion joints, requires higher laborer skill

TABLE 2-2 Pipe joints and their applications

Type of Material	Type of Joint	Application
Ductile iron	Push-on or mechanical	General use where flexibility is required
	Flanged	Where valves or fittings are to be attached in vaults or above grade
	Flexible ball	River crossings or in very rugged terrain
	Restrained	To resist thrust forces and in unstable soils
Concrete	Galvanized steel ring, bell-and-spigot types, or their variations with elastomeric gaskets	All locations
Plastic (PVC)	Bell and spigot type	Most commonly used for typical municipal uses (ASTM F 477)
	Solvent weld	Only for small lines
(HDPE)	Thermal butt-fusion, flange assemblies, or	ASTMD 2657
	mechanical methods recommended by manufacturer	Joining HDPE pipe to valves and ductile-iron fittings
Steel	Mechanical sleeve coupling	All diameters, but especially on pipe too small for a person to enter
	Rubber gasket joints	Low-pressure applications
	Welded joints	High-pressure applications, 24-in. and larger pipes
	Flanged joints	Where valves or fittings are to be attached
	Expansion joints	Allows movement so that expansion or contraction is not cumulative over several lengths

 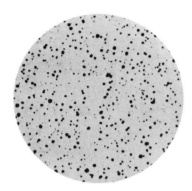

Cast Iron (Graphite Flakes)   Ductile Iron (Graphite Nodules)

FIGURE 2-10 Microphotographs of cast iron and ductile-iron
Courtesy of the Ductile Iron Pipe Research Association

DIP is available in standard pressure classes ranging from 150 to 350 psi (1,034 to 2,413 kPa) and in diameters of 3 to 64 in. (76 to 1,625 mm). The standard lengths of DIP are 18 and 20 ft (5.5 and 6.1 m).

Advantages of DIP include

- good durability and flexural strength,
- smooth interior (*C* value of 140),
- fracture resistance, and
- good exterior corrosion resistance in most soils.

Disadvantages of DIP include

- possible external corrosion in aggressive environments if not protected (see AWWA Standard C105/A21.5, *Polyethylene Encasement for Ductile-Iron Pipe Systems* [most recent edition]).

Ductile-iron pipe joints

The following types of joints are generally used today for connecting DIP and fittings:

- Flanged joints
- Mechanical joints
- Ball-and-socket or submarine joints
- Push-on joints
- Restrained joints
- Grooved and shouldered joints

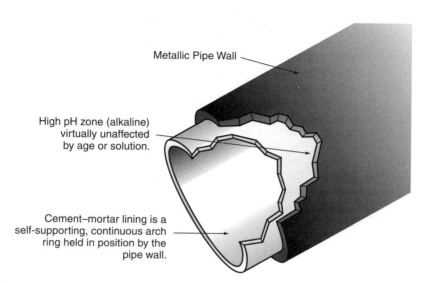

Metallic Pipe Wall

High pH zone (alkaline) virtually unaffected by age or solution.

Cement–mortar lining is a self-supporting, continuous arch ring held in position by the pipe wall.

FIGURE 2-11 Placement of cement-mortar lining on pipe
Courtesy of Mainlining Services, Inc.

Details of a mechanical joint are shown in Figure 2-12. AWWA standards for fittings for DIP are listed in the bibliography at the end of this chapter.

Flanged joints (AWWA C115). Flanged joints consist of two machined surfaces that are tightly bolted together with a gasket between them. They are primarily used in exposed locations where rigidity, self-restraint, and tightness are required, such as inside treatment plants and pump stations. They are also used where valves or other flanged appurtenances need to be installed. Flanged joints should not normally be used for buried pipe because of their lack of flexibility to compensate for ground movement.

Mechanical joints. A mechanical joint is made by bolting a movable follower ring on the spigot to a flange on the bell. The follower ring compresses a rubber gasket to form the seal. Mechanical joints are more expensive than some other joints but they make a very positive seal and require little technical expertise by installers. The joints also allow some deflection of the pipe at installation. They provide considerable flexibility in the event there is ground settlement after the pipe is installed.

Ball-and-socket joints. Ball-and-socket joints are special-purpose joints, most commonly used for intakes and river crossings. Their great advantages are that they provide for a large deflection (up to 15°) and are positively connected so they won't come apart. The large amount of allowable deflection also makes this type of joint useful for pipelines laid across mountainous terrain. Both bolted and boltless flexible pipe joints designed on the ball-and-socket principle are available.

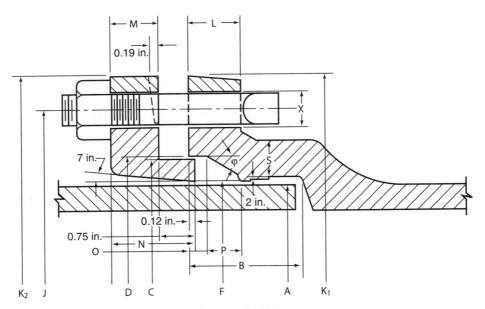

FIGURE 2-12 Mechanical-joint example for sizes 3–48 in.

Push-on joints. The most recently developed (and now the most popular) joints used in water distribution systems are push-on joints. The joint consists of a bell with a specially designed recess to accept a rubber ring gasket. The spigot end must have a beveled edge so it will slip into the gasket without catching or tearing it.

Workers assemble the joint by lubricating the gasket and spigot end with special non-toxic lubricant and pushing the spigot into the bell. Push-on joints are both less expensive to manufacture and quicker to assemble than mechanical joints because there are no bolts to install. When the pipe joint is assembled, the rubber ring gasket is compressed to produce a watertight seal.

Push-on joints are available in several designs and permit considerable flexibility in pipe alignment. Small diameters may be assembled by hand. Larger sizes usually require mechanical assistance, such as pulling the spigot end of the pipe into the bell with a come-along or pushing it home with a backhoe.

Restrained joints. Various types of restrained joints are available from different manufacturers. They are used where it is necessary to ensure that joints do not separate. Some manufacturers offer special versions of the push-on joint with a restraining feature, and others have separate devices for this purpose.

Grooved and shouldered joints. The grooved joint utilizes a bolted, segmental, clamp-type, mechanical coupling with a housing that encloses a U-shaped rubber gasket. The housing locks the pipe ends together and compresses the gasket against the outside of the pipe ends. The ends of the pipe are machine-grooved to accept the housing (Figure 2-13). The shouldered joint is similar except that the pipe ends are shouldered instead of grooved. These joints are covered in AWWA Standard C606, *Grooved and Shouldered Joints* (most recent edition).

An explanation of AWWA standards is provided in appendix A. Information on DIP is available from the Ductile Iron Pipe Research Association at the address listed in appendix B.

Fittings for iron pipe

A wide variety of ductile-iron and cast-iron fittings are available for use with iron pipe as illustrated in Figure 2-14.

Steel Pipe

Steel pipe has been in use in US water systems since 1852. It is frequently used where there is particularly high pressure or where very large-diameter pipe is required. Figure 2-15 shows a picture of riveted steel pipe being installed in 1907.

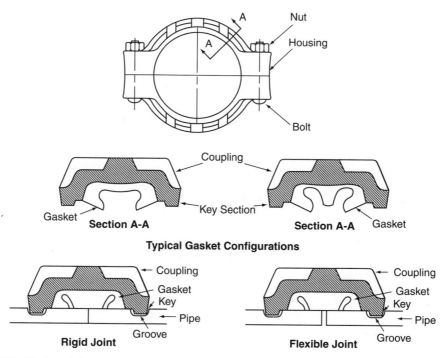

FIGURE 2-13 General coupling and joint configurations

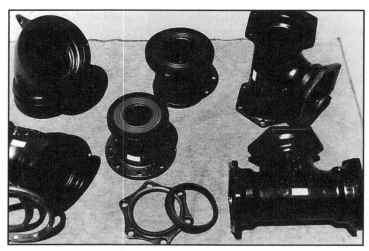

FIGURE 2-14 Typical iron pipe fittings
Courtesy of U.S. Pipe and Foundry Company

FIGURE 2-15 Riveted steel pipe being installed in Philadelphia in 1907
Courtesy of American Iron and Steel Institute

Advantages and Disadvantages

Advantages of steel pipe include

- relatively light weight;
- competitive price, particularly in larger diameters;
- ease of fabricating special configurations;
- designs that will withstand high internal pressures and loads if necessary;
- relative ease of transporting and installing; and
- resistance to shock loads and ability to bend to some degree without buckling.

Disadvantages of steel pipe include

- potential to both internal or external corrosion if not properly protected,
- the need to carefully consider external loads in the installation design, and
- potential for a partial vacuum (caused by rapidly emptying, relatively thin-walled pipe) to cause pipe distortion or complete collapse, as shown in Figure 2-16. (The design for proper vacuum relief should be checked by methods detailed in AWWA Manual M11, *Steel Pipe—A Guide for Design and Installation*.)

Because steel pipe is generally most competitive in sizes larger than 16 in. (400 mm), it is primarily used for feeder mains in water distribution systems and for long-distance transmission mains. In special cases, steel pipe has been fabricated in diameters of 30 ft (9 m). The length of fabricated pipe varies depending on the diameter and shipping restrictions, but 45-ft (14-m) lengths are usually considered optimal for shipping and handling. The thickness of steel plate used to fabricate the pipe varies depending on both the internal water pressure of the pipeline and the external loads that will be exerted on the pipe.

The interior of steel pipe is usually protected with either cement mortar or epoxy, as specified in AWWA Standard C205, *Cement–Mortar Protective Lining and Coating for Steel Water Pipe—4 In. (100 mm) and Larger—Shop Applied*, and AWWA Standard C210, *Liquid-Epoxy Coating Systems for the Interior and Exterior of Steel Water Pipelines* (most recent editions). The exterior of water main pipe must be protected from corrosion and also against abrasion if the pipe is to be buried. AWWA standards provide for a variety of plastic coatings, bituminous materials, and polyethylene tapes depending on the degree of protection required. For the highest degree of protection, a coated pipeline is commonly also provided with cathodic protection.

Steel pipe joints and fittings

Pipe lengths are commonly joined by welding for pipe diameters of 600 mm (24 in.) or larger using butt-welded or lap-welded joints. Steel pipe can also be joined by various types of mechanical joints similar to the joints used for other types of pipe, as illustrated in Figures 2-17

FIGURE 2-16 Collapsed steel pipe
Courtesy of the Los Angeles Department of Water and Power

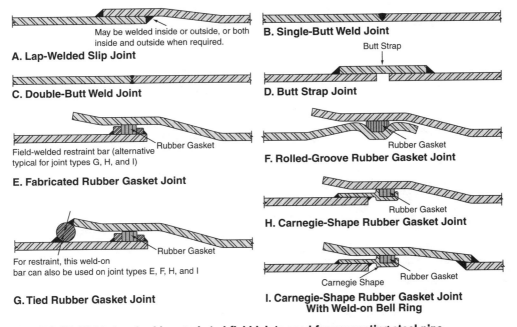

A. Lap-Welded Slip Joint

May be welded inside or outside, or both inside and outside when required.

B. Single-Butt Weld Joint

C. Double-Butt Weld Joint

D. Butt Strap Joint

Butt Strap

E. Fabricated Rubber Gasket Joint

Field-welded restraint bar (alternative typical for joint types G, H, and I)

Rubber Gasket

F. Rolled-Groove Rubber Gasket Joint

Rubber Gasket

G. Tied Rubber Gasket Joint

For restraint, this weld-on bar can also be used on joint types E, F, H, and I

Rubber Gasket

H. Carnegie-Shape Rubber Gasket Joint

Rubber Gasket

I. Carnegie-Shape Rubber Gasket Joint With Weld-on Bell Ring

Carnegie Shape

Rubber Gasket

FIGURE 2-17 Welded and rubber-gasketed field joints used for connecting steel pipe

and 2-18. In installations where steel pipe is expected to experience a high degree of expansion and contraction, expansion joints can be installed, as illustrated in Figure 2-19.

Cast-iron, ductile-iron, fabricated steel, and stainless-steel bends and fittings are usually used for standard changes in either direction or size for smaller-diameter steel pipe.

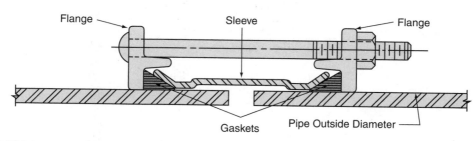

FIGURE 2-18 Detail of a sleeve coupling used for connecting steel pipe sections

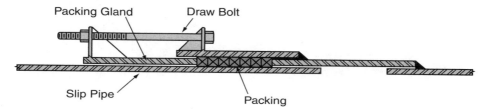

FIGURE 2-19 Detail of one type of expansion joint for steel pipe

Long-radius bends can also be made in smaller-size pipe by the use of "wrinkle bending," which is done by heating the pipe with a welding flame at several points on the inside of the bend. Pipe fabricators can also furnish any form of special fittings desired.

Steel plant piping

Steel piping is frequently used for plant piping because it is inexpensive and relatively light for handling during installation. In addition, special sizes, shapes, and outlets can be fabricated either offsite or onsite by welding (Figure 2-20). External corrosion of piping in a plant can be minimized if the pipe is protected with a properly applied coating and also if the plant is dehumidified.

Information on steel pipe is available from the Steel Tube Institute of North America at the address listed in appendix B.

Plastic Pipe

Plastic pipe is widely used by water utilities. Plastic pipe was first introduced in the United States around 1940. There are many different types of thermoplastic and thermoset plastic materials that can be manufactured to have various properties. Two properties are of particular importance for water supply piping. First, the plastic piping must have a long-term hydrostatic stress rating to withstand both internal and external pressures. Second, there must not be any harmful substances in the plastic that will leach into the water to cause tastes, odors, or adverse health effects.

FIGURE 2-20 Complex piping fabricated from steel pipe
Courtesy of L.B. Foster Co.

Plastic materials

All plastic pipe used to convey potable water must be certified for conformance with the NSF International Standard 61 for potable water use. It must also meet applicable AWWA or other industry standards. More details on standards are included in appendix A.

Plastic materials used for fabricating water main pipe include PVC, polyethylene (PE), and polybutylene (PB). Composite plastic pipe is also available in the form of plastic pipe reinforced with fiberglass. Plastic materials will not usually react with water on the inside or with soil on the outside, so no extra corrosion protection is necessary.

Permeation

Research and actual occurrences have documented that organic compounds can pass through the walls of plastic pipe, even though the pipe is carrying water under pressure. The process by which the molecules pass through the plastic is called permeation. If gasoline, fuel oil, or other organic compounds have saturated the soil around plastic pipe, a disagreeable taste or odor will often be created in the water. More importantly, the contamination can pose a significant health threat to customers using the water. The organic compounds will also soften the plastic pipe, leading to its eventual failure.

Plastic pipe should therefore not be installed in areas where there is known soil contamination by organic compounds or where future contamination is possible, such as close to old petroleum storage tanks.

PVC pipe

PVC pipe is manufactured by an extrusion process and is available in various pressure ratings. It is by far the most widely used type of plastic pipe material for small-diameter water mains. The AWWA C900 standard for PVC pipe in sizes 4 to 12 in. (100 to 300 mm) is based on the same outside diameter as for DIP. In this way, standard DIP bends, tees, and other fittings can be used with PVC pipe. However, AWWA C905 for PVC transmission pipe in sizes 14 to 48 in. (350 to 1,200 mm) is available in outside-diameter sizes based on either iron pipe size or cast-iron pipe size. Information on PVC water main pipe is available in AWWA Manual M23, *PVC Pipe—Design and Installation.*

Advantages of PVC pipe include

- exceptionally smooth interior (*C* value of at least 150);
- chemical inertness;
- generally lower cost and greater ease in shipping;
- greater ease of installation because of lighter weight;
- moderate flexibility, so it will adapt to ground settlement; and
- ease of handling and cutting.

Disadvantages of PVC pipe include

- susceptibility to damage (gouges deeper than 10 percent of the wall thickness can seriously affect pipe strength),
- difficulty in locating because it is nonconductive,
- inability to be thawed electrically,
- need for rigid adherence to the use of proper tools and procedures when service taps are made,
- susceptibility to damage from the ultraviolet radiation in sunlight,
- susceptibility to permeation,
- need for careful bedding during installation to maintain pipe shape, and
- susceptibility to buckling under a vacuum.

PVC joints and fittings. PVC pipe may be joined by a bell-and-spigot push-on joint in large diameters. In the push-on method, either (1) a rubber gasket set in a groove in the bell end or (2) a double bell coupling provides the necessary seal.

Injection-molded PVC fittings are available in sizes up to 8 in. (200 mm), and fabricated fittings are available in larger sizes. Cast-iron, ductile-iron, and cast-steel fittings are also often used. Figure 2-21 illustrates the use of a restrained fitting on PVC pipe.

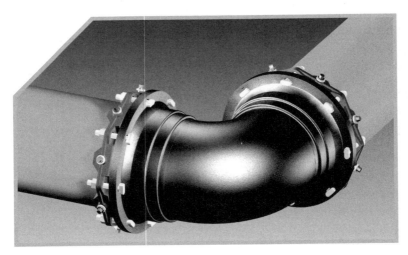

FIGURE 2-21 Restrained fitting
Courtesy of EBAA Iron, Inc.

PE and PB pipe

Extruded PE and PB are primarily used for the manufacture of water service pipe in small sizes. This is covered in detail in chapter 15. The use of PB has decreased remarkably because of structural difficulties causing premature pipe failures.

Details of large-diameter PE pipe are found in AWWA Standard C906, *Polyethylene (PE) Pressure Pipe and Fittings, 4 In. (100 mm) Through 63 In. (1,600 mm), for Water Distribution and Transmission* (most recent edition). Information on PVC, PE, PB, and other types of plastic pipe is available from some of the organizations listed in appendix B.

Fiberglass pressure pipe

Fiberglass pipe is available for potable water use in sizes from 1 in. through 144 in. (25 mm through 3,600 mm) diameter, in five pressure classes ranging from 50 psi through 250 psi (345 kPa through 1,724 kPa). There are also several different stiffness classes available that are incorporated into the design, depending on the exterior loading that will be applied to the pipe. Advantages of the pipe include corrosion resistance, light weight, low installation cost, ease of repair, and hydraulic smoothness. Disadvantages include susceptibility to mechanical damage, low modulus of elasticity, and lack of a standard jointing system.

One method of manufacturing the pipe is called filament winding. A continuous glass-fiber roving saturated with resin is wound around a mandrel in a carefully controlled pattern and under controlled tension. The inside diameter of the pipe is fixed by the mandrel diameter, and the wall thickness is governed by the pressure and stiffness class desired.

The other manufacturing method is centrifugal casting. The resin and fiberglass reinforcement are applied to the inside of a mold that is rotated and heated. The outside diameter of the pipe is determined by the mold and the inside diameter varies depending on the wall thickness. Fittings are made by filament winding, by spraying chopped fiberglass and resin on a mold, or by joining cut pieces of pipe to make mitered fittings.

Several different methods are used by various manufacturers to join pipe sections and fittings. One method is to butt the sections together and wrap the joint with fiberglass material and resin. The other methods include a variety of tapered bell-and-spigot joints that are bonded with adhesives. Pipe, fitting, and adhesive for joining usually are not interchangeable between pipe from different manufacturers. Fiberglass pipe is covered in AWWA Standard C950, *Fiberglass Pressure Pipe* (most recent edition).

Concrete Pipe

The use of concrete pressure pipe has grown rapidly since 1950. The pipe provides a combination of the high tensile strength of steel and the high compressive strength and corrosion resistance of concrete. The pipe is available in diameters ranging from 10 to 252 in. (250 to 6,400 mm) and in standard lengths from 12 to 40 ft (3.7 to 12.2 m).

Concrete pipe is available with various types of liners and reinforcement. The four types in common use in the United States and Canada are

1. Prestressed concrete cylinder pipe
2. Bar-wrapped concrete cylinder pipe
3. Reinforced concrete cylinder pipe
4. Reinforced concrete noncylinder pipe

Prestressed concrete cylinder pipe

Prestressed concrete cylinder pipe has been made in the United States since 1942. Two general types are manufactured: lined-cylinder pipe (Figure 2-22) and embedded-cylinder pipe (Figure 2-23).

For either type, manufacturing starts with a full length of welded steel cylinder. After joint rings are attached to each end, the pipe is hydrostatically tested to verify watertightness. For embedded-cylinder pipe, the steel cylinder is cast within a concrete core.

After the pipe core has cured on embedded-cylinder pipe, hard-drawn steel wire is helically wrapped around the exterior of the cylinder. The wire spacing is accurately controlled to produce a predetermined compression of the concrete and steel core. The core is then covered with a cement slurry and a dense coating of mortar to provide corrosion protection for the steel cylinder and wire. The pipe is then ready for use after the mortar has cured.

For lined-cylinder pipe, the concrete core is cast within the steel cylinder, which is then wrapped with high-strength steel wire and coated with portland-cement mortar.

Lined-cylinder pipe is generally available in diameters from 16 to 60 in. (406 to 1,524 mm). Embedded-cylinder pipe is commonly available in diameters from 24 to 144 in. (610 to 3,658 mm). By variations of design, pipe can be fabricated to withstand more than 400 psi (2,758 kPa) of pressure and more than 100 ft (30 m) of cover. Prestressed concrete cylinder pipe is covered in AWWA Standard C301, *Prestressed Concrete Pressure Pipe, Steel-Cylinder Type*, for manufacture, and by AWWA Standard C304, *Design of Prestressed Concrete Cylinder Pipe* (most recent editions).

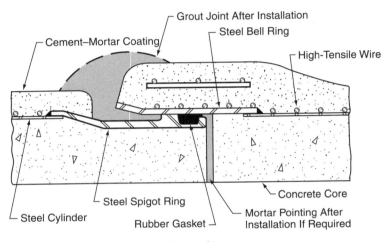

FIGURE 2-22 Prestressed concrete lined-cylinder pipe
Drawing furnished by American Concrete Pressure Pipe Association

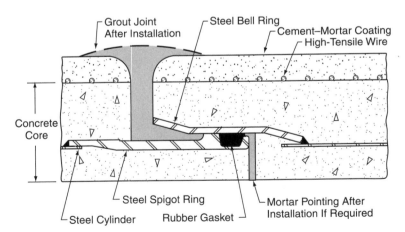

FIGURE 2-23 Prestressed concrete embedded-cylinder pipe
Drawing furnished by American Concrete Pressure Pipe Association

Bar-wrapped concrete cylinder pipe

Bar-wrapped concrete cylinder pipe is manufactured using a hydrostatically tested steel cylinder with welded attached joint rings. The steel cylinder is lined with concrete and wrapped with mild steel bar. The entire assembly is coated with mortar for corrosion protection (Figure 2-24). The core is then protected with a cement–mortar coating.

Bar-wrapped concrete cylinder pipe is manufactured mainly in Canada and the western and southwestern United States. It is normally available in diameters of 10 to 72 in. (250 to 1,800 mm). This pipe design is covered by AWWA Manual M9, *Concrete Pressure Pipe*. Manufacture is covered by AWWA Standard C303, *Concrete Pressure Pipe, Bar-Wrapped, Steel-Cylinder Type* (most recent edition).

Reinforced concrete cylinder pipe

Reinforced concrete cylinder pipe is manufactured by casting mild steel reinforcing cages and a steel cylinder with welded joint rings within a thick concrete core (Figure 2-25). Pipe of this design is generally available in diameters from 24 to 144 in. (610 to 3,658 mm). The design is covered by AWWA Manual M9, *Concrete Pressure Pipe*. Manufacture is covered by AWWA Standard C300, *Reinforced Concrete Pressure Pipe, Steel-Cylinder Type* (most recent edition).

Reinforced concrete noncylinder pipe

Reinforced concrete noncylinder pipe does not contain an internal watertight membrane (steel cylinder). Therefore, its use is limited to internal pressures of less than 55 psi (379 kPa).

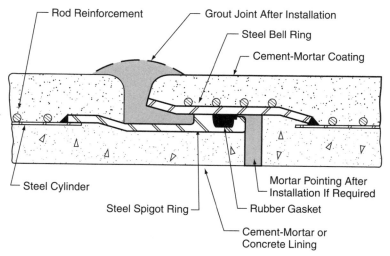

FIGURE 2-24 Pretensioned concrete cylinder pipe
Drawing furnished by American Concrete Pressure Pipe Association

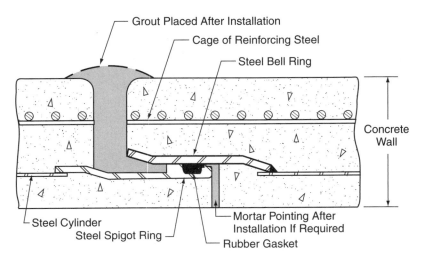

FIGURE 2-25 Reinforced concrete cylinder pipe
Drawing furnished by American Concrete Pressure Pipe Association

The reinforcement consists of one or more steel cages of welded wire fabric or helically wrapped rods welded to longitudinal rods. The pipe is made by placing concrete around the reinforcement by vertical casting. Noncylinder pipe is covered by AWWA Standard C302, *Reinforced Concrete Pressure Pipe, Noncylinder Type* (most recent edition) for manufacture, and AWWA Manual M9 for design.

The general advantages of concrete pipe include the following:

- Can be manufactured relatively inexpensively in larger sizes
- Can be manufactured to withstand relatively high internal pressure or external load
- Resistant to both internal and external corrosion
- Properly manufactured and installed, has a very long, trouble-free life
- Minimal bedding requirements during installation

Disadvantages of concrete pipe include the following:

- Heavy weight makes shipping long distances expensive
- Special handling equipment required due to weight
- Exact pipe lengths and fittings required for installation must be carefully laid out in advance
- More difficult to tap and repair

Joints and bends

As shown in Figures 2-22 through 2-25, pipe joints are sealed with a rubber gasket that is compressed as the steel joint rings are pushed together. The metal parts are then protected from corrosion with a covering of mortar.

Installers can achieve horizontal and vertical deflections by ordering pipe lengths with standard beveled ends. Bends are made with fabricated concrete and steel fittings made up to standard angles or to meet special conditions.

Additional information on concrete pipe is available from the American Concrete Pressure Pipe Association at the address listed in appendix B.

BIBLIOGRAPHY

AWWA Standard for Cement–Mortar Protective Lining and Coating for Steel Water Pipe—4 In. (100 mm) and Larger—Shop Applied. ANSI/AWWA C205. Denver, Colo.: American Water Works Association (latest edition).

AWWA Standard for Concrete Pressure Pipe, Bar-Wrapped, Steel-Cylinder Type. ANSI/AWWA C303. Denver, Colo.: American Water Works Association (latest edition).

AWWA Standard for Design of Prestressed Concrete Cylinder Pipe. ANSI/AWWA C304. Denver, Colo.: American Water Works Association (latest edition).

AWWA Standard for Fiberglass Pressure Pipe. ANSI/AWWA C950. Denver, Colo.: American Water Works Association (latest edition).

AWWA Standard for Grooved and Shouldered Joints. ANSI/AWWA C606. Denver, Colo.: American Water Works Association (latest edition).

AWWA Standard for Liquid-Epoxy Coating Systems for the Interior and Exterior of Steel Water Pipelines. ANSI/AWWA C210. Denver, Colo.: American Water Works Association (latest edition).

AWWA Standard for Polyethylene Encasement for Ductile-Iron Pipe Systems. ANSI/AWWA C105/A21.5. Denver, Colo.: American Water Works Association (latest edition).

AWWA Standard for Polyethylene (PE) Pressure Pipe and Fittings, 4 In. (100 mm) Through 63 In. (1,600 mm), for Water Distribution and Transmission. ANSI/AWWA C906. Denver, Colo.: American Water Works Association (latest edition).

AWWA Standard for Prestressed Concrete Pressure Pipe, Steel-Cylinder Type. ANSI/AWWA C301. Denver, Colo.: American Water Works Association (latest edition).

AWWA Standard for Reinforced Concrete Pressure Pipe, Noncylinder Type. ANSI/AWWA C302. Denver, Colo.: American Water Works Association (latest edition).

AWWA Standard for Reinforced Concrete Pressure Pipe, Steel-Cylinder Type. ANSI/AWWA C300. Denver, Colo.: American Water Works Association (latest edition).

Holsen, T.M., et al. 1991. Contamination of Potable Water by Permeation of Plastic Pipe. *Jour. AWWA* 83(8):53.

Manual M9, Concrete Pressure Pipe. 2008. Denver, Colo.: American Water Works Association.

Manual M11, Steel Pipe—A Guide for Design and Installation. 2004. Denver, Colo.: American Water Works Association.

Manual M23, PVC Pipe—Design and Installation. 2004. Denver, Colo.: American Water Works Association.

Moser, A.P., and K.G. Kellogg. 1994. *Evaluation of Polyvinyl Chloride (PVC) Pipe Performance.* Denver, Colo.: American Water Works Association and American Water Works Association Research Foundation.

Thompson, C., and D. Jenkins. 1987. *Review of Water Industry Plastic Pipe Practices.* Denver, Colo.: American Water Works Association Research Foundation and American Water Works Association.

Water Distribution Operator Training Handbook. 2005. Denver, Colo.: American Water Works Association.

Water Utility Experience With Plastic Service Pipe. 1992. Denver, Colo.: American Water Works Association and American Water Works Association Research Foundation.

Ysusi, M.A. 2000. System Design: An Overview, Chapter 3 in *Water Distribution Systems Handbook*, ed. by L.W. Mays. New York: McGraw-Hill.

Water Storage

Water storage is essential for meeting all of the domestic, industrial, and fire demands of most public water systems. Water may be stored before and/or after treatment. Reservoirs for the storage of raw water are discussed in *Water Sources*, another book in this series. The primary subject of this chapter is distribution storage, which refers to the storage of treated water ready for use by customers.

WATER STORAGE REQUIREMENTS

The type and capacity of water storage required in a distribution system vary with the size of the system, the topography of the area, how the water system is laid out, and various other considerations. The primary types of water storage structures are

- hydropneumatic tanks,
- ground-level reservoirs,
- buried reservoirs, and
- elevated tanks.

Purposes of Water Storage

Water storage in the distribution system may be required for the following purposes:

- Equalizing supply and demand
- Increasing operating convenience
- Leveling out pumping requirements
- Decreasing power costs
- Providing water during power source or pump failure
- Providing large quantities of water to meet fire demands
- Providing surge relief
- Increasing detention times
- Blending water sources

Equalizing supply and demand

As illustrated in Figure 3-1, the demand for water normally changes throughout the day and night. If treated water is not available from storage, the wells or water treatment plant must have sufficient capacity to meet the demand at peak hour (the busiest water-use hour). This high capacity is not generally practical or economical. The peak-hour demand in Figure 3-1 is approximately 175 percent of the average demand for the day. This means that without storage, the plant capacity would have to be almost double the size that is needed to meet average demand.

With adequate storage, water can be treated or supplied to the system at a relatively even rate over a 24-hour period. During midday, when the demand increases, the excess requirement can be made up from storage. Figure 3-1 shows that from 10 p.m. to 7 a.m., the demand on the system is below the average rate. During this time, the reservoirs are being filled. From 7 a.m. to 10 p.m., the demand is greater than the supply. The reservoirs are then being used to feed water back into the system.

A water system that purchases water from another water system is usually offered these three storage options:

1. The purchasers can draw water as needed, in which case they may need no storage of their own. They will, however, have to pay a premium price for the water, because the supplying system must supply the necessary storage or adequate plant capacity to meet demands.
2. The purchasers can take water at a relatively uniform or maximum rate throughout the day. In this case, the water cost is usually substantially less, but the purchasing system will have to provide enough storage of its own to supply peak demands.
3. The purchasers can draw all or most of their water at night, when the treatment plant usually has excess capacity. The water purchased during off-peak hours is usually much less expensive. The rate charged for the water is often called a "dump rate." Under this plan, the purchasing system must furnish adequate storage to supply day-time demand for at least one day.

Increasing operating convenience

In some situations, storage is provided to allow a treatment plant to be operated for only one or two shifts, thereby reducing personnel costs. In this situation, storage provides the water required through the night when the plant is shut down.

Leveling out pumping requirements

The demand for water is continually changing in all water systems. Demand depends on the time of day, day of the week, weather conditions, and many other factors. If there is no storage at all, the utility must continually match the changing demand by selecting pumps of varying sizes. Frequent cycling of pumps (i.e., turning them on and off frequently) causes increased wear on pump controls and motors. It also increases electrical costs.

If a water system has adequate *elevated* storage, pump changes can be minimized. Water will then flow from the tank if a sudden increase in consumer use is greater than what the operating pump can provide. Likewise, water in excess of the current demand will refill the tank when there is a reduction in water use on the system. The pumping rate will not need to be changed immediately.

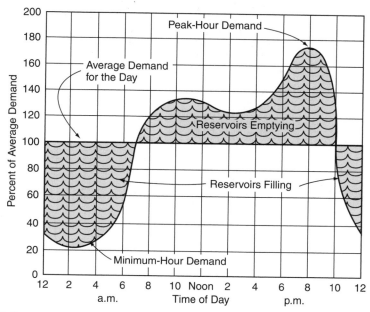

FIGURE 3-1 Daily variation of system demand

However, if a water system has only underground storage, the utility must meet system water demands by constantly adjusting the pumping rate, either from the water source or from a storage facility. In this situation, the utility can not take advantage of the force of gravity to move water through the system.

Decreasing power costs

Storage also allows pumping costs to be reduced if there are special rate structures for electrical power. Many electric power utilities have off-peak rates that are lower at certain times— usually during the night. A system can often achieve substantial savings in power costs if most large pumps are operated during off-peak hours. Water utilities can pump water to elevated storage during the night and do a minimum amount of pumping during the day.

Elevated storage can also help a water system avoid using large pumps to meet occasional peak demands, which usually provides a savings in electricity costs as well.

Providing water during source or pump failure

There are times when power failure, mechanical breakdown, or maintenance work will prevent use of source water pumps or even an entire treatment plant. In addition, sudden extreme increases in demand can be caused by main breaks, broken hydrants, or similar problems. Elevated storage will automatically provide additional water to the system during such emergencies. However, if only pumped storage is available, the utility must be able to begin pumping immediately in order to maintain system pressure.

Another situation in which a relatively large amount of storage may be required involves a water system that depends on a single, long transmission main as its only water source. These systems usually provide storage equal to at least one average day's water use to allow for possible repair or maintenance of the transmission main.

Providing water to meet fire demands

A major purpose of distribution system storage is to meet fire demands. Although fire demand may not occur very often, the rate of water use is usually much greater than for domestic peak demands. Water systems are usually designed to meet fire demand in addition to normal customer needs. Fire demand can account for as much as 50 percent of the total capacity of a storage system.

Providing surge relief

As pumps turn on and off and valves are opened and closed, tremendous pressure changes, known as water hammer, can surge through the distribution system. Excessive water hammer can seriously damage pipes and appurtenances. Elevated storage tanks greatly assist in absorbing the shocks of water hammer by allowing the shock wave to travel up the riser and into the upper tank section.

Increasing detention times

The time that water spends in storage after disinfectants are added, but before the water is delivered to the first customer, can be counted toward the disinfectant contact time required for surface water treatment. Storage reservoirs located at the discharge of a treatment plant provide system storage and help meet disinfection requirements at the same time.

Blending water sources

Some water systems use water from two or more sources, with each source having different water quality. An example is a system that uses water from both ground and surface sources. Blending different sources together in a reservoir will often improve the quality of marginally acceptable water.

Both industrial and domestic customers dislike day-to-day changes in hardness, taste, temperature, or chemical composition. Blending water from different sources in a reservoir allows water of relatively uniform quality to be furnished to the system.

Capacity Requirements

Distribution storage capacity is based on the maximum water demands in the different parts of the system. It varies for different systems and can be determined only by qualified engineers after a careful analysis and study of the system. The storage capacity needed for fire protection should be based on the recommendations of fire underwriters' organizations. Because so many variables are involved, operators should contact the Insurance Services Office or the fire insurance rating office in their state to obtain information.

The amount of emergency storage required is also based on both the reliability of the water source and the availability of backup equipment and standby power sources. A system having two power lines from different substations is usually considered to have a relatively secure power source in an emergency. It is even more secure to have engine generators and/or engine-driven pumps that will supply at least average system demands. A water system should also have backup pumps, so that maximum demands can be met even if major pumping units should fail.

TYPES OF TREATED-WATER STORAGE FACILITIES

Water storage tanks can be classified based on

- type of service,
- configuration, and
- type of construction material.

Type of Service

Water storage tanks can be used for either operating storage or emergency storage. An operating storage tank generally "floats" on the system. In other words, the tank is directly connected to the distribution piping, and the elevation of the water in the tank is determined by the pressure in the system. Water flows into the tank when water demand is low, and it empties from the tank when demand exceeds supply.

Emergency storage is designed to be used only in exceptional situations, such as during high-demand fires. A tank installed by an industry for use with the industry's own fire protection sprinkler system is an example of emergency storage. Because emergency storage water is not constantly being turned over, it stagnates, loses all residual chlorine, and can become contaminated. It is not usable as a potable water source unless it is further treated. Because the water in emergency storage tanks is not circulated, the tanks must be provided with heating equipment in colder climates to prevent the water from freezing.

In establishing storage for primarily standby or seasonal use, one must consider how to keep the stored water from degrading in quality. Water stored for a period of time can, in some circumstances, develop a foul taste or odor or become unsafe.

Some operators of water systems with reservoirs that are primarily designed to meet summer demand have found it best to drain some of the reservoirs during the winter. Otherwise, a procedure must be established to regularly remove some of the water from each reservoir and replace it with fresh water every day or two.

Configuration

Distribution storage facilities can be either located at the ground level or elevated. These facilities are classified as tanks, standpipes, or reservoirs.

Elevated tanks

Elevated tanks generally consist of a water tank supported by a steel or concrete tower. In general, an elevated tank floats on the distribution system. Occasionally, system pressure could become so high that the tank would overflow. In these cases, an altitude valve must be installed on the tank fill line to keep the tank from overflowing.

Standpipes

A tank that rests on the ground with a height that is greater than its diameter is generally referred to as a standpipe. In most installations, only water in the upper portion of the tank will furnish usable system pressure, so that most larger standpipes are equipped with an adjacent pumping system that can be used in an emergency to pump water to the system from the lower portion of the tank.

Standpipes combine the advantages of elevated storage with the ability to store a large quantity of water. Note that a relatively large amount of water must regularly be circulated through the tank to keep the water fresh and to prevent freezing.

Reservoirs

The term *reservoir* has a wide range of meanings in the water supply field. For raw water, reservoirs are generally ponds, lakes, or basins that are either naturally formed or constructed for water storage. For storage of finished water, the term is usually applied to a large aboveground or underground storage facility. Reservoirs may also be referred to as ground-level tanks.

Distribution system reservoirs are usually used where very large quantities of water must be stored or when an elevated tank is objectionable to the public. If a reservoir can be located on a high rise of ground, it can float on the system. This way, all or most of the water is directly available to the system without the need for pumping.

When a ground-level or buried reservoir is located at a low elevation on the distribution system, water is admitted through a remotely operated valve. A pump station is provided to transfer the water into the distribution system. The valves and pumps are usually controlled from the treatment plant or a central control center.

Completely buried reservoirs are often used where an aboveground structure is objectionable, such as in a residential neighborhood. In some cases, the land over a buried reservoir can be used for recreational facilities such as a ball field or tennis court.

The initial cost of a ground-level reservoir is generally less than providing equal elevated storage. However, the water must regularly be circulated in and out of a ground-level reservoir to keep it fresh. Thus, there is a continuing cost for power and the maintenance of pumping equipment. The cost of constructing a completely buried reservoir will be higher than for one at ground level because of the considerable amount of excavation required.

Hydropneumatic systems

For very small water systems that cannot afford to install an elevated or ground-level tank, a hydropneumatic pressure tank will often provide adequate continuity of service for domestic use. As illustrated in Figure 3-2, a steel pressure tank is kept partially filled with water and partially filled with compressed air. Some tanks also have a flexible membrane that separates the air and water.

The compressed air maintains water pressure when use exceeds the pump capacity. The tank also reduces the frequency with which the pump needs to be turned on and off. It provides water for a limited time in the event of pump failure.

Type of Construction Material

Over the years, reservoirs have been constructed from a variety of materials. Materials of construction and surfaces in contact with potable water should comply with all applicable standards (ANSI/AWWA D100, D102, D103, D104, D110, D115, D120, D130, and ANSI/NSF Standard 61). Early reservoirs were constructed with earth-embankment techniques. Today, steel and concrete are the most widely used construction materials.

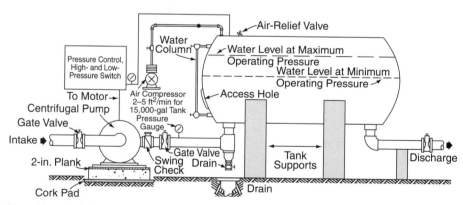

NOTE: Use special rubber hose fitting between pump and pressure tank for quiet operation.

FIGURE 3-2 Hydropneumatic water pressure system
This material is used with permission of John Wiley & Sons, Inc., from Environmental Engineering and Sanitation, 4th ed., Joseph A. Salfvato, © 1992 by John Wiley & Sons, Inc.

Earth-embankment reservoirs

Early distribution storage facilities were often earth-embankment reservoirs that were built on higher ground, if available, or near the service area. These reservoirs were usually constructed partly by excavation and partly by embankment. They were paved with stone riprap, brick, or concrete on the slopes and bottoms and were almost always open at the top.

These installations were effective and valuable in system operation at the time, but problems included leakage, freezing, contamination by birds and animals, and algae growth. State and federal regulations now prohibit the storage of potable water in open reservoirs.

Steel tanks

Most elevated tanks and standpipes, as well as many ground-level tanks, are constructed out of steel. The thickness of the steel used in constructing a tank varies with the pressure exerted on the tank wall. The upper walls may be relatively thin, but the lower walls of a tall tank may have a thickness of 2 in. (50 mm) or more.

Early steel tanks were constructed of panels that were riveted together, but this method was gradually replaced by welding starting around 1930. Welded tanks are much easier to maintain because of their smooth surface. Welding also allows tanks to be designed in new, more pleasing designs that would not be possible with riveted construction. Standards for constructing potable water tanks from welded steel are covered in AWWA Standard D100, *Welded Carbon Steel Tanks for Water Storage* (most recent edition).

Figure 3-3 shows an example of the unusual types of designs that can be achieved with welded steel construction. Most newer elevated tanks are of the single-pedestal design, which is more visually pleasing and easier to maintain because of its smooth surfaces.

Bolted tanks are also available in reservoir and standpipe configurations. The tanks are constructed of uniformly sized panels that are assembled with gaskets or a sealant to achieve a watertight seal at the bolted joints. The panels are factory coated for long-term corrosion protection. The coating systems presently available include hot-dipped galvanized, fused glass, thermoset liquid epoxy, and thermoset dry epoxy types. Figure 3-4 shows a typical bolted tank. Guidelines for bolted steel tanks are found in AWWA Standard D103, *Factory-Coated Bolted Carbon Steel Tanks for Water Storage* (most recent edition).

Concrete tanks and reservoirs

The first concrete reservoirs were open, but later designs were roofed with wood or concrete. A properly ventilated reservoir will generally not have sanitation problems. Concrete reservoirs may be constructed using several different techniques.

Cast-in-place concrete tanks are constructed much the same as the basement for a house. However, much more reinforcing is needed so the tank can resist the weight of the water. In addition, more care is needed to reduce leakage. The shape of a cast-in-place concrete reservoir is generally square or rectangular.

FIGURE 3-3 A 1,000,000-gal single-pedestal spheroidal elevated tank painted to look like a hot air balloon
Courtesy of CB&I

FIGURE 3-4 Typical bolted steel tank
Courtesy of Engineered Storage Products Company

It is difficult to prevent some cracking of cast-in-place concrete tanks, but the cracks can usually be repaired with new types of flexible caulking materials. Otherwise, membrane liners can be added to prevent water loss.

Circular, prestressed concrete tanks are constructed by first installing an inner concrete core wall that establishes the reservoir's circular form. Steel wire is then wrapped around the core wall under tension (Figures 3-5 and 3-6). The steel wrapping is then protected by layers of hydraulically applied concrete (gunite).

The walls of prestressed tanks can be made thinner and with less reinforcing steel than those of cast-in-place concrete tanks. However, the tanks must be built only by qualified contractors. Prestressed tanks are constructed in accordance with AWWA Standard D110, *Wire- and Strand-Wound, Circular, Prestressed Concrete Water Tanks* (most recent edition).

Hydraulically applied concrete-lined reservoirs consist of an earth embankment that is covered with reinforced, hydraulically applied concrete. This process is similar to that used for constructing many swimming pools. Reservoirs can be constructed at relatively low cost with this method, but they are usually shallow and difficult to cover.

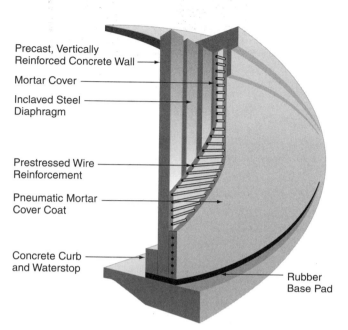

Precast, Vertically
Reinforced Concrete Wall

Mortar Cover

Inclaved Steel
Diaphragm

Prestressed Wire
Reinforcement

Pneumatic Mortar
Cover Coat

Concrete Curb
and Waterstop

Rubber
Base Pad

FIGURE 3-5 Sectional view of a prestressed concrete tank
Courtesy of Preload Inc.

FIGURE 3-6 Circumferential wire prestressing operation
Courtesy of Preload Inc.

LOCATION OF DISTRIBUTION STORAGE

The location of distribution storage facilities is closely associated with the system hydraulics and water demands in various parts of the system. It is also affected by the availability of appropriate land and by public acceptance of the structure.

Elevated Storage

Some concerns linked to the location of elevated storage include the quality of the system hydraulics, pumping and transmission costs, and aesthetic considerations.

Relationship to system hydraulics

When elevated storage is placed adjacent to the pump station, as illustrated in Figure 3-7A, the head loss to the farthest portion of the system may be excessive through normal size piping. As a result, additional transmission mains may have to be added to the system to provide enough pressure to remote areas.

To avoid the cost of increased main size, the elevated tank can be located beyond the service area (Figure 3-7B), so that the pressure will be significantly improved with existing main size. In this situation, however, there must be adequate main capacity to the remote location to refill the tank during off-peak periods.

An even better solution is shown in Figure 3-7C. Here the tank is located in the part of the service area that originally had the lowest pressure. In this case, the pressure is slightly improved over the situation before the tank was installed. In addition, smaller-diameter mains can be connected to the tank because the flow from the tank is split into two directions.

It is often desirable to provide several smaller storage units in different parts of the system rather than one large tank of equal capacity at a central location. Smaller pipelines are used for decentralized storage. In addition, other things being equal, the smaller units don't need as high a flow-line elevation or pumping head. Of course, the cost of building and maintaining several smaller tanks must be compared with the cost of larger water mains that would accomplish the same objective.

An elevated tank should be located on the highest point of ground that also meets other hydraulic criteria. A tank at a higher elevation will not need to be as tall as a tank at a lower elevation. Eliminating 50 to 75 ft (15 to 23 m) in the height of a tank can provide a considerable cost savings.

Minimizing pumping and transmission costs

As illustrated in Figure 3-8, there can be a very substantial pressure loss when water is pumped over a long distance. The situation worsens drastically under high-demand conditions. The location of storage near the load center, as shown in Figure 3-8, would allow the pumping station to operate near average-day conditions most of the time.

A. Adjacent to Pump Station

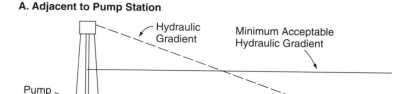

B. Beyond Service Area

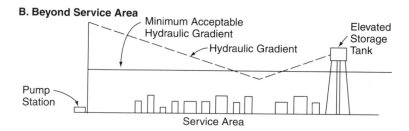

C. Adjacent to Area With Lowest Pressure

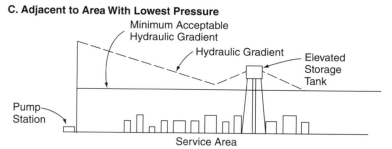

FIGURE 3-7 Different locations of elevated storage
Courtesy of Public Works Magazine

Aesthetic concerns

Although existing elevated tanks rarely bother anyone, property owners will usually object to a new tank being built near their homes. Although many designs are very pleasing and colors can be selected to minimize the visual impact, objections are usually still made. Another complaint is that an elevated tank might cause television interference.

As a result, the best hydraulic location and the most economical design are not always the deciding factors in the location of an elevated tank. In some cases, the only acceptable location will be in an industrial area or public park. In cases where public feeling is very strong, a water utility may have to construct ground-level storage, which is more aesthetically acceptable.

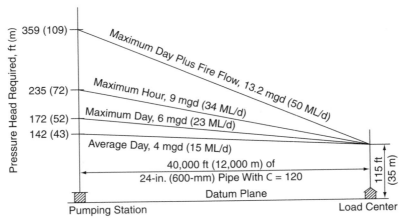

FIGURE 3-8 Sufficient pumping head must be provided when there is no storage

Ground-Level Storage

Ground-level storage is generally used where a large quantity of water must be stored or where topography or community objections do not permit the economical location of an elevated tank. A relatively large parcel of land is required to accommodate both the reservoir and an accompanying pump station. If the distribution system has several pressure zones, ground-level storage and booster pumping are often located at the pressure zone boundary. This way, water from the lower zone flows into the reservoir and is pumped to the zone with higher pressure.

WATER STORAGE FACILITY EQUIPMENT

This section discusses the types of equipment associated with elevated and ground-level storage facilities.

Elevated Storage Tanks

Elevated storage tanks (Figure 3-9) are often constructed of steel. The tanks contain a variety of equipment options including inlet and outlet pipes, overflow piping, drain connections, monitoring devices, valving, air vents, access hatches, ladders, protective coatings, cathodic protection, and obstruction lighting.

Inlet and outlet pipes

The same pipe is generally used as both the inlet and outlet pipe on an elevated tank. This pipe is called a riser. In cold climates, any exposed risers on multicolumn tanks are generally 6 ft (1.8 m) in diameter or larger to allow for some freezing around the edge and for expansion when water turns to ice. In extremely cold climates, an exposed riser may have

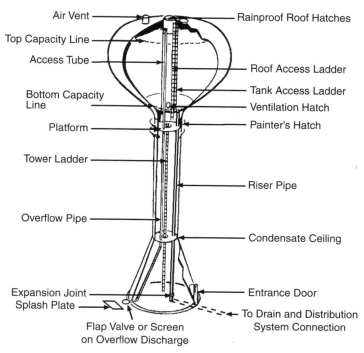

Air Vent
Top Capacity Line
Access Tube
Bottom Capacity Line
Platform
Tower Ladder
Overflow Pipe
Expansion Joint
Splash Plate

Rainproof Roof Hatches
Roof Access Ladder
Tank Access Ladder
Ventilation Hatch
Painter's Hatch
Riser Pipe
Condensate Ceiling
Entrance Door
To Drain and Distribution System Connection
Flap Valve or Screen on Overflow Discharge

FIGURE 3-9 Principal accessories for an elevated storage tank
Courtesy of CB&I

to be insulated or heated. An advantage of the single-pedestal support design is that heat can be provided within the support column to keep the temperature around the riser pipe above freezing. In locations where freezing temperatures do not occur, a small riser pipe with just enough capacity to carry maximum flow may be used.

Overflow pipe

An overflow pipe is necessary on all tanks to safeguard the tank in the event that the water-level controls fail. The overflow pipe should be brought down from the maximum tank level to a point within about 1 ft (0.3 m) of the ground surface. It should discharge to a splash plate or drainage inlet structure to prevent soil erosion at the tank foundation. The overflow pipe should never be directly connected to a sewer or storm drain.

The overflow pipe discharge should be covered by a weighted flap valve or screen to exclude bugs and animals, but it must positively open or break away so that the discharge is unobstructed if overflow takes place.

Drain connection

An elevated tank must be furnished with a drain connection to empty the tank for maintenance and inspection. A commonly used method is to install a fire hydrant on the main connection to the distribution system, with a valve on either side of the hydrant tee (Figure 3-10). Utility personnel can empty the tank by draining it through the hydrant after closing one of the valves. After draining is complete, the valve positions can be reversed so that the hydrant can be used for tank maintenance work.

If the water drained from a tank is discharged onto the ground, the utility must make provisions to carry away the considerable flow created without eroding or flooding the surrounding area. If the drained water is discharged to a storm sewer, the design must be such that a potential cross-connection is not created.

Monitoring devices

The water level in a tank can be measured either by a pressure sensor at the base of the tank or by a level sensor inside the tank. The level is then usually indicated on an instrument at the site, and the data transmitted to a central location. In many systems, water-level information is used to manually or automatically operate pumps that maintain system pressure. The level indicator is usually also furnished with alarms to alert the operator of an unusually high or low water level in the tank.

Valving

An elevated tank must be furnished with a valve at the connection to the distribution system. This valve is used to shut the tank off for maintenance and inspection. When a tank floats on the system, this valve is left open for water to pass in and out of the tank as pressure fluctuates.

If a tank is not tall enough to accept full system pressure without overflowing, an altitude valve is installed on the distribution system connection. This valve automatically shuts off flow to the tank when the water level approaches the overflow point. Flow out of the tank, however, is usually unrestricted.

Air vents

Air vents must be installed to allow air to enter and leave the tank as the water level falls and rises. If the vent capacity is insufficient to admit air when water is draining from the tank, the vacuum created could collapse the tank.

Vents should be screened to keep out birds and animals that might contaminate the water. A screen with ¼-in. (6-mm) mesh openings is required by most state regulations. Insects seldom contaminate stored water, and insect screening is not recommended because it may become clogged or covered with ice. Some vent designs have flap valves that will operate to relieve excess pressure or vacuum if the screen should become blocked.

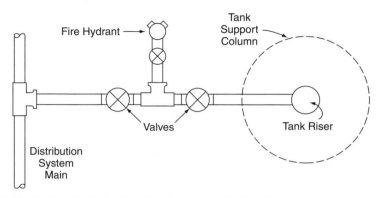

FIGURE 3-10 Fire hydrant installed for draining an elevated tank

Access hatches

Water storage tanks must have hatches installed for both entry and ventilation during maintenance and inspection. Hatches on the roof of a tank must be installed with rims under the cover. These rims are designed to prevent surface runoff from entering the tank.

The hatch at the bottom of a large-diameter wet riser must be designed to withstand the pressure of the water column. Some designs use a large exterior flange with multiple bolts. Others have a hatch with an interior cover that is held in place by a crossbar and large bolt. All hatches and access holes must be securely locked when not in use to prevent vandalism and reduce the security risk. Figure 3-11 illustrates how all access to a single-column tank is through a secure doorway.

Ladders

Multicolumn tanks generally have three ladders. The tower ladder runs up one leg of the tower from the ground to the balcony. The bottom of most ladders begins about 8 ft (2.5 m) from the ground to deter unauthorized persons from climbing the tank. On other tanks, the ladder is installed with the base at the ground surface but with a heavy metal shield locked in place over the bottom of the ladder to prevent unauthorized entry.

The second ladder, or tank ladder, runs from the balcony to the roof. The third ladder, called the roof ladder, is installed on the top of the roof from the tank ladder to the roof access hatch. Tanks located in warm climates may also have an inside ladder for the convenience of those working inside the tank. However, in climates where ice may form in the tank, an inside ladder cannot be permanently installed because it would be destroyed by the moving ice.

Pedestal tanks such as those illustrated in Figures 3-9 and 3-11 are furnished with similar ladders. One extends from the ground to the bottom of the tank, another goes up through the access tube to the top of the tank, and another is located inside the tank if icing is not a problem.

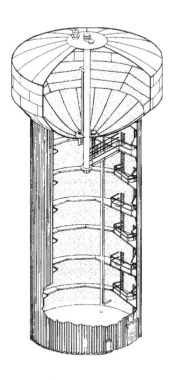

FIGURE 3-11 Exterior view and cut-away drawing of a fluted column elevated tank
Courtesy of CB&I

All ladders and safety devices including safety cage, safety cable, or safety rail, must meet Occupational Safety and Health Administration (OSHA) requirements. Ladders may also need to have rest platforms or roof-ladder handrails.

Coatings

Steel, when exposed to the environment, oxidizes and deteriorates, particularly when the environment includes both oxygen and moisture. It is therefore necessary to protect both the interior and exterior surfaces of steel tanks from corrosion. Several types of paints or coatings can be used for this purpose.

Interior coatings must be able to withstand the following conditions:

- Constant immersion in water
- Varying water temperatures
- Alternate wetting and drying periods
- Ice abrasion
- High humidity and heat
- Chlorine and mineral content in the water

The exterior tank coating must endure similar conditions. In addition, it must maintain a good appearance over a reasonable period of time. A recent concern that has emerged concerning older tanks is that many were originally coated with lead-based paint. If the old coating is to be removed by sand blasting, it is important to observe state and federal regulations that restrict the levels of both lead and silica that can be released into the atmosphere. Special blast abrasive and methods of containing the lead must be used if a lead coating is to be removed. In addition, at the conclusion of the work, the removed material must be tested. It may have to be disposed of as a potentially hazardous waste.

It is important that the new coating be effective and safe. It must not cause taste or odor problems. A reliable manufacturer should be consulted in selecting a coating system that meets NSF International standards (see appendix A).

Inspecting the work of a painting contractor is both dangerous and beyond the knowledge of most water system operators. It is generally a good idea to employ a qualified third-party firm to inspect the work. The inspectors selected should be trained and experienced in tank painting. They must be able to climb and use rigging to reach the work areas for inspection. Competing painting contractors should not be used as inspectors because of possible conflicts of interest.

Cathodic protection

In addition to a good inside coating system, cathodic protection can further reduce corrosion of the interior walls below the water line of steel tanks. Cathodic protection reverses the flow of current that tends to dissolve iron from the tank surface and cause rust and corrosion. If electrodes are placed in the water and a direct current is impressed on them, the electrodes will corrode (or "be sacrificed") instead of the tank.

In warm climates, the electrodes can be suspended from the tank roof and left in place until they have corroded to the point of requiring replacement (Figure 3-12). In climates where ice is formed in the tank, the ice will quickly pull down suspended electrodes. A submerged system can be installed that resists damage by icing (Figure 3-13). Although the anodes will generally last up to 10 years, it is recommended that cathodic protection systems be inspected annually to ensure they are operating properly.

Although cathodic protection systems can be operated on bare steel or in a tank with a badly deteriorated coating, a large amount of power will be used and the anodes will quickly

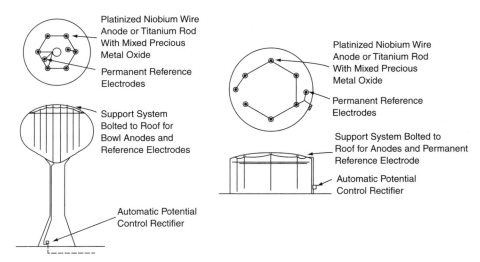

FIGURE 3-12 Typical methods of suspending cathodic protection anodes in steel tanks where icing conditions do not exist

Courtesy of Corrpro Waterworks, a Corrpro Company, 1055 West Smith Road, Medina, Ohio 44256

disintegrate. The usual approach is to install a good coating system and a cathodic system. This combination extends the useful life of the coating and anodes with minimal power consumption. Details on the installation and operation of cathodic protection for steel water tanks are covered in AWWA Standard D104, *Automatically Controlled, Impressed-Current Cathodic Protection for the Interior of Steel Water Tanks* (most recent edition).

Obstruction lighting
Depending on the height and location of an elevated tank, the Federal Aviation Administration (FAA) may require the installation of obstruction lights or strobe lights to warn aircraft in the vicinity. In particularly hazardous locations, orange and white obstruction painting may also be required. It is even possible that the FAA will not allow the construction of an elevated tank in some locations, so an application for FAA clearance should be made before a site location is finalized.

Ground-Level Storage Facilities
The equipment required for a ground-level storage facility includes inlet and outlet pipes; overflow pipes, vents, and hatches; drains; and corrosion protection.

Inlet and outlet pipes
Ground-level and buried storage tanks usually have a single pipe that serves as both inlet and outlet. Separate inlet and outlet pipes may be used if there is a need to improve the

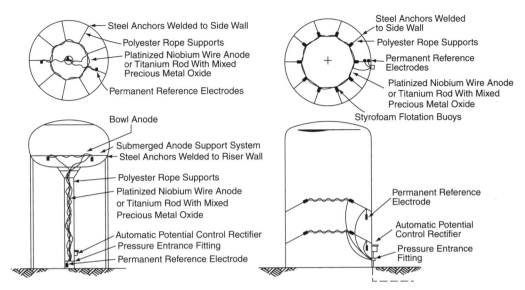

FIGURE 3-13 Typical methods of suspending cathodic protection anodes in steel tanks where icing conditions exist

Courtesy of Corrpro Waterworks, a Corrpro Company, 1055 West Smith Road, Medina, Ohio 44256

circulation of water within the tank. Improved circulation helps the quality of the water remain uniform and reduces freezing. The opening of the inlet–outlet pipe is usually located a short distance above the floor. Alternatively, a silt stop may be installed to keep sediment on the floor from being drawn into water leaving the tank.

Overflow pipes, vents, and hatches

The same general comments for overflows, vents, and hatches on elevated tanks also apply to ground-level tanks. However, for ground-level tanks, this equipment is generally even more accessible to vandals. In particular, if the area over or around a buried reservoir is used for a play area or other public use, all exposed equipment must be especially sturdy and designed with child safety in mind.

Drains

If the location and topography are appropriate, it is convenient to have a gravity drain on a ground-level tank. However, this should not be done if a possible cross-connection will be created. If a drain cannot be installed, and if there is a pump station at the tank, the usual procedure for emptying the tank is to pump most of the water to the system. It is not generally wise to pump all the way to the outlet opening because any sediment on the bottom will be disturbed and pumped into the system as cloudy water. The remaining water in the tank must then be removed by portable pumps.

Corrosion protection

Steel ground-level tanks should be painted and protected by cathodic protection in the same manner as elevated steel tanks. Aboveground steel tanks are generally set on a concrete ring-wall foundation. The section of steel wall immediately above the foundation is a particularly vulnerable point for rusting. Care should be taken to keep this area free of weeds and dirt all the way around the tank.

Concrete tanks are not generally painted, but occasionally there is a demand to paint the exterior of an aboveground tank for aesthetic reasons. An exterior coating may also be used for corrosion protection in areas near the ocean with salty air or where there are industrial gases in the air. The utility must take extreme care in selecting the proper time and correct painting system. If the paint does not stick properly and begins to flake or peel, removing the coating could be extremely costly. There may also be a need to coat the interior or to place a membrane liner in an older concrete tank. This serves as a means of protecting concrete that is disintegrating.

OPERATION AND MAINTENANCE OF WATER STORAGE FACILITIES

This section discusses some of the normal concerns in the daily operation of water storage facilities.

Cold-Weather Operation

In cold climates, freezing water in tanks can be a serious problem. Freezing is more likely in systems using surface water sources because the water entering the tank may, at times, be just above freezing temperature. Even though groundwater is usually somewhat warmer, icing problems in storage tanks can also occur during extremely cold weather.

If ice forms inside a tank, it can damage the paint, cathodic protection system, ladders, and overflow pipes that are exposed. If it is expected that ice may form in a tank, there should be nothing exposed in the interior that can be damaged.

Freezing in ground-level tanks is not usually serious because of the way in which the tanks are operated. As long as some water is pumped from the tank to the system and new, warmer water is filled back in every day, there may only be nominal freezing around the walls.

Multicolumn elevated tanks constructed in areas with freezing temperatures are furnished with greatly oversized risers. It is expected that some ice will form around the walls in very cold weather. The water circulating in and out of the tank will normally be able to pass through the riser. However, if the riser is allowed to freeze solid, the upper tank may then also freeze. If this happens, severe structural damage can result because water expands as it turns to ice.

It is impossible to keep ice from forming around the walls of an elevated tank under extremely cold conditions. This ice is not normally of concern. What is most important is to prevent thick ice from forming on the water surface. If ice is allowed to become so thick that it will resist the pressure of the water, the tank will become unusable. The tank could also be damaged by the expansion of the ice.

To keep ice from getting too thick, the utility must make a special effort in extremely cold weather to provide frequent changes in the water level. This will both discharge cold water and bring in warmer water. It will also break up any surface ice that has formed. If automatic controls normally maintain an almost constant water level in a tank, they should be overridden in very cold weather to provide pressure fluctuations. If a tank should become frozen, it is possible to use a steam generator or electric heaters to thaw it, but this process can be very costly.

Basic Maintenance

It is recommended that all storage structures be completely inspected and cleaned every 3 to 5 years.

Elevated tanks

Elevated tanks should periodically be drained, cleaned, inspected, repaired (if necessary), and painted. Because of the specialized nature of the work and the many dangers involved in working in high and confined places, this work should be performed by a competent contractor.

Ground-level tanks

Ground-level tanks can generally be drained, cleaned, and inspected by water system workers as long as careful attention is given to the dangers of working in confined spaces and other potentially dangerous conditions. The surfaces of the walls and floor should be cleaned thoroughly with a high-pressure water jet or by sweeping or scrubbing. All water and dirt must then be flushed from the tank. If workers notice pitting of steel tank walls or disintegration of concrete, it is usually best to obtain professional advice on how to prevent further damage.

If painting is necessary, it is extremely important that adequate measures be taken to exhaust the paint fumes. Paint fumes in a confined space can cause injury or even death.

Disinfection

After cleaning and/or painting are complete, water storage tanks must be disinfected before being placed in service. Either (1) liquid sodium hypochlorite solution or (2) calcium hypochlorite granules or tablets may be used. Alternate methods of chlorination are listed in AWWA Standard C652, *Disinfection of Water-Storage Facilities* (most recent edition). The state water supply agency should also be consulted for any specific requirements.

In the first method, the volume of the entire tank is chlorinated, so that the water will have a free chlorine residual of at least 10 mg/L after the proper detention time. The detention time is 6 hours if the disinfecting water is chlorinated before entering the tank, and 24 hours if the water is mixed with hypochlorite in the tank. The chlorine level must be reduced to acceptable levels before the water is used or discharged.

The second method involves spraying or painting all interior tank surfaces with a solution of 200 mg/L available chlorine. This procedure requires special precautions and should be done only by trained, experienced, and properly equipped personnel.

In the third method, 6 percent of the tank volume is filled with a solution of 50 mg/L available chlorine for at least 6 hours. Then the tank is completely filled and the solution held for 24 hours. An advantage of this method is that if the test results are satisfactory, the chlorine levels of the water are generally acceptable for immediate use of the water in the distribution system.

The amount of chlorine needed can be determined from Table 3-1. If the chlorine solution must be neutralized before being discharged, Table 3-2 indicates the amount of chemical required. Check with the local sewer department before discharging any highly chlorinated wastewater into a sanitary sewer. Consult with the state or local environmental agency (regulatory agency) before discharging highly chlorinated water anywhere. The agency may have special disposal requirements.

After the chlorination procedure is completed, and before the tank is placed in service, water from the full tank must be tested for bacteriological safety (coliform test). If the tests show the sample to be bacteriologically safe according to the requirements of the state regulatory agency, the tank can be placed in service. If the samples test unsafe, further disinfection must be performed and repeat samples taken until two consecutive samples test safe. Details of bacteriological testing are included in *Water Quality*, another book in this series.

Inspection

Water storage facilities must be inspected periodically to find any structural problems and correct them before they become serious. Cracks or holes in tanks can cause a loss of water, contamination of the water supply, and possible eventual total failure of the structure.

Tanks should be inspected for corrosion and cracks on both the inside and outside. This requires draining the tank to check the surfaces and the operation of the cathodic protection equipment. Overflows and vents should be examined to make sure that they are not blocked and that screens are clean and in place. AWWA Manual M42, *Steel Water Storage Tanks*, (most recent edition), includes maintenance procedure information.

If an altitude valve is used, it should be checked to make sure it is allowing water into the tank and stopping the flow when the tank is full. It should also be checked for speed of opening and closing to make sure it does not cause water hammer. Inspectors should check level sensors and pressure gauges by changing the water level in the tank. Controls should be examined to see if they are operating at the desired pressures.

TABLE 3-1 Amounts of chemicals required to give various chlorine concentrations in 100,000 gal (378.5 m³) of water*

Desired Chlorine Concentration in Water, mg/L	Liquid Chlorine Required		Sodium Hypochlorite Required						Calcium Hypochlorite Required	
			5% Available Chlorine		10% Available Chlorine		15% Available Chlorine		65% Available Chlorine	
	lb	(kg)	gal	(L)	gal	(L)	gal	(L)	lb	(kg)
2	1.7	(0.77)	3.9	(14.7)	2.0	(7.6)	1.3	(4.9)	2.6	(1.2)
10	8.3	(3.76)	19.4	(73.4)	9.9	(37.5)	6.7	(25.4)	12.8	(5.8)
50	42.0	(19.05)	97.0	(367.2)	49.6	(187.8)	33.4	(126.4)	64.0	(29.0)

*Amounts of sodium hypochlorite are based on concentrations of available chlorine by volume. For either sodium hypochlorite or calcium hypochlorite, extended or improper storage of chemicals may have caused a loss of available chlorine.

TABLE 3-2 Amounts of chemicals required to neutralize various residual chlorine concentrations in 100,000 gal (378.5 m³) of water

Residual Chlorine Concentration, mg/L	Chemical Required, lb (kg)			
	Sulfur Dioxide (SO_2)	Sodium Bisulfite ($NaHSO_3$)	Sodium Sulfite (Na_2SO_3)	Sodium Thiosulfate ($Na_2S_2O_3 \cdot 5H_2O$)
1	0.8 (0.36)	1.2 (0.54)	1.4 (0.64)	1.2 (0.54)
2	1.7 (0.77)	2.5 (1.13)	2.9 (1.32)	2.4 (1.09)
10	8.3 (3.76)	12.5 (5.67)	14.6 (6.62)	12.0 (5.44)
50	41.7 (18.91)	62.6 (28.39)	73.0 (33.11)	60.0 (27.22)

Water storage tanks should be checked frequently for vandalism and signs of forced entry. Access doors and their locks should be kept in good repair. Ladders must provide safe access for authorized personnel, but the entry should be locked to exclude unauthorized access. Security must be provided to prevent unauthorized access and protect the water from possible contamination.

Aviation warning lights, if installed, should be maintained so they will provide adequate warning to aircraft. Lights should be checked regularly to make sure they are working and clean enough that they do not have reduced light output. FAA regulations require that the bulbs be replaced before they reach 75 percent of their normal life expectancy.

Records

Most water systems have a relatively small number of distribution storage tanks, so the record-keeping system does not have to be extensive. Basic information to be recorded should include the tank location, the dates of inspection, conditions noted during the inspection, the dates maintenance was performed, and notes on the maintenance performed.

If maintenance is performed by outside contractors, a copy of the contracts, reports, and lists of repairs that were made should be maintained in a separate file for each structure. The phone numbers of equipment suppliers and contractors equipped to make repairs should also be kept available for use in an emergency.

WATER STORAGE FACILITY SAFETY

Only trained and experienced operators should be allowed to work on elevated and ground-level storage tanks and standpipes. This is hazardous work and dangerous for untrained workers. Special precautions are also needed for work on or in tanks. In these confined working areas, workers must guard against slipping and falling from dangerous heights.

In addition to the safety precautions listed in chapter 13, the following guidelines apply to work in and around storage facilities:

- The security of ladders must be checked frequently. Required safety cages or safety cable equipment must be provided.

- Workers must be provided with boots and clothing for working in wet and slippery conditions.

- Workers performing disinfection must be provided with special protective goggles and gloves.

- Special fans or other ventilation equipment must be provided inside tanks while work is being done there.

- Adequate light must be provided inside tanks for workers to perform their work properly and safely. Special care must be taken to use waterproof wiring and light units to prevent shocks in a wet environment.

BIBLIOGRAPHY

ANSI/NSF Standard 61: Drinking Water System Components—Health Effects. Ann Arbor, Mich.: NSF International.

AWWA Standard for Automatically Controlled, Impressed-Current Cathodic Protection for the Interior of Steel Water Tanks. ANSI/AWWA D104. Denver, Colo.: American Water Works Association (latest edition).

AWWA Standard for Coating Steel Water-Storage Tanks. ANSI/AWWA D102. Denver, Colo.: American Water Works Association (latest edition).

AWWA Standard for Disinfection of Water-Storage Facilities. ANSI/AWWA C652. Denver, Colo.: American Water Works Association (latest edition).

AWWA Standard for Factory-Coated Bolted Carbon Steel Tanks for Water Storage. ANSI/AWWA D103. Denver, Colo.: American Water Works Association (latest edition).

AWWA Standard for Flexible-Membrane Materials for Potable Water Applications. ANSI/AWWA D130. Denver, Colo.: American Water Works Association (latest edition).

AWWA Standard for Tendon-Prestressed Concrete Water Tanks. ANSI/AWWA D115. Denver, Colo.: American Water Works Association (latest edition).

AWWA Standard for Thermosetting Fiberglass-Reinforced Plastic Tanks. ANSI/AWWA D120. Denver, Colo.: American Water Works Association (latest edition).

AWWA Standard for Welded Carbon Steel Tanks for Water Storage. ANSI/AWWA D100. Denver, Colo.: American Water Works Association (latest edition).

AWWA Standard for Wire- and Strand-Wound, Circular, Prestressed Concrete Water Tanks. ANSI/AWWA D110. Denver, Colo.: American Water Works Association (latest edition).

Distribution System Maintenance Techniques. 1987. Denver, Colo.: American Water Works Association.

Elevated Water Storage Tanks: Maintenance. 2001. Video. Denver, Colo.: American Water Works Association.

Elevated Water Storage Tanks: Safety and Security. 2001. Video. Denver, Colo.: American Water Works Association.

Manual M3, Safety Practices for Water Utilities. 2002. Denver, Colo.: American Water Works Association.

Manual M25, Flexible-Membrane Covers and Linings for Potable-Water Reservoirs. 2000. Denver, Colo.: American Water Works Association.

Mays, L.W. 1999. *Water Distribution Systems Handbook.* New York: McGraw-Hill.

Salfvato, J.A. 1992. *Environmental Engineering and Sanitation,* 4th ed. New York: John Wiley & Sons.

Water Distribution Operator Training Handbook. 2005. Denver, Colo.: American Water Works Association.

CHAPTER 4

Pumps and Pumping Stations

Pumping facilities are required wherever gravity cannot be used to supply water to the distribution system under sufficient pressure to meet all service demands. When used, pumping accounts for most of the energy consumed in water supply operations.

Pumping equipment also represents a major part of a utility's investment in equipment and machinery. If properly operated and maintained, quality pumps, electric motors, and engines can give decades of efficient and reliable service. Typical uses of pumps in a water system are listed in Table 4-1 and diagrammed in Figure 4-1.

PUMP STATIONS

Pump stations that supply water to the distribution system are referred to as main pump stations. They are located near a treatment facility or a potable water storage facility, and they pump directly into the piping system. High lift pumps are used to pump directly into transmission lines. Booster pumps are used to increase the pressure in the pipelines and may be located anywhere in the water distribution system. Booster pump stations typically are located remotely from the main pump station. One example is in hilly topography where pressure zones are required. Booster pump stations may be needed for only handling peak flows. Generally there are two types of pumps used for potable water pumping applications. These are the vertical turbine type (line-shaft and submersible) and the centrifugal horizontal or vertical split case pump (ANSI/AWWA Standards E102 and E103). Another excellent reference on pumping station design is by Jones (2008).

TABLE 4-1 Pump applications in water systems

Application	Function	Pump Type
High service	Discharge water under pressure to distribution system	Centrifugal
Booster	Increase pressure in the distribution system or to supply elevated storage tanks	Centrifugal
Well	Lift water from shallow or deep wells and discharge it to the treatment plant, storage facility, or distribution system	Centrifugal or ejector (jet)

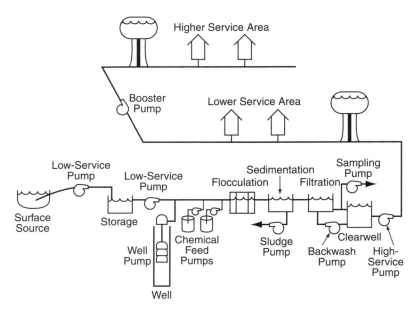

FIGURE 4-1 Uses of pumps in a water system

Booster Pump Stations

Booster pumps may be located above or below ground as shown in the piping schematics in Figure 4-2. The schematic in Figure 4-2A shows a booster pump station with a centrifugal pump housed in a building above ground. Figure 4-2B shows an in-line booster pump that could be an underground horizontal submersible turbine pump or a centrifugal pump in a vault. Note the valve arrangement with two gate valves and a check valve. The gate valves would be used to isolate the pump for repairs, maintenance, or replacement. The check valve is to control the direction of flow.

Pump Station Supplying Distribution System

Figure 4-3 shows pump applications that are supplying the distribution system. Figure 4-3A is an application that uses an underground reservoir and pump station with a vertical turbine pump with the supply from a well or treatment plant. Figure 4-3B shows the application of a ground storage tank with centrifugal pumps in the pump station.

Pump Station Layout

A typical pump station layout is shown in Figure 4-4, and the pumping station sections are shown in Figure 4-5. Notice in Figure 4-4 that the piping layout includes suction piping (for water entering the pump station on the suction side of the pumps) and discharge

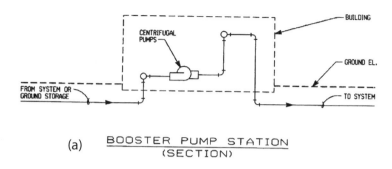

(a)   BOOSTER PUMP STATION
(SECTION)

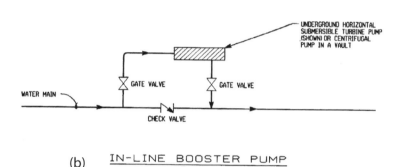

(b)   IN-LINE BOOSTER PUMP
(PLAN)

FIGURE 4-2 Booster pump stations (Department of Army)

piping (for water leaving the station on the discharge side of the pumps). The design of the suction piping is important to minimize pressure losses and allow sufficient flow into the pump. The suction piping must not have air leaks. For larger-size suction piping the pipe joints are flanged, and for smaller pipes the joints are screwed or flanged.

Pumping stations also use various types of valves in the piping system. These include gate valves, globe valves, angle valves, cone valves, butterfly valves, ball valves, check valves, and relief valves. For flow modulation to provide desired flows and/or pressures, globe, ball, cone, and butterfly valves are best suited. Gate valves are used in the suction piping in order to isolate the pump, with the valve stem installed horizontally to avoid air pockets. In the discharge piping for each pump, both a check valve and a gate valve (or butterfly valve) are used for each pump as shown in Figure 4-4. A check valve is located between the pump and the gate valve. Check valves protect the pump from excessive back-pressure and from backflows in the case of power failure. The gate valve is used to isolate the pump and check valve for maintenance. Also shown in Figure 4-5 are surge (pressure)

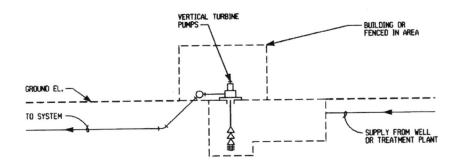

(a) <u>UNDERGROUND RESERVOIR AND PUMP STATION</u>
(SECTION)

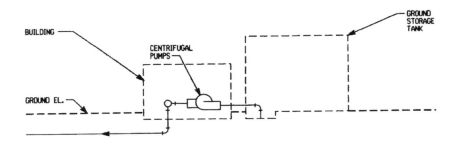

(b) <u>GROUND STORAGE TANK AND PUMP STATION</u>
(SECTION)

FIGURE 4-3 Alternative pump applications (Department of Army)

relief valves in the discharge piping for the purpose of protecting the pump and piping system from excessive surge pressures. They are also used for flow control and/or pressure regulation.

Pump Performance Curves for Pump Stations

Basic Science Concepts and Applications presents a section on pump curves, which are graphs showing the characteristics of a particular pump. They are commonly referred to as pump characteristic curves. The four characteristics are discharge (capacity), head,

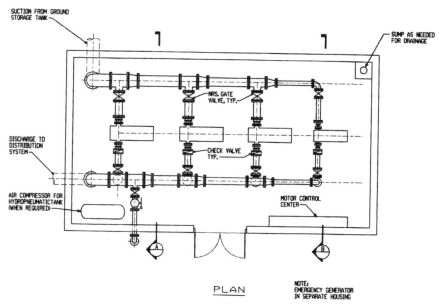

FIGURE 4-4 Typical layout of a pumping station (Department of Army)

power, and efficiency, which are related to each other, allowing them to be plotted on the same graph. Three curves are used that relate head to discharge (capacity), efficiency to discharge, and power to discharge. In pump stations where there are pumps placed in parallel as discussed above, the pump curves for two or more pumps operated in parallel can also be determined.

The design of a pump station requires the determination of the head requirements, which can be expressed as a system head curve, including the static or pressure difference and the system friction (head loss). In pumping from one location of the piping system to another, there is a friction (head loss) through the piping system and possibly an elevation or pressure difference. A simplified head curve in Figure 4-6 illustrates the concept of a system head curve.

In a simple pump station with two identical constant speed pumps operating in parallel, the pump curves are shown in Figure 4-7. The head-discharge curve (pump performance curve) with one pump operating and the head-discharge curve with two pumps operating are shown. The curve with two pumps operating is determined by adding the capacity of both pumps for each respective head. Also shown in Figure 4-7 is a system head curve. The point of intersection of the pump performance and the system head curve represents the discharge at which the pump operates. For the operation of one pump, the discharge is 60 percent of the rated station capacity, and for two pumps the discharge is 100 percent of the rated capacity.

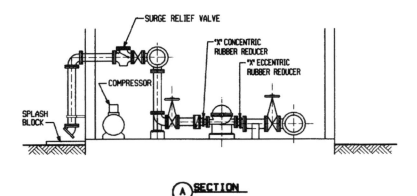

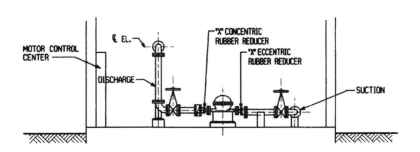

FIGURE 4-5 Pumping station sections (Department of Army)

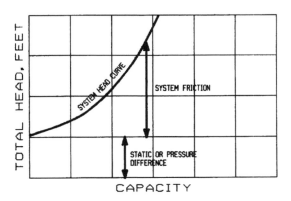

FIGURE 4-6 System head curve (Department of Army)

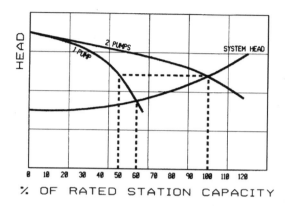

FIGURE 4-7 Operation of two constant speed pumps operating in parallel (Department of Army)

TYPES OF PUMPS

Two basic categories of pumps are used in water supply operations: velocity pumps and positive-displacement pumps. Velocity pumps, which include centrifugal and vertical turbine pumps, are used for most distribution system applications. Positive-displacement pumps are most commonly used in water treatment plants for chemical metering.

Velocity Pumps

Velocity pumps use a spinning impeller, or propeller, to accelerate water to high velocity within the pump casing. The high-velocity, low-pressure water leaving the impeller can then be converted to high-pressure, low-velocity water if the casing is shaped so that water moves through an area of increasing cross section. This increasing cross-sectional area may be achieved in two ways:

1. The volute (expanding spiral) casing shape, as in the common centrifugal pump (Figure 4-8A), may be used.
2. Specially shaped diffuser vanes or channels may be used, such as those built into the bowls of vertical turbine pumps (Figure 4-8B).

Velocity pump design characteristics

Two designs of velocity pumps are widely used in water systems: centrifugal pumps (volute pumps) and turbine pumps. A feature distinguishing velocity pumps from positive-displacement pumps is that velocity pumps will continue to operate undamaged, at least for a short period, when the discharge is blocked. When this happens, a head builds up that is typically greater than the pressure generated during pumping, and water recirculates within the pump impeller and casing. This flow condition is referred to as *slip*.

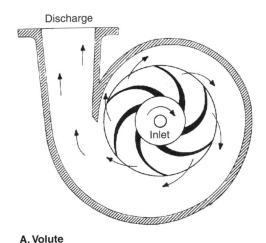

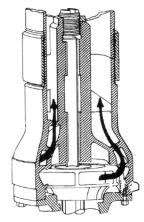

A. Volute   **B. Duffuser**

FIGURE 4-8 Centrifugal pump casings
This material is used with permission of John Wiley & Sons, Inc. from Centrifugal Pump Design, John Tuzson, ©2000 by John Wiley & Sons, Inc.

Depending on the casing shape, impeller design, and direction of flow within the pump, velocity pumps can be manufactured with a variety of operating characteristics.

Radial-flow designs. In the radial-flow (or centrifugal) pump, shown in Figure 4-8, water is thrown outward from the center of the impeller into the volute or diffusers that convert the velocity to pressure. The type of centrifugal pump commonly used in water supply practice is a radial-flow, volute-case type. A cutaway of a typical single-stage pump is shown in Figure 4-9. Centrifugal pumps of this type generally develop very high heads and have correspondingly low-flow capacities.

In general, any centrifugal pump can be designed with a multistage configuration. Each stage requires an additional impeller and casing chamber in order to develop increased pressure, which adds to the pressure developed in the preceding stage.

Although the pressure increases with each stage, the flow capacity of the pump does not increase beyond that of the first stage. There is no theoretical limit to the number of stages that are possible. However, mechanical considerations such as casing strength, packing leakage, and input power requirements do impose practical limitations.

Axial-flow designs. The axial-flow pump, shown in Figure 4-10, is often referred to as a propeller pump. It has neither a volute nor diffuser vanes. A propeller-shaped impeller adds head by the lifting action of the vanes on the water. As a result, the water moves parallel to the axis of the pump rather than being thrown outward as with a radial-flow pump. Axial-flow pumps handle very high volume but add limited head. Pumps of this design must have the impeller submerged at all times because they are not self-priming.

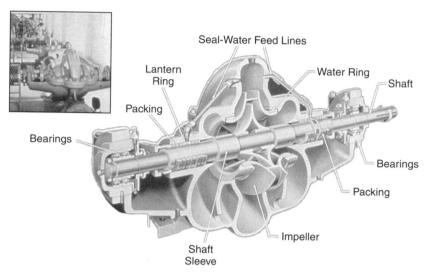

FIGURE 4-9 Cutaway of a single-stage pump

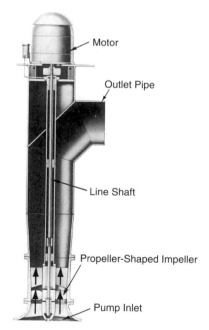

FIGURE 4-10 Axial-flow pump
Courtesy of Ingersoll-Dresser Pump Company

Mixed-flow designs. The mixed-flow pump, illustrated in Figure 4-11, is a compromise in features between radial-flow and axial-flow pumps. The impeller is shaped so that centrifugal force will impart some radial component to the flow. This type of pump is useful for moving water that contains solids, as in raw-water intakes.

Centrifugal pumps

The volute-casing type of centrifugal pump, shown in Figure 4-12, is used in most water utility installations. A wide range of flows and pressures can be achieved by varying the width, shape, and size of the impeller, as well as by varying the clearance between the impeller and casing. The pumps can develop a head up to 250 ft (76 m) per stage and efficiencies up to 75 or 85 percent.

Initial cost is relatively low for a given pump size, and relatively little maintenance is required. However, periodic checks are advised to monitor impeller wear and packing condition.

Advantages and disadvantages vary with the type of centrifugal pump used. Some of the advantages are

- wide range of capacities (Available capacities range from a few gpm to 50,000 gpm [190,000 L/min]. Heads of 5–700 ft [1.5–210 m] are generally available);
- uniform flow at constant speed and head;
- simple construction;
- small amounts of suspended matter in the water will not jam the pump;
- low to moderate initial cost for a given size;
- ability to adapt to several drive types—motor, engine, or turbine;
- moderate to high efficiency at optimal operation;
- no need for internal lubrication;
- little space required for a given capacity;
- relatively low noise level; and
- ability to operate against a closed discharge valve for short periods without damage.

Some of the disadvantages are

- an efficiency that is at best limited to a narrow range of discharge flows and heads,
- flow capacity that is greatly dependent on discharge pressure,
- generally no self-priming ability,
- potential for running backward if stopped with the discharge valve open, and
- potential for impeller to be damaged by abrasive matter in water or become clogged by large quantities of particulate matter.

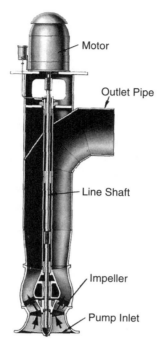

Motor

Outlet Pipe

Line Shaft

Impeller

Pump Inlet

FIGURE 4-11 Mixed-flow pump
Courtesy of Ingersoll-Dresser Pump Company

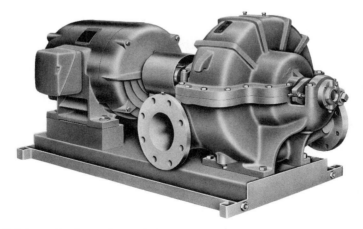

FIGURE 4-12 Volute-casing type of centrifugal pump

Vertical turbine pumps

Vertical turbine pumps have an impeller rotating in a channel of constant cross-sectional area, which imparts mixed or radial flow to the water. As liquid leaves the impeller (Figure 4-13), velocity head is converted to pressure head by diffuser guide vanes. The guide vanes form channels that direct the flow either into the discharge or through diffuser bowls into succeeding stage inlets.

Turbine pumps are manufactured in a wide range of sizes and designs, combining efficiency with high speeds to create the highest heads obtainable from velocity pumps. The clearance between the diffuser and the impeller is usually very small, limiting or preventing internal backflow and improving efficiency. Efficiencies in the range of 90–95 percent are possible for large units. However, the closely fitting impeller prohibits pumping of any solid sediment, such as sand, fine grit, or silt. Turbine pumps have a higher initial cost and are more expensive to maintain than centrifugal volute pumps of the same capacity.

The major advantages of turbine pumps are

- uniform flow at constant speed and head;
- simple construction;
- individual stages capable of being connected in series, thereby offering multiple head capacities for a single pump model;
- adaptability to several drive types—motor, engine, or turbine;
- moderate to high efficiency under the proper head conditions;
- little space occupied for a given capacity; and
- low noise level.

The main disadvantages are

- high initial cost,
- high repair costs,
- the need to lubricate support bearings located within the casing,
- inability to pump water containing any suspended matter, and
- an efficiency that is at best limited to a very narrow range of discharge flow and head conditions.

Deep-well pumps. For deep-well service, a shaft-type vertical turbine pump requires a lengthy pipe column housing, a drive unit, a drive shaft, and multiple pump stages. In this type of pump, a drive unit is located at the surface, with the lower shaft, impeller, and diffuser bowls submerged (Figure 4-14). This type of pump requires careful installation to ensure proper alignment of all shafting and impeller stages throughout its length. Deep-well turbines have been installed in wells with lifts of over 2,000 ft (610 m).

FIGURE 4-13 Turbine impeller
Courtesy of Ingersoll-Dresser Pump Company

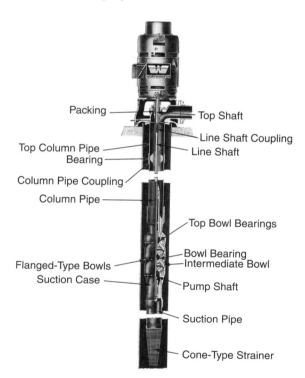

FIGURE 4-14 Deep-well pump
Courtesy of Ingersoll-Dresser Pump Company

Submersible pumps. Multistage mixed-flow centrifugal pumps or turbine pumps with an integral or close-connected motor may be designed for operation while completely submerged, in which case they are termed submersible pumps. As shown in Figure 4-15, the entire pump and motor unit is placed below the water level in a well.

Booster pumps. Vertical turbine pumps are often used for in-line booster service to increase pressure in a distribution system. The unit is actually a turbine pump that has the motor and pumps mounted close together and is installed in a sump. As shown in Figure 4-16, this type of unit is commonly called a "can" pump. The sump receives fluid and maintains an adequate level above the turbine pump suction.

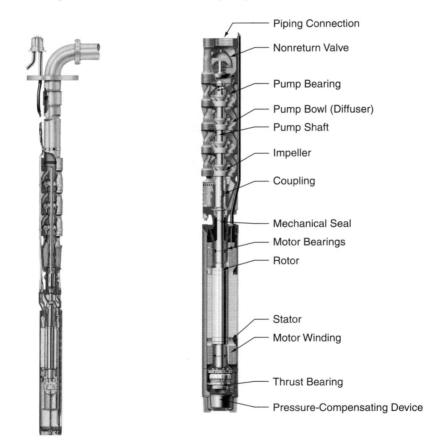

FIGURE 4-15 Vertical turbine pump driven by a submersible motor (left), Cross-sectional view of a submersible pump (right)
Images provided courtesy of Flowserve Corporation

Centrifugal-jet pump combination

Figure 4-17 illustrates a centrifugal-jet pump combination at the ground surface that generates high-velocity water that is directed down the well to an ejector. Jet pumps are widely used for small, private wells because of their low initial cost and low maintenance. They are rarely used for public water systems because of their relatively low efficiency.

Positive-Displacement Pumps

Early water systems used reciprocating positive-displacement pumps powered by steam engines to obtain the pressure needed to supply water to customers. These pumps have essentially all been replaced with centrifugal pumps, which are much more efficient. The only types of positive-displacement pumps used in current water systems are some types of portable pumps used to dewater excavations, as well as chemical feed pumps.

Reciprocating pumps

As illustrated in Figure 4-18, reciprocating pumps have a piston that moves back and forth in a cylinder. The liquid is admitted and discharged through check valves. Flow from reciprocating pumps generally pulsates, but this can be minimized by the use of multiple cylinders or pulsation dampeners. Reciprocating pumps are particularly suited for applications where very high pressures are required or where abrasive or viscous liquids must be pumped.

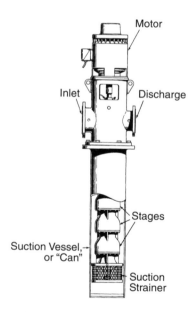

FIGURE 4-16 Turbine booster pump

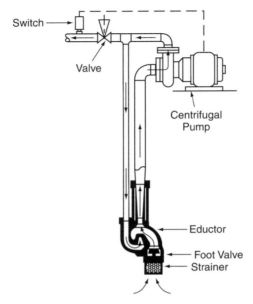

FIGURE 4-17 Centrifugal-jet pump combination
Reproduced with permission of McGraw-Hill Companies from Pump Handbook by Karassik et al., 2001.
Published by McGraw-Hill.

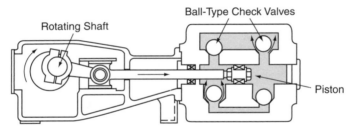

FIGURE 4-18 Double-acting reciprocating pump
Courtesy of the Hydraulic Institute

Rotary pumps

Rotary pumps use closely meshed gears, vanes, or lobes rotating within a close-fitting chamber. The two most common types, which use gears or lobes, are shown in Figure 4-19.

OPERATION OF CENTRIFUGAL PUMPS

The procedures for centrifugal pump operation vary somewhat from one brand of pump to another. The manufacturer's specific recommendations should be consulted before

Gear Type

Lobe Type

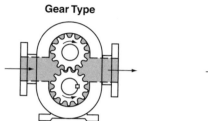
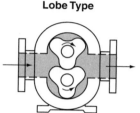

FIGURE 4-19 Rotary pumps
Courtesy of the Hydraulic Institute

operating any unit. The procedures described in this chapter are typical and will serve as a guide if manufacturer's instructions are not available.

Additional details on the theory of pump operations—including pumping rates, pump heads, horsepower and efficiency, and reading pump curves—are contained in *Basic Science Concepts and Applications*, another book in this series.

Pump Starting and Stopping

A major consideration in starting and stopping large pumps is the prevention of excessive surges and water hammer in the distribution system. Large pump-and-motor units have precisely controlled automatic operating sequences to ensure that the flow of water starts and stops smoothly. General procedures for starting pumps are listed below.

- Check pump lubrication.
- Prime the pump:

    - Where head exists on the suction side of the pump, open valve on the suction line and allow any air to escape from the pump casing through the air cocks.
    - Where no head exists on the suction side and a foot valve is provided on the suction line, fill the case and suction pipe with water from any source, usually the discharge line.
    - Where no head exists on the suction side and no foot valve is provided, the pump must be primed by a vacuum pump or ejector operated with steam, air, or water.

- After priming, start the pump with the discharge valve closed.
- When the motor reaches full speed, open the discharge valve slowly to obtain the required flow. To avoid water hammer, do not open the valve suddenly.
- Avoid throttling the discharge valve; this wastes energy.
- Before shutting down the pump, close the discharge valve slowly to prevent water hammer in the system.

Pump starting

Centrifugal pumps do not generate any suction when dry, so the impeller must be submerged in water for the pump to start operating. If a pump is located above water level, a foot valve is often provided on the suction piping to hold the pump's prime (i.e., to keep some startup water in the pump). The foot valve, which is a type of check valve, prevents water from draining out of the pump when the pump is shut down.

Pump prime can also be maintained by placing a vacuum connection connected to both the pump suction and the high point of the pump, as illustrated in Figure 4-20. The priming valve automatically removes any air that accumulates and keeps the pump completely full of water at all times. (See the details of an air-and-vacuum relief valve in chapter 6, Figure 6-15.)

Controlling water hammer is important when a pump is being started. Large pumps are furnished with a valve on the discharge that is opened slowly after the pump gets up to speed. As a result, the surge of water does not produce a serious shock in the distribution system.

Pump stopping

A check valve is usually installed in the discharge piping of small pumps to stop flow immediately after the pump stops. This will prevent reverse flow through the pump. However, the sudden shutdown of a pump may cause water hammer. Relief valves or surge chambers may be installed to absorb the pressure shock.

On large pumps, smooth shutdown is ensured by closing the discharge valve slowly while the pump is still running and then shutting off the pump just as the valve finally closes. In this manner, the pumping unit is eased off the system. Some form of power-activated

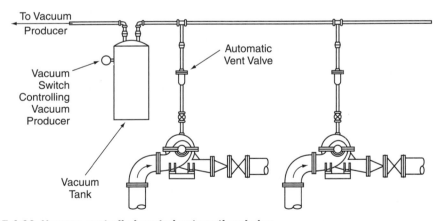

FIGURE 4-20 Vacuum-controlled central automatic priming
Reproduced with permission of McGraw-Hill Companies from Pump Handbook by Karassik et al., 2001. Published by McGraw-Hill.

valve is necessary to obtain slow valve closure. Figure 4-21 shows power-operated discharge valves installed on large pumps. The valve operators are equipped with handwheels for manual operation in the event of a power failure.

Whenever a power failure occurs, the motor will stop while the discharge valve is still open, so there must be a way for the valve to close very rapidly before the pump reverses itself and begins to run backward. Battery power or an emergency hydraulic system is usually provided to operate the valves in an emergency.

Whenever a pump must be shut down for more than a short period in freezing weather, the pump and exposed suction and discharge piping must be drained of water to prevent freezing. If the pump will be out of service for an extended period of time, the pump and motor bearings should be flushed and regreased, and packing should be removed from the stuffing box. The units should also be covered to prevent moisture damage to the motor windings and bearings.

Flow control

Pumps are usually operated at constant speed. System pressure is controlled by having various size pumps start or stop as necessary. Throttling the discharge valve or using variable-speed motors or pump drives are other ways to control the flow rate.

A major disadvantage of cycling pumps on and off as a means of controlling output is excessive motor wear. Medium-size motors should not be cycled (i.e., started and stopped) more frequently than every 15 minutes, and larger motors should be cycled even less frequently. Frequent starting also increases power costs. Frequent cycling of pumps is an indication that the system probably does not have adequate distribution system storage.

FIGURE 4-21 Pump discharge valves
Courtesy of Henry Pratt Company

Throttling the discharge valve in an attempt to approximate the required system flow should be done only when elevated storage is not available or when other, smaller pumping units are out of operation. In general, throttling should be avoided because it wastes energy. It is also necessary to make sure that the valves used for throttling are appropriate for this purpose. Gate valves should not be used for throttling because the gate is loose in its guides and will vibrate when it is not fully open or shut. The best valves for throttling are plug, ball, self-actuating, or altitude valves. Butterfly valves can be used for throttling for short periods of time, but extended use may damage them.

If the system design requires continually varying pump discharge rates, variable-speed drives should be provided. Numerous variable-speed package drives are on the market, including continuously variable and stepped-speed motors, as well as constant-speed motors driving variable-speed electrical, hydraulic, and mechanical speed reducers coupled to the pump. Pumps can also be driven by variable-speed motors, which have either variable-voltage or variable-frequency controls.

Monitoring Operational Variables

A primary requirement at every pumping station is to measure the amount of water pumped and provide a record of water delivered to the system. It is also usually necessary to monitor pressure in the system and elevated tank levels as a way to control pump operation. Pumping station production records also provide the basis for the plant maintenance schedule. Past records are usually reviewed to determine the need for equipment replacement.

Suction and discharge heads

Pressure gauges should be connected to both the suction and discharge sides of a pump at the pressure taps supplied on the pump. The gauges should be mounted in a convenient location so the operator can frequently check pump performance. The pressure readings can also be electronically transmitted to a control room.

Bearing and motor temperature

The most common way to check bearing and motor temperatures in a small- to medium-size plant is by feel. Experienced operators check pump operation by putting a hand on the motor and the pump bearing surfaces. They know how warm the surfaces should be. If a surface is substantially hotter than normal, the unit should be shut down and the cause of excessive heat investigated.

Special thermometers or temperature indicators are also available for monitoring the temperature at critical points in the pump and motor. It is particularly wise to have these monitors installed on equipment at unattended pump stations. The monitors will sound an alarm and automatically shut down the unit if the temperature gets too high.

Vibration

As with temperature, experienced operators get to know the normal feel and sound of each pump unit. They should investigate any change they notice. Vibration detectors are sometimes used on large pump and motor installations to sense equipment malfunctions, such as misalignment and bearing failure, that will cause excessive vibration. The detectors can also be used to shut down the unit if vibration increases beyond a preset level.

Speed

Monitoring the pump speed of variable-speed pumps is important because these pumps may experience cavitation (the creation of vapor bubbles) at low speeds. Centrifugal-speed switches can be installed on the pumping unit or contacts can be provided on a mechanical speed-indicating instrument to sound alarms or shut off the system if the speed goes too high or too low. Other systems use a tachometer generator that generates a voltage in proportion to speed. This voltage is used to drive a standard indicator near the pump or at a remote location. Underspeed and overspeed alarms can be activated by the speed-sensing device.

General observations

An operator should also monitor surge-tank air levels, recording meters, and intake-pipe screens. Pumps with packing seals should be adjusted so that there is always a small drip of water leaking around the pump shaft. Idle pumps should be started and run weekly. All operations and maintenance should be recorded in log books.

Finally, operators must remain attentive to the general condition of the pump on a day-to-day basis. Unusual noises, vibrations, excessive seal leakage, hot bearings or packing, or overloaded electric motors are all readily apparent to the alert operator who is familiar with the normal sound, smell, sight, and feel of the pump station. Reporting and acting on such problems immediately can prevent major damage that might occur if the problem were allowed to remain until the next scheduled maintenance check.

CENTRIFUGAL PUMP MAINTENANCE

A regular inspection and maintenance program is important in maintaining the condition and reliability of centrifugal pumps. Bearings, seals, and other parts all require regular adjustment or replacement because of normal wear. General housekeeping is also important in prolonging equipment life.

Mechanical Details of Centrifugal Pumps

Size and construction may vary greatly from one volute-type centrifugal pump to another, depending on the operating head and discharge conditions for which the pumps are designed. However, the basic operating principle is the same. Water enters the impeller eye from the pump suction inlet. There it is picked up by curved vanes, which change the flow direction from axial to radial. Both pressure and velocity increase as the water is impelled outward and discharged into the pump casing. The major components of a typical volute-type centrifugal pump are described in the following paragraphs.

Casing

Water leaving the pump impeller travels at high velocity in both radial and circular directions. To minimize energy losses due to turbulence and friction, the casing is designed to convert the velocity energy to additional pressure energy as smoothly as possible. In most water utility pumps, the casing is cast in the form of a smooth volute, or spiral, around the impeller. Casings are usually made of cast iron, but ductile iron, bronze, and steel are usually available on special order.

Single-suction pumps

Single-suction pumps are designed with the water inlet opening at one end of the pump and the discharge opening placed at right angles on one side of the casing. Single-suction pumps, also called end-suction pumps, are used in smaller water systems that do not have a high volume requirement. These pumps are capable of delivering up to 200 psi (1,400 kPa) pressure if necessary, but for most applications they are usually sized to produce 100 psi (700 kPa) or less.

The impeller on some single-suction pump units is mounted on the shaft of the motor that drives the pump, with the motor bearings supporting the impeller (Figure 4-22A). This arrangement is called the close-coupled design. Single-suction pumps are also available with the impeller mounted on a separate shaft, which is connected to the motor with a coupling (Figure 4-22B). In this design, known as the frame-mounted design, the impeller shaft is supported by bearings placed in a separate housing, independent of the pump housing.

The casing for a single-suction pump is manufactured in two or three sections or pieces. All housings are made with a removable inlet-side plate or cover, held in place by a row of bolts located near the outer edge of the volute. Removing the side plate provides access to the impeller. The pump does not have to be removed from its base in order to have the side plate removed. However, all suction piping must be removed to provide sufficient access.

Some manufacturers cast the volute and the back of the pump as a single unit. Other manufacturers cast them as two separate pieces, which are connected by a row of bolts, similar to the inlet side plate. In units with separate backs, the impeller and drive unit can be removed from the pump without having to disturb any piping connections.

Double-suction pumps

Water enters the impeller of a double-suction pump from two sides and discharges outward from the middle of the pump. Although water enters the impeller from each side, it enters the housing at one location (usually on the opposite side of the discharge opening).

Internal passages in the pump guide the water to the impeller suction and control the discharge water flow.

The double-suction pump is easily identified because of its casing shape (Figure 4-23). The motor is connected to the pump through a coupling, and the pump shaft is supported by ball or roller bearings mounted external to the pump casing.

The double-suction pump is usually referred to as a horizontal split-case pump. The term *horizontal* does not indicate the position of the pump. It refers to the fact that the housing is split into two halves (top and bottom) along the center line of the pump shaft, which is normally set in the horizontal position. However, some horizontal split-case pumps are designed to be mounted with the drive shaft in a vertical position, with the drive motor placed on top. Double-suction pumps can pump over 10,000 gpm (38,000 L/min), with heads up to 350 ft (100 m). They are widely used in large systems.

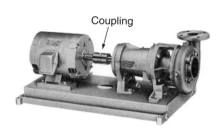

A. Close-Coupled Pump   B. Frame-Mounted Pump

FIGURE 4-22 Single-suction pumps

FIGURE 4-23 Double-suction pump casing shape
Courtesy of Ingersoll-Dresser Pump Company

Removing the bolts that hold the two halves of the double-suction casing together makes it possible to remove the casing's top half. Most manufacturers place two dowel pins in the bottom half of the casing to ensure proper alignment between the halves when they are reassembled. It is important that the machined surfaces not be damaged when the halves are separated.

Impeller

Most pump impellers for water utility use are made of bronze, although a number of manufacturers offer cast iron or stainless steel as alternative materials. The overall impeller diameter, width, inlet area, vane curvature, and operating speed affect impeller performance and are modified by the manufacturer to attain the required operating characteristics. Impellers for single-suction pumps may be of the open, semiopen, or closed design, as shown in Figure 4-24. Most single-suction pumps in the water industry use impellers of the closed design, although a few have semiopen impellers. Double-suction pumps use only closed-design impellers.

Wear rings

In all centrifugal pumps, a flow restriction must exist between the impeller discharge and suction areas to prevent excessive circulation of water between the two. This restriction is made using wear rings. In some pumps, only one wear ring is used, mounted in the case. In others, two wear rings are used, one mounted in the case and the other on the impeller. The wear rings are identified in Figure 4-25.

The rotating impeller wear ring (or the impeller itself) and the stationary case wear ring (or the case itself) are machined so that the running clearance between the two effectively restricts leakage from the impeller discharge to the pump suction. The clearance is usually 0.010–0.020 in. (0.25–0.50 mm). Rings are normally machined from bronze or cast iron, but stainless-steel rings are available. The machined surfaces will eventually wear to the point that leakage occurs, decreasing pump efficiency. At this point, the rings need to be replaced or the wearing surfaces of the case and impeller need to be remachined.

Shaft

The impeller is rotated by a pump shaft, usually machined of steel or stainless steel. The impeller can be secured to the shaft on double-suction pumps using a key and a very tight fit (also called a shrink fit). Because of the tight fit, an arbor press or gear puller is required to remove an impeller from the shaft.

In end-suction pumps, the impeller is mounted on the end of the shaft and held in place by a key nut. The end of the shaft may be machined straight or with a slight taper. However, removing the impeller usually will not require a press. Several other methods are also used for mounting impellers.

Semiopen   Closed

FIGURE 4-24 Types of impellers
Courtesy of Goulds Pumps, ITT Industries

Shaft sleeves

Most manufacturers provide pump shafts with replaceable sleeves for the packing rings to bear against. If sleeves are not used, the continual rubbing of the packing can eventually wear out the shaft, which would require replacement. A shaft could be ruined almost immediately if the packing gland were too tight. Where shaft sleeves are used, operators can repair a damaged surface by replacing the sleeve, a procedure considerably less costly than replacing the entire shaft. The sleeves are usually made of bronze alloy, which is much more resistant than steel to the corrosive effects of water. Stainless-steel sleeves are usually available for use where the water contains abrasive elements.

Packing rings

To prevent leakage at the point where the shaft protrudes through the case, either packing rings or mechanical seals are used to seal the space between the shaft and the case. Packing consists of one or more (usually no more than six) separate rings of graphite-impregnated cotton, flax, or synthetic materials placed on the shaft or shaft sleeves (Figure 4-26). Asbestos material, once common for packing, is no longer used on potable water systems. The section of the case in which the packing is mounted is called the stuffing box. The adjustable packing gland maintains the packing under slight pressure against the shaft, stopping air from leaking in or water from leaking out.

To reduce the friction of the packing rings against the pump shaft, the packing material is impregnated with graphite or polytetrafluoroethylene to provide a small measure of lubrication. It is important that packing be installed and adjusted properly.

Lantern rings

When a pump operates under suction lift, the impeller inlet is actually operating in a vacuum. Air will enter the water stream along the shaft if the packing does not provide an effective seal. It may be impossible to tighten the packing sufficiently to prevent air from entering

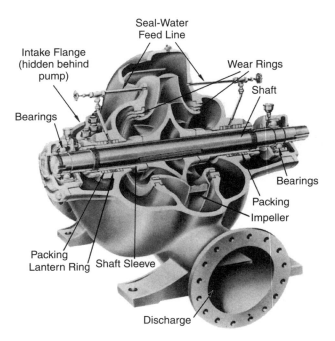

FIGURE 4-25 Double-suction pump
Courtesy of Ingersoll-Dresser Pump Company

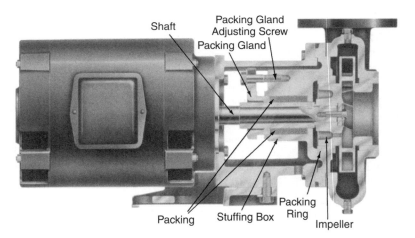

FIGURE 4-26 Pump packing locations
Courtesy of Aurora Pump

without causing excessive heat and wear on the packing and shaft or shaft sleeve. To solve this problem, a lantern ring (Figure 4-27) is placed in the stuffing box. Pump discharge water is fed into the ring and flows out through a series of holes leading to the shaft side of the packing. From there, water flows both toward the pump suction and away from the packing gland. This water acts as a seal, preventing air from entering the water stream. It also provides lubrication for the packing.

Mechanical seals

If the pump must operate under a high suction head (60 psig [400 kPa (gauge)] or more), the suction pressure itself will compress the packing rings, regardless of operator intervention. Packing will then require frequent replacement. Most manufacturers recommend using a mechanical seal under these conditions, and many manufacturers use mechanical seals for low-suction-head conditions as well. The mechanical seal (Figure 4-28) is provided by two machined and polished surfaces—one is attached to and rotates with the shaft, the other is attached to the case. Contact between the seal surfaces is maintained by spring pressure.

The mechanical seal is designed so that it can be hydraulically balanced. The result is that the wearing force between the machined surfaces does not vary regardless of the suction head. Most seals have an operating life of 5,000 to 20,000 hours. In addition, there is little or no leakage from a mechanical seal—a leaky mechanical seal indicates problems that should be investigated and repaired. A major advantage of mechanical seals is that there is no wear or chance of damage to shaft sleeves.

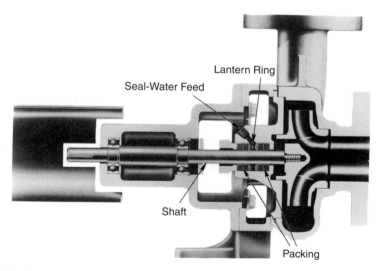

FIGURE 4-27 Lantern ring placed in the stuffing box
Courtesy of Aurora Pump

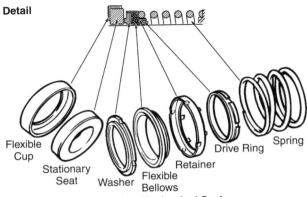

Detail

Flexible Cup
Stationary Seat
Washer
Flexible Bellows
Retainer
Drive Ring
Spring

Parts of a Mechanical Seal

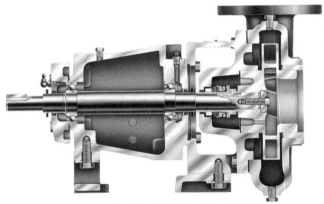

Mechanical Seal Installed in a Pump

FIGURE 4-28 Mechanical seal parts and placement
Courtesy of Aurora Pump

A major disadvantage of mechanical seals is that they are more difficult to replace than packing rings. Replacing the mechanical seal often requires removing the shaft and impeller from the case. Another disadvantage is that failure of a mechanical seal is usually sudden and accompanied by excessive leakage. Packing rings, by contrast, normally wear gradually, and the wear can usually be detected long before leakage becomes a problem. Mechanical seals are also more expensive than packing.

Bearings

Most modern pumps are equipped with ball-type radial and thrust bearings. These bearings are available with either grease or oil lubrication and provide good service in most water utility applications. They are reasonably easy to maintain when manufacturer's recommendations are followed, and new parts are readily available if replacement is required. Ball bearings will usually start to get noisy when they begin to fail, enabling operators to plan a shutdown for replacement.

Couplings

Frame-mounted pumps have separate shafts connected by a coupling. The primary function of couplings is to transmit the rotary motion of the motor to the pump shaft. Couplings are also designed to allow slight misalignment between the pump and motor and to absorb the startup shock when the pump motor is switched on. Although the coupling is designed to accept a little misalignment, the more accurately the two shafts are aligned, the longer the coupling life will be and the more efficiently the unit will operate (Figure 4-29).

Various coupling designs are supplied by pump manufacturers. Couplings may be installed dry or lubricated. Most couplings are of the lubricated style and require periodic maintenance, usually lubrication at 6-month or annual intervals. Dry couplings using rubber or elastomeric membranes do not require any maintenance, except for periodic visual inspection to make sure they are not cracking or wearing out. The rubber or elastomer used for the membrane must be carefully selected for the pump, because the corrosive chemicals used in water treatment plants could affect the life and operation of the coupling.

Inspection and Maintenance

A well-defined inspection and maintenance schedule is necessary to ensure long and reliable pump service and to preserve warranty rights on new units. It is important to maintain complete records of inspections and any service performed. An operator can best evaluate pump condition by comparing a pump's current performance to its performance when it was first installed. Therefore, complete testing immediately following installation is important.

The following should be checked periodically to ensure maximum operating efficiency and minimum maintenance expenditures:

* Priming
* Packing and seals
* Bearings
* Vibration
* Alignment
* Sensors and controls
* Head (pressure gauges)
* Cavitation

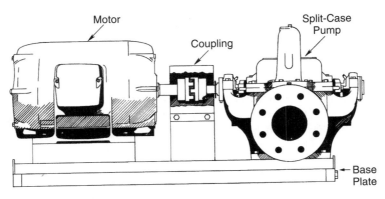

FIGURE 4-29 Alignment of motor and pump at coupling

Priming

Pumps must be checked before startup to be sure they are primed. Capacity will be reduced and water-lubricated internal wear rings may be damaged if a pump is only partially primed. If a pump is primed from an overhead tank or other gravity supply, it may be started as soon as water shows at the top vent cocks. If it is primed by a vacuum pump, the action of the device itself will indicate when the casing is filled with water. Some pumps are equipped with a float switch that will not complete the circuit for starting the motor unless the pump is completely primed.

If the pump does not have a vacuum priming system, petcocks on top of the pump case should be opened routinely during operation to bleed off any air that might collect there. Continuous bleeds or air-release valves should be installed on pumps at unattended pump stations. All valves in the suction line must be fully open.

Packing

A correctly packed and adjusted stuffing box should be trouble-free. Packing should be inspected annually if a pump is run on a regular basis. The easiest way to do this is to remove the top of the casing or the packing gland. When the packing wears or is compressed to the point where it is impossible to tighten the gland further, a new set of rings should be installed. It is generally not considered good practice to make up for wear and compression by adding new rings on top of old ones. However, manufacturer's maintenance specifications may allow the addition of one more ring.

New packing must be installed with care to extend its life. The following guidelines should be observed:

- During disassembly, keep parts in the order in which they are removed, including the number of rings before and after the lantern ring.
- Clean the stuffing box.

- Check the shaft sleeve. Replace it if it is badly worn.
- Use packing and sleeve material that is the proper size and compatible with the expected service. For a severely abrasive or corrosive service, consult the pump manufacturer. Although precut packing is available, most operators purchase bulk packing and cut what they need for each job.
- Replace parts in the opposite order that they were removed.
- Replace all packing rings. The ends of the strips used to form rings may be cut diagonally or square. The ends should be carefully butted, with joints staggered at least 90°. Four or more rings are usually placed with their ends at 90° intervals. Be sure to replace the lantern ring in its original sequence.
- Never overtighten the packing. Draw each ring down firmly. After all rings are installed, back off on the gland nuts about one full turn before the pump is started. Before its initial startup with new packing, the pump should rotate fairly easily by hand, unless it is very large. If it does not turn over, the packing should be loosened until it does, provided there are no other problems.
- At startup, the packing should be allowed to leak freely until conditions stabilize and it is evident that there are no hot spots in the area of the packing or bearings. Then gradually tighten the packing gland until the packing allows a slow drip.
- While the pump is running, adjust the gland by tightening the gland-bolt nuts. Tighten the nuts evenly, and tighten each nut no more than ⅙ of a turn every 20–30 minutes.
- Never tighten packing glands to the point where there is no leakage. Doing so will cause premature packing wear and scored shaft sleeves.
- After the initial installation and adjustment, check the packing regularly and adjust the gland whenever leakage increases. The leakage rate should be checked daily if possible.
- If cooling or sealing water is injected into the box, set its flow pressure to 15–20 psi (100–140 kPa) more than the pressure on the inboard end of the stuffing box. Excessive pressure will cause increased wear of the packing and sleeve.

Mechanical seals

The operating temperature of a mechanical seal should never exceed 160°F (71°C). If there is a possibility that a seal will exceed this limit, it should be water cooled. The water can be supplied by the pump discharge or from an external source. The water must not contain dirt, grit, or other abrasive materials that could damage the seal. If, for instance, the pump is used for pumping raw water, seal water must be supplied from the filtered water system with adequate precautions taken to prevent a cross-connection.

It is important that the mechanical seal is designed for the stuffing-box pressure at which it will operate. A pump that develops a partial vacuum on the suction side must be fitted with a close-clearance bushing between the seal and the suction passage of the pump in order to maintain lubrication, cooling, and flushing fluid at the seal.

Bearings

Regular inspection and lubrication of bearings is essential to efficient pump and motor operation. Lubrication points should be checked and lubricated at the intervals prescribed by the manufacturer. The following checks are important:

- Oil level in bearing housings
- Free movement and proper operation of oil rings
- Proper oil flow for pressure-feed systems
- Proper type of grease for grease-lubricated bearings
- Proper amount of grease in the housing
- Bearing temperature

Oil-lubricated bearings. The bearing housing of oil-lubricated bearings should be kept filled with a good grade of filtered mineral oil. The oil should be changed after the first month of operation of a new pump. After the first oil change, future oil changes should be performed every 6 to 12 months, depending on operating frequency and environmental conditions. Whenever oil is changed, and especially after the first oil change, the oil should be inspected for signs of bearing wear or excessive dirt. The following viscosities are recommended for use with various antifriction bearings:

- Ball and cylindrical roller bearings—70 Saybolt standard units (SSU) oil rated at operating temperature.
- Spherical roller bearings—100 SSU oil rated at operating temperature.
- Spherical roller and thrust bearings—150 SSU oil rated at operating temperature.

It is important that the bearing housing not be overfilled with oil. Most housings have the correct oil level indicated on a sight glass.

Grease-lubricated bearings. For grease-lubricated bearings, similar maintenance is required. During the initial run-in period, bearing temperatures must be closely monitored. The housing of bearings that are within an acceptable temperature range can be touched with the bare hand. After about 1 month of initial full-service operation, all bearings should be regreased.

Bearings should not be overgreased. Because of the internal friction caused by the churning grease, a bearing will run hotter if the grease pocket is packed too tightly. The grease used for lubricating antifriction pump bearings should be a sodium soap-base type that meets Anti-Friction Bearing Manufacturer's Association group 1 or 2 classifications. Bearings should be regreased every 3 to 6 months according to the following procedures:

1. Open the grease drain plug at the bottom of the bearing housing.
2. Fill the bearing with new grease until grease flows from the drain plug.

3. Run the pump with the drain plug open until the grease is warm and no longer flows from the drain.
4. Replace the drain plug.

Grease does not need to be added to bearings between intervals unless the bearing seals are bad and grease has been lost. If bearings run hot, it is usually because the grease is packed in too tightly. In this case, the grease drain plug should be opened and some grease allowed to drain out. If the bearing has been disassembled, cleaned, and flushed, the housing should be refilled to one third of its capacity.

Bearing replacement. The life of a ball bearing will vary with the conditions of load and speed. Most pump bearings have a minimum operating life of 15,000 hours (about 1 ½ years), but they may last much longer. As a general rule, if a pump has been operating continuously for 1 or 2 years, the bearings should be replaced if the pump is taken out of service for any other repairs. It is more economical to replace bearings while other work is being performed than to wait until excessive heat or noise warns that the bearings are about to fail.

Operators should be extremely careful when removing or installing bearings. Appropriate bearing pullers and hydraulic presses should be used, especially if there is a chance that bearings will have to be reused. It is often good practice to leave serviceable bearings mounted until replacements have been obtained. Proper installation procedures, as directed by the manufacturer, should be followed to prevent damage to new bearings during installation.

Vibration monitoring

On high-speed or large pump units, operators may use instruments to periodically monitor vibration in the vertical, lateral, and axial planes. This will give an early indication of possible future problems. When problems occur, an experienced person will usually be able to detect undesirable vibration merely by listening to or touching the unit. In general, on units for which periodic vibration checks will be performed, a vibration test should be conducted immediately after the pump is installed to establish a baseline condition. This test should be followed by periodic measurements of vibration at intervals recommended by the manufacturer. For critical equipment that is operated continuously, monthly checks may be required. Less critical equipment may be checked annually.

Vibration sensors may be installed on pumping units or portable equipment may be used to make the measurements. Vibration readings should be taken on the shafts in or near the bearings. Any change from the baseline measurements of vibration magnitude or frequency indicates potential problems.

When operators observe vibration or overheating of a bearing or coupling, they should inspect the pump's shaft coupling and impeller immediately for possible imbalance. The imbalance could result from the presence of foreign material caught in the impeller, the accumulation of scale, mechanical breakage, or loss of metal due to corrosion or cavitation.

Alignment

Excessive vibration may also be caused by misalignment. Alignment should be checked in pumping units after they are first installed and brought up to operating temperature (often called a "hot alignment" check). This is particularly important for frame-mounted pumps. A record of the initial readings should be made using a dial indicator gauge, as shown in Figure 4-30. Alignment should then be checked again periodically to ensure that the initial readings have not changed. Vibration due to misalignment is a common cause of bearing failures.

Sensors and controls

Operating personnel should know how and why each part of an automatic control system functions. They should understand how each part affects the operation of the system. The normal sequence and timing of each operation should be determined and any deviations identified for correction. Technicians familiar with each type of equipment should repair and adjust all of the more complex control devices. Pressure and float sensors may require adjustment seasonally or with changes in water demand.

Head

Pump discharge and suction heads should be checked and compared with baseline performance figures for when the pump was first installed. Some wear will necessarily reduce performance, but major reductions in capacity should be identified and corrected before the inefficient unit begins to waste large amounts of energy. Strip or circular chart recordings of pump performance can be helpful in identifying reduced pump capacity.

Operators can check an installed centrifugal pump for wear by closing the discharge valve and then reading the discharge pressure. This pressure can be compared with the original pump characteristics, after appropriate deductions are made for suction pressure. If the shutoff head is close to the original value, the pump is not greatly worn.

Cavitation

Under certain circumstances, a pump can pull water so hard that some of the water turns to small bubbles of vapor. This is called cavitation. The bubbles explode against the impeller, making the sound as if there are marbles in the pump. The consequences of cavitation may be only the bothersome noise, but over time the impeller will usually be damaged and the metal eroded.

To avoid cavitation, a pump should not be continuously operated at a rate much higher or lower than it was designed for. In addition, the suction requirements for the pump must be met. Turbine and submersible pumps must be provided with a deep enough sump to keep them submerged. Intake screens should also be routinely cleaned if necessary to avoid suction restrictions.

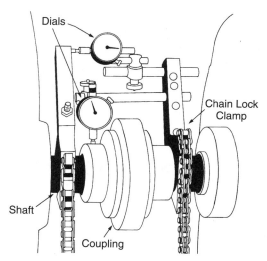

FIGURE 4-30 Dial indicator alignment gauge

Major repairs

Major repair jobs that may be required for pumps at infrequent intervals include replacement of bearings, wearing rings, shaft sleeves, and impellers. These jobs require that the pump be removed from service for a period of time. Planning is essential to ensure that the work goes smoothly and quickly. If a pump is not too large, it can be returned to the manufacturer or the utility shop for repair. If the repair must be made in the field, it is especially important to have all parts on hand before the pump is removed from service.

Care should be taken to ensure that the correct replacement parts are used. If the identical part is not available, a substitute part may be used if it meets or exceeds the standards of the original part. The parts themselves or accurate specifications can be obtained from the pump manufacturer before the repairs are begun.

In addition to replacement parts, any necessary special tools should be on hand before the job is started. For example, bearing removal will usually require a bearing puller or a hydraulic press, possibly fitted with special collars.

RECORD KEEPING

An adequate record of equipment specifications and maintenance performed will assist the operator in scheduling inspections and needed service work, evaluating pump equipment, and assigning personnel. An appropriate system could be based on data cards. Each card should list the make, model, capacity, type, date and location installed, and other information for both driver and driven unit. The remarks section should include the serial or part numbers of special components (such as bearings) that are likely to require eventual replacement.

A separate operating log should be kept that lists all pumping units along with a record of the operating hours. This record is an essential feature of any reasonable periodic service or maintenance schedule. In addition, a daily work record should be kept on each piece of equipment.

Many water systems put records such as those described here on a computer database, where the information is more easily stored and accessed. See chapter 16 of this book for more information.

PUMP SAFETY

Specific pump safety precautions must be recognized and observed with respect to

- moving machinery,
- material handling,
- personal protection equipment,
- hand tools,
- electrical devices, and
- fire safety.

Machinery should always be stopped before it is cleaned, oiled, or adjusted. The controlling switchgear should be locked out before any work begins, so that another person cannot start the machinery. A conspicuous tag should also be posted on or over the control panel, giving notice that the equipment is under repair and should not be restarted. The name of the person who locked the equipment out should also be listed.

Before a machine is restarted, check to be sure that all personnel are clear of danger and that working parts are free to move without damage. When disconnecting equipment, follow the manufacturer's instructions for disconnecting and securing drive and rotating equipment.

Guards over rotating parts should be secured in place to protect workers who are near equipment when it is in operation. Guards with hinged or movable sections should be provided where it is necessary to change belts, make adjustments, or add lubricants. If a guard or enclosure is within 4 in. (100 mm) of a moving part, the maximum opening in the screen should not exceed ½ in. (13 mm) across. (This will prevent an operator's fingers from getting too close to moving parts.) Guards placed more than 4 in. (100 mm) but less than 15 in. (380 mm) from a moving part can have openings no more than 2 in. (50 mm) across. The guard should be strong enough to provide complete safety, and guard structures should be constructed so that they cannot be pushed or bent against moving parts. Guards should be removed for maintenance only when the machinery is not in operation. Other aspects of safety (such as for working with tools) are discussed in other chapters of this book.

BIBLIOGRAPHY

ANSI/NSF Standard 61: Drinking Water System Components—Health Effects. Ann Arbor, Mich.: NSF International.

AWWA Standard for Submersible Vertical Turbine Pumps. ANSI/AWWA E102. Denver, Colo.: American Water Works Association (latest edition).

AWWA Standard for Horizontal and Vertical Line-Shaft Pumps. ANSI/AWWA E103. Denver, Colo.: American Water Works Association (latest edition).

Basic Science Concepts and Applications. 2003. Denver, Colo.: American Water Works Association.

Centrifugal Pumps and Motors: Operation and Maintenance. 1992. Denver, Colo.: American Water Works Association.

Headquarters, Department of the Army, Water Supply: Pumping Stations, TM 5-813-6, 1992.

Hydraulic Institute Pump Standards—Centrifugal Pumps. ANSI/HI. Latest edition. Parsippany, N.J.: Hydraulic Institute.

Hydraulic Institute Pump Standards—Vertical Pumps. ANSI/HI. Latest edition. Parsippany, N.J.: Hydraulic Institute.

Karassik, I.J., J.P. Messina, P. Cooper, and C.C. Heald. 2001. *Pump Handbook,* 3rd ed. New York: McGraw-Hill.

Rishel, J.B. 2002. *Water Pumps and Pumping Systems.* New York: McGraw-Hill.

Jones, G.M., ed. , Sanks, R.L., G.Tchobanoglous, and B.E. Bosserman II. 2008. *Pumping Station Design.* 3rd ed. Woburn, Mass.: Butterworth-Heinemann.

Tuzon, J. 2000. *Centrifugal Pump Design.* New York: John Wiley & Sons.

Water Distribution Operator Training Handbook. 2005. Denver, Colo.: American Water Works Association.

CHAPTER 5

Hydraulics of Water Distribution Systems

The purpose of this chapter is to provide some of the concepts of the hydraulics of water distribution systems beyond that provided in *Basic Science and Concepts and Applications*, which is in the same series of AWWA books. That book introduces some elementary concepts of the hydraulics of water distribution systems. These include piezometric surface and hydraulic grade line, hydraulic head, head loss (including friction losses and minor losses), pumping heads, and pump curves. This chapter discusses the hydraulics of pipe flow, the hydraulics of a water system including topics of water age and system performance, the use of network analysis, and the concepts of hydraulic transients.

HYDRAULICS OF PIPE FLOW

Pipe Flow

The Hazen–Williams formula for the velocity of flow in a pipe is expressed as

$$V = 1.318 C R^{0.63} S_f^{0.54}$$

where V is the flow velocity in feet per second, C is the Hazen–Williams coefficient (Table 5-1), R is the hydraulic radius in ft, and S_f is the friction slope in feet per feet.

TABLE 5-1 Typical C values (Hazen–Williams coefficients) for different pipe materials

Pipe Material	C Value
Asbestos–cement	140
Brass	130–140
Cast iron	
New, unlined	130
Old, unlined	40–120
10 years old	110
20 years old	90
Cement lined	120–140
Tar coated	110–130
Concrete/cement lined	
New	140
Old	100–130

Table continued on next page

TABLE 5-1 Typical C values (Hazen–Williams coefficients) for different pipe materials (Continued)

Pipe Material	C Value
Copper	140
Ductile iron	
New, lined	140
Fire hose (rubber)	135
Lead	130–140
Plastic (PVC)	130–140
Steel	
Coal-tar enamel	140
New, lined	140

The hydraulic radius is the circular area (A) of the inside diameter of the pipe divided by the wetted perimeter (P), which is the circumference of the inside diameter of the pipe for a full pipe flow. Then the hydraulic radius is $R = A/P$. For a 15 in. (1.25 ft) diameter pipe, the area is $\pi D^2/4$ or $3.14 D^2/4$, where D is the inside diameter in feet, so that the area A is expressed as $A = 0.785D^2$ for full flow in a circular pipe of diameter D. The wetted perimeter for a full pipe flow is πD. The hydraulic radius for a full pipe flow is $R = (\pi D^2/4)/(\pi D) = D/4$. For the 15 in. pipe, the hydraulic radius is $1.25/4 = 0.3125$.

The expression for flow rate ($Q = AV$) in a circular pipe presented in *Basic Science Concepts and Applications* is

$$Q = 0.785D^2V$$

For the 15 in.-diameter pipe, if the velocity of flow was 2 feet per second, the flow rate would be $Q = 0.785(1.25\ \text{ft})(1.25\ \text{ft})(2\ \text{ft/sec}) = 2.45\ \text{ft}^3/\text{sec}$.

The friction slope, S_f, is the friction head loss (h_L) over a length of pipe divided by the length (L) of the pipe, so that $S_f = h_L/L$. The discharge in a pipe is expressed as $Q = AV$, which can be used to define the flow rate by computing the velocity using the Hazen–Williams formula, then

$$Q = AV = 1.318ACR^{0.63}S_f^{0.54}$$

From this equation for flow rate and the fact that the friction slope is $S_f = h_L/L$, the head loss can be expressed as

$$h_L = 4.727C^{-1.852}D^{-4.87}LQ^{1.852}$$

or

$$h_L = KQ^{1.852}$$

where

$$K = 4.272C^{-1.852}D^{-4.87}L$$

Consider the 15-in. diameter pipe with a flow of 2.45 ft³/sec discussed above. We want to determine the head loss due to friction in a 5,000 ft-long pipe. The pipe is a new ductile-iron pipe with a C value of 140 (see Table 5-1). The K value is computed as

$$K = 4.727C^{-1.852}D^{-4.87}L = 4.727(140)^{-1.852}\left(\frac{15}{12}\right)^{-4.87}(5000) = 0.845$$

The head loss due to friction is then computed as

$$h_L = KQ^{1.852} = 0.845(2.45)^{1.852} = 4.44 \text{ ft}$$

The hydraulic grade line was defined as the line that connects all the piezometric surfaces along a pipeline in *Basic Science Concepts and Applications*. If the hydraulic grade line is 1,000 ft at the upstream end of the pipe, the hydraulic grade line at the end of the 5,000-ft pipe would be 1,000 ft – 4.439 ft = 995.561 ft.

Minor Head Losses

Minor head losses (also called local losses) are caused by the added turbulence that occurs at bends and fittings in a water distribution network. They can be accounted for by assigning the fitting a minor loss coefficient. The minor head loss is the product of this coefficient and the velocity head of the pipe ($v^2/2g$), i.e.,

$$h = K\frac{v^2}{2g}$$

where K = minor loss coefficient, V = flow velocity, and g = acceleration of gravity. The importance of including such losses depends on the layout of the network and the degree of accuracy required. Table 5-2 provides minor loss coefficients for several types of fittings.

HYDRAULICS OF THE SYSTEM

Water Demand Patterns

As discussed in chapter 3, the demand for water changes throughout the day and night (refer to Figure 3-1), so that water demand variations occur on a daily basis. Daily demand variations can be shown on a diurnal demand curve, which plots the percentage of daily demand versus time. Figure 5-1 shows an average day diurnal flow that is expressed in terms of the peaking factor as a function of the 24-hr day.

TABLE 5-2 List of minor loss coefficients (Rossman 2000)

Fitting	Loss Coefficient
Globe valve, fully open	10.0
Angle valve, fully open	5.0
Swing check valve, fully open	2.5
Gate valve, fully open	0.2
Short-radius elbow	0.9
Medium-radius elbow	0.8
Long-radius elbow	0.6
45 degree elbow	0.4
Closed return bend	2.2
Standard tee - flow through run	0.6
Standard tee - flow through branch	1.8
Square entrance	0.5
Exit	1.0

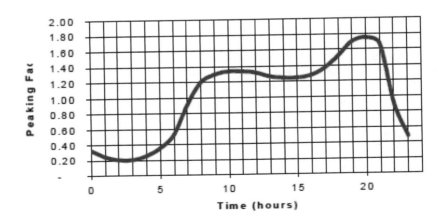

FIGURE 5-1 Average day flow diurnal curve

Water Age

Review of composite usage patterns suggests that typically, water age due to storage in the water distribution system is highest in the early morning hours and lowest in the late evening. Several indicators may suggest high water age (USEPA 2002), which include aesthetic considerations that may be identified by consumers, and results of distribution system monitoring efforts. Indicators can be triggered by factors other than water age, such as insufficient source water treatment, pipe materials, and condition/age of the distribution system. Water demand changes or use pattern changes, such as those caused by the relocation of an industrial water user, annexation of a neighboring system, or consolidation of multiple systems, can have a significant impact on water age.

The following aesthetic indicators may be identified during water consumption (USEPA 2002):

- Poor taste and odor: Aged, stale water provides an environment conducive to the growth and formation of taste- and odor-causing microorganisms and substances.
- Discoloration: Water in low-flow areas and dead-ends often accumulate settled deposits over time. During a demand period, these deposits are entrained and degrade the clarity and color of the water.
- Water temperature: Stagnant water will approach the ambient temperature.

The following monitoring indicators require sample collection and analysis (USEPA 2002):

- Depressed disinfectant residual: Chlorine and chloramines undergo decay over time.
- Elevated DBP levels: The reaction between disinfectants and organic precursors occur over long periods.
- Elevated bacterial counts (i.e., heterotrophic plate count).
- Elevated nitrite/nitrate levels (nitrification) for chloraminating systems.

Table 5-3 is a summary of water quality problems associated with water age.

Mixing Time in Cylindrical Tanks

The following relationship has been developed for cylindrical tanks under fill and draw operation (Rossman and Grayman 1999):

$$\text{Mixing time (in seconds)} = 10.2 \, (V_T)^{2/3} / (Qv)^{1/2}$$

where V_T is the volume of water in tank at start of fill (cubic feet), Q is the inflow rate (cubic feet per second), and v is the inflow velocity in ft/sec. Mixing time is highly dependent on the inflow rate and quantity of water flowing into the reservoir. Often, mixing can be increased

TABLE 5-3 Water quality problems associated with water age (USEPA 2002)

Chemical Issues	Biological Issues	Physical issues
Disinfection by-product Formation*	Disinfection by-product Biodegradation*	Temperature increases
Disinfectant decay	Nitrification*	Sediment Deposition
Corrosion control effectiveness*	Microbial regrowth/recovery/ shielding*	Color
Taste and odor	Taste and odor	

*Denotes water quality problem with direct potential public health impact.

by operating a reservoir so that the inflow rate is increased. Additionally, daily turnover rates (inflow/outflow quantity) can be increased to reduce overall water age.

Hydraulics of Dead Ends

Dead ends of pipes in water distribution systems hydraulically create a space where the velocity is zero, resulting in the water remaining in the pipe for an extended period of time. Water quality deterioration is often a function of the time that water is resident in the distribution system. The longer the water is in contact with pipe walls and reacting with constituents in the water, the greater the opportunity for water quality changes. In other words, as the water stagnates, metal concentrations may increase as a result of the pipes leaching metals into the water. Bacteriological growth in stagnant areas is also a concern. Network analysis (hydraulic models) can be used to determine water age throughout a water distribution system and to evaluate methods and modifications to reduce the water age. Automatic flushing hydrants, automated blow-off valves, and flush valves can be used to improve water quality at the end of lines.

Hydraulics of System Performance

The following scenarios are for the purpose of describing the hydraulics of system performance. Each of these types of analysis can be generated using network analysis described in the next section. These analyses and results are based on the work of Tom Walski.

Pumping Capacity

One way to evaluate pumping capacity is to consider the effect on tank levels for three alternative pumps: a properly sized pump (good pump), an undersized pump (small pump), and an oversized pump (large pump) that feed a tank. Figure 5-2 shows the fluctuations in tank water level for a system comparing the three pumps. The properly sized pump results show the tank

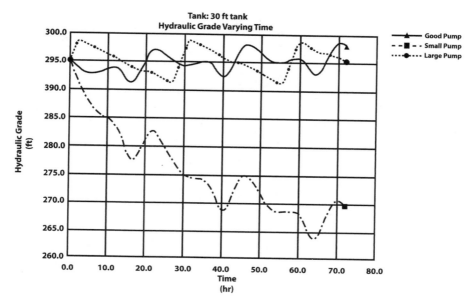

FIGURE 5-2 Tank level vs. time for three alternative pumps
Courtesy of T. Walski

level fluctuations for a pump that matches well with demands and tank levels remain in an acceptable range. The undersized pump would only work when the water level dropped to an elevation where the discharges matched the demand. The oversized pump cycles on and off several times, filling the tank in a few hours, and then remaining off for several hours.

Pipe Capacity

Consider three different pipe sizes for a main that feeds a tank and compare the hydraulic grade versus time for each to compare their performances. The objective is to keep the tank level (hydraulic grade) above a target level of 290 ft. The hydraulic grade verses time presented in Figure 5-3 for a 10-in. pipe shows that the tank levels stay above the desired tank level. The 8-in. pipe results in the tank not filling properly and excessive draining of the tank. If a 12-in. pipe were used, there would be only a slight improvement of capacity. It may be that the 12-in. pipe would be satisfactory for future expansion of the system.

Tank Volume

Consider three different sizes (volumes) of tanks and compare the hydraulic grade versus time for each to compare their performances. Figure 5-4 shows that the 30-ft diameter tank results in a reasonable fluctuation in hydraulic grade typical of a diurnal demand. The smaller, 20-ft diameter tank results in more cycling of the pumps but could be

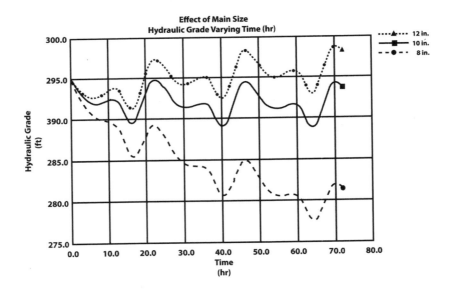

FIGURE 5-3 Tank level vs. time for different main sizes

Courtesy of T. Walski

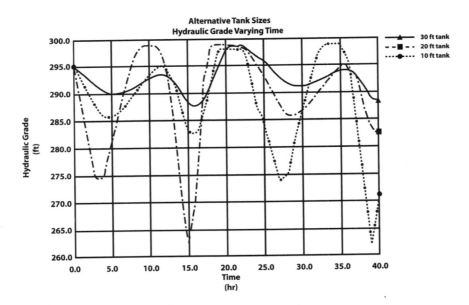

FIGURE 5-4 Performance of different tank sizes

Courtesy of T. Walski

satisfactory. The 10-ft diameter tank results in too much cycling of the pumps, such that the system cannot keep up with peak demands.

Tank Elevation

Figure 5-5 illustrates the hydraulic grade versus time for three tanks of different elevations having the same overflow elevation of 300 ft. The desired range of the hydraulic grade is between 290 and 300 ft. Tank Low is the tank with too low of an elevation for the system and pumps. This tank fills quickly and shuts off the pumps, resulting in pressure drops. As a result, the pumps have to keep cycling to keep the hydraulic grade in a desirable range. Tank High is a tank at too high of an elevation so that it is difficult to keep the tank water level in a desired range without the possibility of overpressurizing parts of the system. Results for the base is a tank at a proper elevation so that the pumps do not have to keep cycling to keep the tank water levels in a desired range of elevations.

Low Pressures in System

There may be some locations in a pressure zone that have problems with low pressures. Figure 5-6, which shows the pressure versus time plot for three different situations in a network, is used to further illustrate this problem. The base situation is compared with two others, one with low pressures due to high elevation (High Elv) and one with low pressures due to excessive head loss (High Loss) due to inadequate piping capacity. Customers at a high elevation (High Elv) have pressures that are consistently low over time. Customers with low pressures (High Loss) have inadequate piping capacity only during peak demand periods.

Pressure-Reducing Valves

Consider a pressure-reducing valve (PRV) that is used as a backup feed (sleeper valve). Figures 5-7 and 5-8 show the flow versus time and pressure versus time for the downstream side of a PRV. When the valve is open the pressure is constant at the PRV setting, and when it is closed the downstream-side pressure is dependent on other sources.

Hydraulic Performance During a Fire Event

Figures 5-9 and 5-10 illustrate the tank level (hydraulic grade) as a function of time and the pump discharge as a function of time during a fire event. Figure 5-9 illustrates the extended length of time it takes for the tank to fill after the fire event is over. Figure 5-10 illustrates that the pump discharge increases not only during the fire (for the fire demand) but also after the fire (to refill tanks). This situation causes the pump to move further out on the pump head-discharge curve, and if the discharge is too large could cause the pump to shut down.

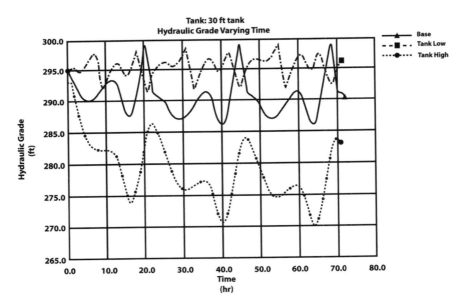

FIGURE 5-5 Tank elevations

Courtesy of T. Walski

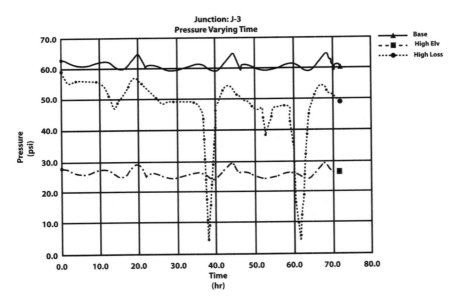

FIGURE 5-6 Low pressure areas

Courtesy of T. Walski

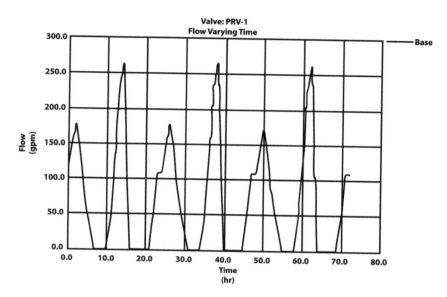

FIGURE 5-7 Flow vs. time for a PRV (sleeper valve)
Courtesy of T. Walski

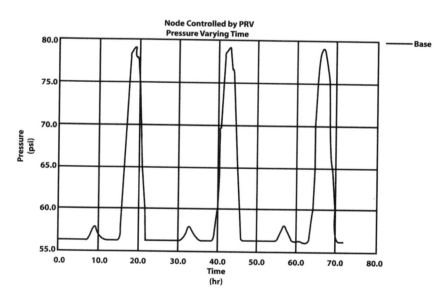

FIGURE 5-8 Pressure vs. time for a PRV (sleeper valve)
Courtesy of T. Walski

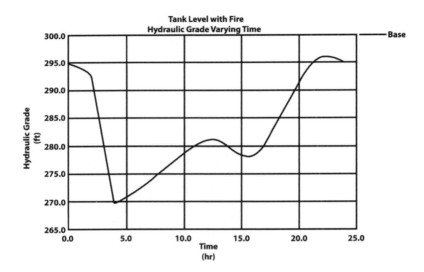

FIGURE 5-9 Tank level during a fire
Courtesy of T. Walski

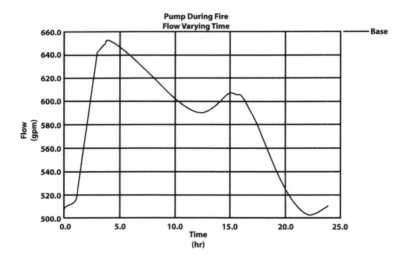

FIGURE 5-10 Pump discharge during a fire
Courtesy of T. Walski

Hydraulic Performance During a Pipe Break

Pipe breaks can result in large flows from the network while the break is being located and isolated. During that time tanks are used to provide the extra water needed. Figure 5-11 shows a plot of the hydraulic grade of a tank during a pipe break.

NETWORK ANALYSIS

What is Network Analysis?

Pipe network flow analysis began with solving problems that asked what flows and pressures are in a network of pipes subject to a known set of inflows and outflows. Such analyses required two sets of equations: conservation of flow at each pipe junction and a relation between the head loss and the discharge in each pipe such as the Hazen–Williams equation presented previously. The set of equations for the pipe network are solved using an iterative numerical technique for the discharge in each pipe and the heads at each of the pipe junctions. Over the years, these types of network analysis have advanced from simple skeletonized models with few pipes to sophisticated computer models such as the EPANET model that can handle very large networks containing thousands of pipes.

In fact, not only can these models handle steady-state flow but now have the capability to perform extended-period simulation (EPS). Steady-state models only consider a single demand at nodes, whereas the extended-period simulation considers the variation of demand as a function over the time period of simulation. Extended-periods simulation

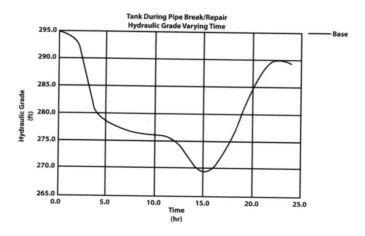

FIGURE 5-11 Tank level during a pipe break
Courtesy of T. Walski

uses a sequence of steady-state simulations over long periods of time. After each simulation period of steady-state flow, the tank levels are updated and demand and operational (e.g., pump operation) changes are made. Water quality can also be modeled along with the hydraulics using EPS. Water quality capabilities of these models for network analysis are discussed further.

Uses of Network Analysis

There are a number of uses for network analysis (network computer models) in planning, design, operations, and management of water distribution systems by water utilities. Cesario (1991) presented some examples as follows:

1. Network models are run to analyze what capital improvements will be needed to serve additional customers and maintain existing services in future years.
2. Network models are used to locate and size network components, such as new mains, storage tanks, pumping stations, and regulator valves.
3. Pump scheduling, tank turnover analysis, energy optimization, and operator training are some of the ways in which network models can be used to improve system operations.
4. EPS can be used to analyze a host of questions related to water quality.
5. Fire-flow studies are used to determine if adequate flow and pressure are available for firefighting purposes, as required for fire insurance ratings.
6. Vulnerability studies are used to test a system's susceptibility to unforeseen occurrence, such as loss of power, major main breaks, extended drought periods, and intrusion of waterborne contamination.

The following discussion includes more specific uses of network computer models for water quality purposes (Rossman 2000):

1. Assessing alternative management strategies for improving water quality throughout a system including
   - altering source utilization within multiple source systems,
   - altering pumping and tank filling/emptying schedules,
   - use of satellite treatment, such as rechlorination at storage tanks, and
   - targeted pipe cleaning and replacement.
2. Study such water quality phenomena as
   - blending water from different sources,
   - age of water throughout a system,
   - loss of chlorine residuals,
   - growth of disinfection by-products, and
   - tracking contaminant propagation events.

3.   Other uses dealing with water quality include
   *   calculating travel time, residence time, or water age in a network,
   *   defining zones in a network served by a particular source and/or assessing the degree of blending with water from other sources,
   *   determining the impacts of accidental or intentional contamination, and
   *   identifying appropriate sampling locations within the water distribution network.

EPANET Capabilities (from Rossman 2000)

EPANET is a computer program that performs extended-period simulation of hydraulic and water quality behavior within pressurized pipe networks. A network consists of pipes, nodes (pipe junctions), pumps, valves, and storage tanks or reservoirs. EPANET tracks the flow of water in each pipe, the pressure at each node, the height of water in each tank, and the concentration of a chemical species throughout the network during a simulation period comprised of multiple time steps. In addition to chemical species, water age and source tracing can also be simulated.

EPANET is designed to be a tool for improving our understanding of the movement and fate of drinking water constituents within distribution systems. It can be used for many different kinds of applications in distribution systems analysis. Sampling program design, hydraulic model calibration, chlorine residual analysis, and consumer exposure assessment are some examples in which EPANET can help. Running under Windows, EPANET provides an integrated environment for editing network input data, running hydraulic and water quality simulations, and viewing the results in a variety of formats. These include color-coded network maps, data tables, time series graphs, and contour plots.

EPANET models a water distribution system as a collection of links connected to nodes. The links represent pipes, pumps, and control valves. The nodes represent junctions, tanks, and reservoirs. Figure 5-12 illustrates how these objects can be connected to one another to form a network.

Hydraulic Modeling Capabilities

Full-featured and accurate hydraulic modeling is a prerequisite for doing effective water quality modeling. EPANET contains a state-of-the-art hydraulic analysis engine that includes the following capabilities:

*   Places no limit on the size of the network that can be analyzed
*   Computes friction head loss using the Hazen–Williams, Darcy–Weisbach, or Chezy–Manning formulas
*   Includes minor head losses for bends, fittings, etc.
*   Models constant or variable speed pumps

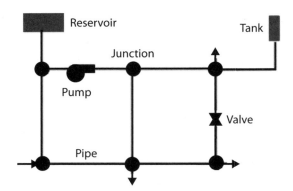

FIGURE 5-12 Physical Components in a water distribution system (Rossman 2000)

- Computes pumping energy and cost
- Models various types of valves including shutoff, check, pressure regulating, and flow control valves
- Allows storage tanks to have any shape (i.e., diameter can vary with height)
- Considers multiple demand categories at nodes, each with its own pattern of time variation
- Models pressure-dependent flow issuing from emitters (sprinkler heads)
- Can base system operation on both simple tank level or timer controls and on complex rule-based controls

Water Quality Modeling Capabilities

In addition to hydraulic modeling, EPANET provides the following water quality modeling capabilities:

- Models the movement of a nonreactive tracer material through the network over time
- Models the movement and fate of a reactive material as it grows (e.g., a disinfection by-product) or decays (e.g., chlorine residual) with time
- Models the age of water throughout a network
- Tracks the percent of flow from a given node reaching all other nodes over time
- Models reactions both in the bulk flow and at the pipe wall
- Uses nth order kinetics to model reactions in the bulk flow
- Uses zero or first-order kinetics to model reactions at the pipe wall
- Accounts for mass transfer limitations when modeling pipe wall reactions
- Allows growth or decay reactions to proceed up to a limiting concentration

- Employs global reaction rate coefficients that can be modified on a pipe-by-pipe basis
- Allows wall reaction rate coefficients to be correlated to pipe roughness
- Allows for time-varying concentration or mass inputs at any location in the network
- Models storage tanks as being either complete mix, plug flow, or two-compartment reactors

Output of Network Analysis

Figure 5-13 illustrates some of the types of results data provided by the EPANET model. For each node (junction), the demand in gal/min (gpm), head (ft), pressure (psi), and chlorine (mg/L) are listed. There are also many other types of output that are available.

HYDRAULIC TRANSIENTS: WATER HAMMER

What is Water Hammer?

Water distribution systems are pressure pipe systems subject to a wide range of physical loads and operational requirements. Underground piping systems must be able to withstand mechanical forces caused by fluid pressure, differential settlement, and concentrated loads. The internal pressure requirement of pipes is also very important because it directly influences the pipe wall thickness of larger pipes, and because pipe manufacturers characterize the mechanical strength of a pipe by the pressure rating. Hydraulic transient flow occurs when flow changes suddenly from one steady-state flow to another steady-state flow. Transient flow is the intermediate stage flow during the transition. This hydraulic transient is also referred to as water hammer. By definition, water hammer is a pressure (acoustic) wave phenomenon created by relatively sudden changes in liquid velocity (Martin 2000). The total force acting within a pipe is the sum of the steady-state pressure and the water hammer pressure in the pipe.

Node ID	Demand GPM	Head ft	Pressure psi	Chlorine mg/L
Junc 10	0.00	1010.67	130.28	1.00
Junc 11	210.00	992.42	122.37	0.85
Junc 12	210.00	990.17	121.40	0.79
Junc 13	140.00	977.08	122.23	0.30
Junc 21	210.00	977.24	120.13	0.74
Junc 22	280.00	976.29	121.88	0.49
Junc 23	210.00	975.76	123.82	0.30
Junc 31	140.00	970.32	117.13	0.53

FIGURE 5-13 Example output for network nodes (Rossman 2000)

Water hammer occurs when a valve such as depicted in Figure 5-14 is closed too quickly. Consider the steady-state flow of water from the tank or reservoir through the pipe with velocity V as shown in Figure 5-14. The pipe has a valve at the end of the pipe that is instantaneously closed, causing a pressure wave to develop upstream of the valve. This fast closure causes a pressure wave having a much greater pressure than the steady-state flow pressure to develop just upstream of the valve. This pressure wave travels upstream toward the reservoir at approximately the speed of sound in water, v_c. Water in the pipe between the valve and the front of pressure wave is at rest and the water between the pressure wave and the reservoir has the initial velocity V. The pressure in the water between the pressure wave and the valve is the initial pressure, p_o, plus the increased pressure, Δp, due to the sudden valve closure. When the pressure wave reaches the reservoir, the pressure imbalance causes water to flow from the pipe into the reservoir at velocity V. This causes a new pressure wave that travels back to the valve.

As the pressure wave travels to the valve, the pressure in the pipe between the reservoir and the pressure wave returns to the initial pressure p_o, and the velocity V is toward the reservoir. When this pressure wave reaches the valve, water in the pipe is flowing toward the reservoir at velocity V. When the pressure wave reaches the valve, a rarefied pressure wave moves back toward the reservoir with a much smaller pressure, less than the pressure in the reservoir. The velocity in the pipe between the pressure wave and valve is now zero. When the pressure wave reaches the reservoir, the water velocity in the entire length of the pipe is zero, and the pressure in the pipe is less than in the reservoir at the pipe end. This pressure imbalance causes flow once again in the pipe toward the valve. This process of the pressure wave moving back and forth from the valve to the reservoir continues until frictional forces dampen out the pressure wave.

Methods that determine the magnitude of the pressure wave consider an instantaneous closure of the valve and can also include the expansion and contraction of the pipes. The increased pressure, Δp, due to the sudden valve closure is computed by multiplying the density of water times the initial velocity of the flow (V) times the pressure wave velocity (v_c), expressed as

$$\Delta p = \rho V v_c$$

This equation is for a rigid pipe, i.e., assuming no expansion or contraction of the pipe with the increased water hammer pressure.

Water hammer is important in water distribution systems because it can cause (1) rupture of pipe and pump casings; (2) pipe collapse; (3) vibration; (4) excessive pipe displacement, pipeline fitting, and support deformation and/or failure; and (5) vapor cavity formation (cavitation, column separation) (Bosserman and Hunt 1998).

How Water Hammer Causes Damage

To give an idea of the increased pressure due to the water hammer created by the sudden closure of the valve, consider that the increased pressure (water hammer pressure) is $\Delta p = \rho V v_c$

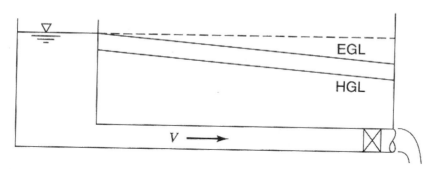

FIGURE 5-14 Steady-state flow in a pipe

for a rigid pipe. For a 2-mile-long, 24-in. diameter pipe, as depicted in Figure 5-14, determine the water hammer pressure for the assumption of an instantaneous closure of the downstream valve if the initial velocity of flow is 4.0 ft/sec. The initial static pressure in the pipe is 86.4 psi. Assuming the pipe is rigid so that $v_c = 4718$ ft/sec and the density of water is 1.94 slugs/ft^3, then

$$\Delta p = 1.94(4.0)(4718) = 36,611 \text{ lb/ft}^2 = 254 \text{ psi}$$

If we considered this a nonrigid pipe, we would take into account the expansion and contraction of the pipe, which would consider the pipe wall thickness, pipe diameter, modulus of elasticity of the pipe, and elasticity of the water. This would reduce the water hammer pressure some, but it would still be very high as compared to the initial pressure in the pipe of around 86.4 psi.

The time for the pressure wave to travel from the valve to the reservoir and back to the valve for the situation in Figure 5-14 is $2L/v_c$. If at the time of closure t_c the valve is less than $2L/v_c$, then the maximum pressure developed at the valve is essentially the same as for instantaneous closure, and the valve closure is considered as a rapid closure. The critical time of closure is then defined as $t_c = 2L/v_c$. For the 2-mile-long pipeline, $t_c = 2(2(5,280))/4,718 = 4.48$ seconds. For closure times greater than this, the full pressure wave does not occur at the valve. The greater the closure time, the less is the pressure wave at the valve.

Causes of hydraulic transients in water distribution systems include: (1) valve opening and closing (frequently occurs), (2) flow demand changes (rarely occurs), (3) controlled pump shutdown (rarely occurs), (4) pump failure (often occurs), (5) pump start-up (rarely occurs), (6) air-venting from lines (frequently occurs), (7) failure of flow on pressure regulation (rarely occurs), (8) pipe rupture (rarely occurs) (Bosserman and Hunt 1998). Pipelines and distribution systems can be subject to surges very frequently and in many cases almost daily. Over time, this can cause damage to the pipeline and equipment.

Surge Control in Pumping Stations and Pipelines

Power failures resulting in the sudden stoppage of pumps is the most frequent cause of surges. An unprotected pump system can result in damage to the equipment and to the pipeline. Surges can cause damages ranging from the loosening of pipe joints to damaging pumps and valves and even pipeline breaks and the shifting of pipes. Even loosened pipe joints can result in contamination to the distribution system from groundwater and back-flow conditions. Pumps do not need to stop quickly nor does a valve need to close suddenly (instantaneously) to cause damaging surges. Basically, almost all pumping systems need surge control equipment to prevent water hammer surges resulting from power failure.

A typical water pumping/distribution system is shown in Figure 5-15. This system has two parallel pumps that pump water from a wet well into a pump header and then the distribution system. Both a relief valve and a surge tank are used for relieving and preventing surges. Figure 5-16 shows a well service air valve for a well pump (vertical turbine pump) that is used to relieve the pressurized air in the pump column during the pump start-up. Figure 5-17 is a schematic of a well service air valve for a vertical turbine pump. This valve is a float-operated valve that is normally open. As water enters the valve, the float rises and closes to prevent the discharge of water.

Pump control valves are used to control surges associated with the starting and stopping of pumps. Pump control valves are wired to the pump electrical circuit to provide an adjustable opening and safe closing times. These closing times are in excess of the critical times for closure. Various types of valves can be used such as butterfly, plug, ball, and others, such as globe-pattern control valves.

When a pump control valve is used with a vertical pump, an air-release valve equipped with a vacuum breaker, as shown in Figure 5-18, can be used. When a pump is started, the opening of the control valve is delayed a few seconds. This allows the air-release valve to expel air slowly through an orifice.

Surge relief valves in many cases are a practical means of surge control. Essentially, a pressure surge lifts a disc that allows the valve to release water to the air or into a wet well. Figure 5-19 shows a globe-pattern control valve that is equipped with surge relief and anticipator controls. The surge anticipator valve opens upon sensing a high- or low-pressure event.

Air valves can also be used to reduce surges in pipelines. These valves essentially prevent the formation of air pockets in pipelines during normal operation. Figure 5-20 shows a surge-suppression air valve that can be used when column separation is expected at an air valve location. A vacuum-breaker valve is another type of air valve that is used where column separation may occur in large pipelines. These valves have components similar to the regulated exhaust device, except the vacuum-breaker disc is held closed by a spring, while the regulated exhaust disc is normally open. The vacuum-breaker valve cannot expel air as it only admits air to prevent the formation of a vacuum pocket, keeping a positive pressure in the pipeline and reducing the surge associated with column separation.

Various other methods are used to control the effects of water hammer in water distribution systems including:

- Air chambers
- Surge tanks (Figure 5-21)
- Air arrestor vacuum tanks (Figure 5-22)
- High-pressure-rated piping, valves, and equipment

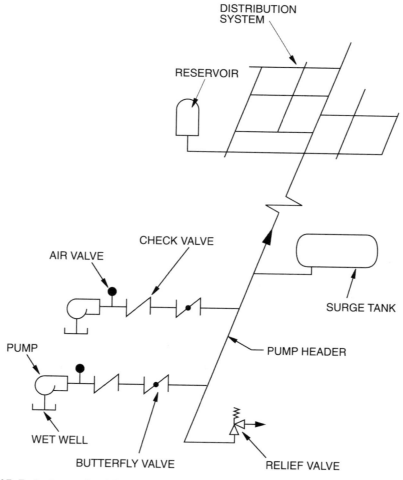

FIGURE 5-15 Typical pumping/distribution system
Courtesy of Val-Matic Valve and Manufacturing Corporation

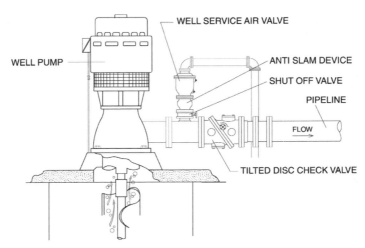

FIGURE 5-16 Vertical turbine well pump with a well service air valve
Courtesy of Val-Matic Valve and Manufacturing Corporation

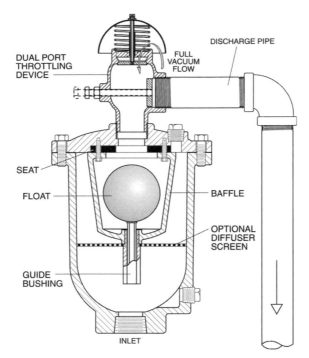

FIGURE 5-17 Well service air valve
Courtesy of Val-Matic Valve and Manufacturing Corporation

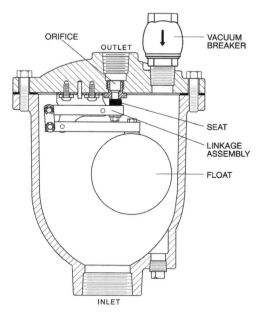

FIGURE 5-18 Air release valve
Courtesy of Val-Matic Valve and Manufacturing Corporation

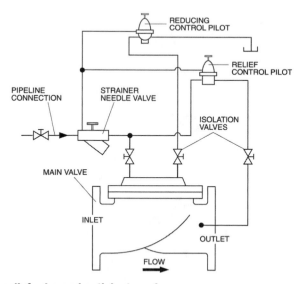

FIGURE 5-19 Surge relief valve and anticipator valve
Courtesy of Val-Matic Valve and Manufacturing Corporation

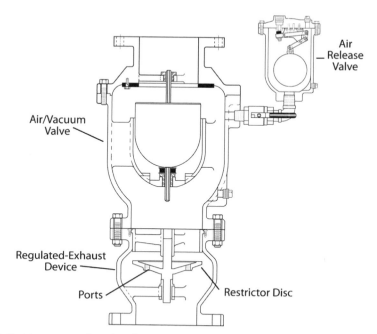

FIGURE 5-20 Surge-suppression air valve
Courtesy of Val-Matic Valve and Manufacturing Corporation

FIGURE 5-21 Surge arrestor tank
Courtesy of AA Tanks

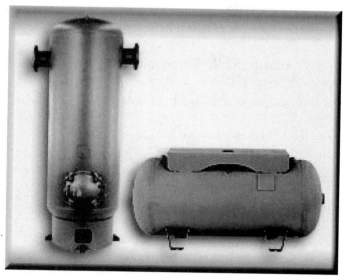

FIGURE 5-22 Air arrestor vacuum tanks
Courtesy of AA Tanks

BIBLIOGRAPHY

Bosserman, B.E. II and Hunt, W.A. 1998. Fundamentals of hydraulic transients, in *Pumping Station Design,* R.L. Sanks, ed. Woburn, Mass.: Butterworth-Heinemann.

Cesario, A.L., 1991. Network analysis from planning , engineering, operations, and management perspectives. *Journal AWWA* 83: 38–42.

Manual M32, Distribution Network Analysis for Water Utilities. 2004. Denver, Colo.: AWWA.

Martin, C.S. 2000. Hydraulic transient design for pipeline systems, in *Water Distribution Systems Handbook,* L.W. Mays, ed. New York: McGraw-Hill.

Mays, L.W., ed. 2000. *Water Distribution Systems Handbook.* New York: McGraw-Hill.

Mays, L.W., ed. 2004. *Water Supply Systems Security.* New York: McGraw-Hill.

Rossman, L. 2000. *EPANET2 User's Manual,* EPA/600/R-00/057. Cincinnati, OH: US Environmental Protection Agency.

Rossman, L. and W.M. Grayman. 1999. Scale-model studies of mixing in drinking water storage tanks. ASCE *Journal of Environmental Engineering* 125 (8):755–761.

US Environmental Protection Agency (USEPA). 2002. Effects of water age on distribution system water quality. Prepared by AWWA and Economic and Engineering Services, Inc. Washington, D.C.: USEPA Office of Ground Water and Drinking Water.

US Environmental Protection Agency (USEPA). 2005. *Water Distribution Systems Analysis: A Reference Guide for Utilities*, EPA Report 600/R-06/028, Cincinnati, OH: USEPA.

Val-Matic Valve and Manufacturing Corporation. 2007. Surge Control in Pumping Stations, *Pumps & Systems*, March.

Val-Matic Valve and Manufacturing Corporation. 2009. Surge Control in Pumping Stations, Elmhurst, Illinois.

Walski, T. Interpreting Extended Period Simulation Models, Vol. 1, No. 1, *Current Methods*, Exton, Penn.: Bentley Haestad Methods.

CHAPTER 6

Valves

Numerous types of valves are required in every water system. They are necessary to operate the distribution system as well as to control treatment processes, pumps, and other equipment. The correct size and type of valves must be selected for each use, and most valves require periodic checking to ensure they are operating properly.

USES OF WATER UTILITY VALVES

Water utility valves are designed to perform several different functions. The principal uses are to

- start and stop flow,
- isolate piping,
- regulate pressure and throttle flow,
- prevent backflow, and
- relieve pressure.

Valves to Start and Stop Flow

Most valves installed in a water system are intended to start and stop flow. They are designed to be either fully open or fully closed under normal conditions. These valves are generally not intended for throttling flow and may be damaged if used in a partially open position for an extended period of time.

Distribution System Isolation Valves

Isolation valves are installed at frequent intervals in distribution piping so that small sections of water main may be shut off for maintenance or repair. The closer the valves are spaced, the fewer the number of customers who will be inconvenienced by having their water turned off while repairs are being made. When a system is laid out in a grid pattern, normal practice is to install at least two valves at each intersection, as illustrated in Figure 6-1. Distribution system valves should normally be maintained in a fully open position.

Hydrant auxiliary valves

An isolation valve is usually installed on the section of pipe leading to each fire hydrant. This allows pressure to be turned off during hydrant maintenance without disrupting service to customers. Fire hydrant design and installation are detailed in chapter 7.

Pump control valves

As discussed further in chapter 4, each water system pump must normally be furnished with at least two control valves. A discharge valve is essential to stop water from flowing backward through a pump that is not in operation. A valve is also usually installed on the suction side of each pump for use during pump repair.

Water service valves

The valves used to connect water services to water mains are usually referred to as corporation stops. They are ordinarily buried, so they are used only when service is being initiated or discontinued.

In service lines, the valves used for temporarily shutting off water to a building are commonly called curb stops. Additional valves are also installed on each side of a water meter for use when the meter is removed for repair. Water service equipment is discussed further in chapter 15.

Valves for Regulating Pressure and Throttling Flow

Throttling the flow of water requires special valve designs that are durable over a long period of time. The two principal types of throttling valves used in a water system are pressure-reducing valves and altitude valves.

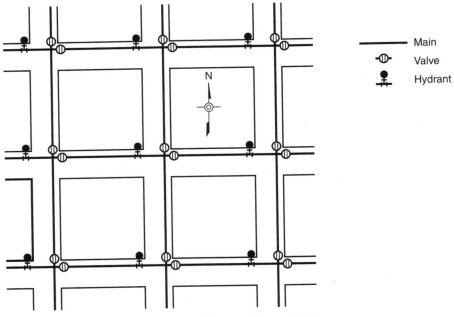

FIGURE 6-1 Valves installed at intersection of mains in a grid pattern

Pressure-reducing valves

As illustrated in Figure 6-2, it is sometimes necessary to create two or more separate pressure zones in a distribution system in order to furnish all customers with water at an adequate (but not excessive) pressure. The most common way of accomplishing this is to take water from a higher-pressure zone and reduce the pressure for the lower zone by using pressure-reducing valves. These valves operate automatically to throttle flow and maintain a lower pressure in the lower distribution system zone.

Altitude valves

Ground-level reservoirs are usually filled through an altitude valve. This type of valve is designed to let the reservoir fill at a regulated rate and to stop flow completely when the tank is full. Altitude valves are also used to control flow to an elevated tank when the tank is not high enough to accept full system pressure. The valve will automatically shut off flow to the tank to keep the tank from overflowing.

Single-acting altitude valves allow water to flow in only one direction. Double-acting valves allow flow in both directions. Figure 6-3 illustrates two types of typical water tank installations.

Valves for Preventing Backflow

There are numerous places in a water distribution system where unsafe water may be drawn into the potable water mains if a temporary vacuum should occur in the system. In addition, contaminated water from a higher-pressure source can be forced through a water system connection that is not properly controlled. Several types of valves are available specifically for installation wherever backflow might occur. This equipment is detailed in chapter 11.

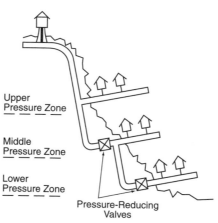

FIGURE 6-2 Pressure-reducing valve installed on a system that has three pressure zones

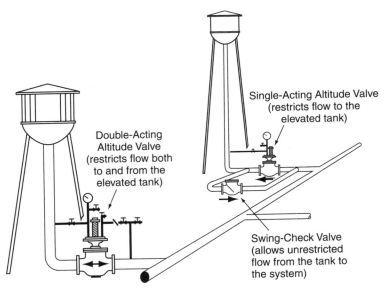

FIGURE 6-3 Altitude valves
Courtesy of GA Industries, Inc.

Valves for Relieving Pressure

Special valves are necessary to protect water systems and plumbing from excessive pressure. They are designed to remain tightly shut under normal pressure but will open when excessive pressure occurs.

Pressure-relief valves

A rapid increase in pressure in a water distribution system is usually called water hammer. The pressure wave that moves rapidly down a pipe can damage valves, burst pipes, or blow pipe joints apart. A common cause of water hammer is opening or closing a valve too fast. Pressure-relief valves can be installed to release some of the energy created by water hammer.

In household plumbing systems, small pressure-relief valves must be installed on all hot water heaters and boilers on customer services. Their purpose is to allow steam and excess pressure to blow off in the event of equipment overheating. If the excess pressure is not allowed to vent, either the tank or boiler will burst or hot water will be forced back through the water service into the water distribution system. A typical pressure-relief valve is shown in Figure 6-4.

Air-relief valves

One of the more common uses of air-relief valves is to automatically vent air that accumulates at high points in transmission pipelines. Air pockets in water mains can substantially

FIGURE 6-4 Pressure-relief valve
Courtesy of Watts Regulator Co.

reduce the effective area of flow through the pipe. This reduction will result in increased pumping costs and restricted flow to parts of the system.

Air-relief valves must also be installed on the discharge of most well pumps to vent air that has accumulated in the well column while the well is not in use. A typical air-relief installation is shown in Figure 6-5.

CLASSIFICATION OF WATER UTILITY VALVES

Most valves used in water systems fall into one of the following general classifications (Figure 6-6):

- Gate valves
- Globe valves
- Needle valves
- Pressure-relief valves
- Air-and-vacuum relief valves
- Diaphragm valves

FIGURE 6-5 Air-relief valve on a well pump discharge
Courtesy of Henry Pratt Company

- Pinch valves
- Rotary valves
- Butterfly valves
- Check valves

The designs of various types of valves are covered in several American Water Works Association (AWWA) standards.

Gate Valves

The most common type of valve found in a water distribution system is the gate valve. The gate, or disk, of the valve is raised and lowered by a screw, which is operated by a handwheel or valve key. When fully open, gate valves provide almost unrestricted flow because the gates are pulled fully up into the bonnet. When closed, the gate seats against the two faces of the valve body, and closes relatively tightly unless the faces have become worn or something becomes lodged under the gate.

Gate valves are not designed to be used to regulate or throttle flow. If used for throttling, the gate mechanism will vibrate and eventually be damaged.

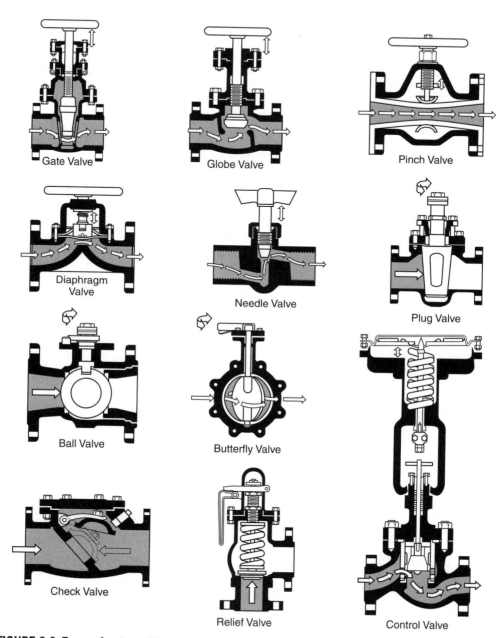

Gate Valve

Globe Valve

Pinch Valve

Diaphragm Valve

Needle Valve

Plug Valve

Ball Valve

Butterfly Valve

Check Valve

Relief Valve

Control Valve

FIGURE 6-6 Types of water utility valves

Courtesy of the Valve Manufacturers Association of America, Washington, D.C.

Generally, gate valves are not installed where they will need to be operated frequently because they require too much time to operate from the fully open to closed positions. Sizes used in water distribution systems generally range from ¾ in. (19 mm) in diameter for water services up to 72 in. (1,830 mm) used on transmission mains. To prevent water from leaking past the stem of a gate valve, a seal is provided either by O-rings or conventional packing.

Nonrising-stem gate valves

The most common type of gate valve installed for buried service is the iron-body, bronze-mounted, nonrising-stem, double-disk gate valve (Figure 6-7). The body is made of cast or ductile iron, and the sealing and operating parts are made of bronze. When double-disk valves are closed, the disks are pushed apart by a wedging mechanism as they reach the bottom of the valve. This mechanism pushes the disks outward to seal against the seat rings. The valve stem itself rotates but does not move up and down. The screw mechanism is located within the valve body, so valves can be directly buried. There is no problem with dirt affecting the mechanism.

Rising-stem gate valves

In situations where a valve is not buried and the water system operator will need to know from inspection whether the valve is open or closed, a rising-stem valve with an outside screw and yoke is often used (Figure 6-8). Outside screw and yoke valves are frequently used in treatment plants and pumping stations.

Horizontal gate valves

Gate valves of the double-disk type, in sizes 16 in. (400 mm) or larger, are often designed to lie horizontally. This way, the valve-operating mechanism does not have to lift the weight of the gate to open the valve (Figure 6-9). These valves may also be used where a large main is not buried very deeply and a vertical valve would extend above ground level.

Horizontal valves are equipped with special tracks in the valve body and bonnet. Rollers on the disks ride in the tracks throughout the length of travel to support the weight of the disks. Bronze scrapers are also provided to move ahead of the rollers in both directions to remove any foreign matter from the tracks. In rolling-disk valves, the disks themselves serve as rollers.

Because of the weight of the disk and the internal friction in the valve, large gate valves are furnished with a geared operator. Several turns of the operating nut may therefore be required to make one turn of the operating shaft. It may take 30 minutes or more to fully open or close a large gate valve. If the valve is old, it usually requires two operators to turn the valve key by hand. Most water utilities with large valves now use a portable power-operated valve operator for operating valves. These units greatly reduce the time and effort required in operating valves.

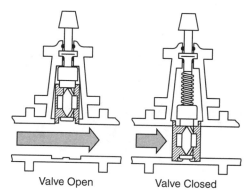

FIGURE 6-7 Operation of nonrising-stem gate valve

FIGURE 6-8 Valve with an outside screw and yoke
Courtesy of Mueller Company, Decatur, Ill.

Bypass valves

Horizontal thrust against the disk of a closed gate valve increases rapidly with valve size. For example, the thrust against a closed valve that is under 75 psi (517 kPa) water pressure is approximately 8,480 lbf (37,700 N) for a 12-in. (305-mm) valve. For the same pressure, the thrust is 15,080 lbf (67,000 N) for a 16-in. (406-mm) valve, and 76,340 lbf (339,000 N) for a 36-in. (914-mm) valve. These thrusts cause a lot of friction on the valve guides, which

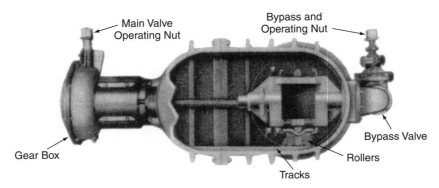

FIGURE 6-9 Horizontal gate valve
Courtesy of Mueller Company, Decatur, Ill.

makes large gate valves extremely difficult to open. For this reason, bypass valves may be needed. Bypass valves are smaller-diameter valves that will allow water to pass around the larger valve. This reduces the differential pressure across the closed disk and makes the main valve easier to open and close. Bypass valves also provide for low-volume flow without the need for opening the main valve.

Tapping valves

As illustrated in Figure 6-10, a tapping valve is a specially designed gate valve used to connect a new water main to an existing main under pressure. One end of the valve is a flange designed to attach to a tapping sleeve, and the other end is usually a regular mechanical joint pipe bell. Tapping valves have a slightly larger inside diameter to allow the tapping machine cutter to pass. Details of tapping procedures are included in chapter 12.

Cutting-in valves

As illustrated in Figure 6-11, a cutting-in valve has one oversized end connection designed to be used with a cutting-in sleeve. The valve and sleeve used together greatly facilitate installation of a new valve in an existing main. Main pressure must be shut off for a period of time to make the installation. This approach is a relatively inexpensive way of adding a valve in a distribution system.

Inserting valves

When it is necessary to install a valve in a water main without shutting off pressure, a special inserting valve can be used (Figure 6-12). Water utilities normally have an inserting valve installed by a company that has the special equipment required to make the installation and specializes in this work.

FIGURE 6-10 Tapping valve
Courtesy of Mueller Company, Decatur, Ill.

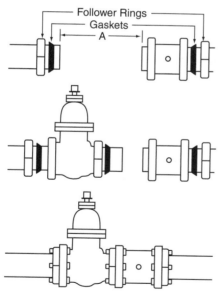

1. Cut out a section of pipe (A) to the length specified by the valve manufacturer.

   Slip a follower ring and gasket over each of the cut ends.

   Slide the sleeve on the one cut end.

2. Mount the valve on the opposite cut end.

   Place a follower ring and gasket on the valve spigot.

3. Push the sleeve up to the valve.

   Tighten all bolts.

FIGURE 6-11 Installation of a cutting-in valve
Courtesy of U.S. Pipe and Foundry Company

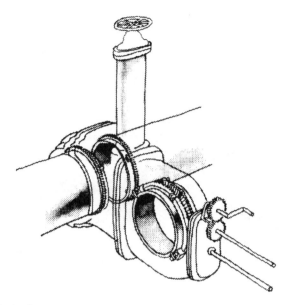

FIGURE 6-12 Inserting valve

Resilient-seated gate valves

New designs of gate valves are available that use a resilient seat of rubber or synthetic material. When the valve is closed, the seat provides a seal between the valve body and the disk. The design offers easier operation and leak-tight shutoff. Designs vary, but in general the resilient material is bonded to either the valve body or the disk. Figure 6-13 shows a cutaway of a resilient-seated valve.

Slide valves

Where the pressure to be regulated is relatively low and tight shutoff is not important, knife gate or sluice gate valves may be used. These valves use a relatively thin gate or blade that slides up and down in a recess, as illustrated in Figure 6-14. Slide valves may be square, oblong, or round in shape.

Globe Valves

The globe valve principle is commonly used for water faucets and other household plumbing. As illustrated in Figure 6-6, the valves have a circular disk that moves downward into the valve port to shut off flow. Although they seat very tightly, globe valves produce high head loss when fully open. They are not suited for distribution mains where head loss is critical.

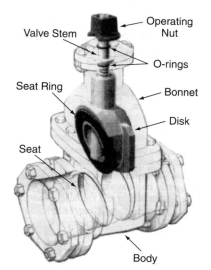

FIGURE 6-13 **Parts of a resilient-seated gate valve**
Courtesy of Mueller Company, Decatur, Ill.

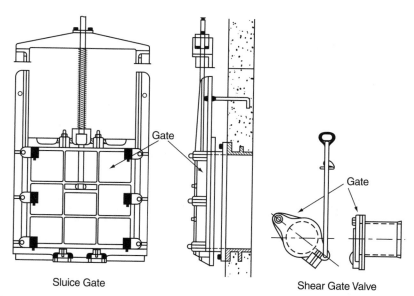

Sluice Gate   Shear Gate Valve

FIGURE 6-14 **Examples of slide valves**

Needle Valves

Needle valves are similar to globe valves except that a tapered metal shaft fits into a metal seat when the valve is closed (see Figure 6-6). Needle valves are available only in small sizes and are primarily used for precise throttling of flow. A common water utility practice is to install needle valves on the hydraulic lines to valve actuators. The needle valves then allow the opening and closing speeds of the actuator to be adjusted.

Pressure-Relief Valves

Pressure-relief valves are similar to globe valves, but their disks are normally maintained against the seat by a spring. The tension on the spring can be adjusted to allow the valve to open if the desired preset pressure is exceeded.

Air-and-Vacuum Relief Valves

As illustrated in Figure 6-15, air-and-vacuum relief valves consist of a float-operated valve that will allow air to escape as long as the float is down. When water fills the container, the float rises and closes the valve. If trapped or entrained air subsequently enters the unit, the float drops long enough to vent the air and then recloses. If a vacuum occurs in a pipeline, the relief valve will admit air to prevent collapse or buckling of the pipe.

Diaphragm Valves

Diaphragm valves operate similarly to globe valves. A manually operated diaphragm valve is illustrated in Figure 6-6. In other types of diaphragm valves, water pressure exerted on a diaphragm is used to assist in closing the valve. Figure 6-16 illustrates the principal parts of an altitude valve, which is the most common type of diaphragm valve used in a water system. Figure 6-3 illustrates how altitude valves are connected to control water to an elevated tank.

Pinch Valves

Pinch valves are closed by pinching shut a flexible interior liner. This type of valve is available only in relatively small sizes, but it is particularly useful for throttling the flow of liquids that are corrosive or might clog other types of valves. The cross section of a pinch valve is illustrated in Figure 6-6.

Rotary Valves

The two principal types of rotary valves used in water systems are plug valves and ball valves. The movable element in a plug valve (Figure 6-6) is a cylinder-shaped or cone-shaped plug that has a passageway or port through it. It requires a one-quarter turn to move from fully open to fully closed. Small plug valves are used as curb stops and corporation stops.

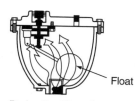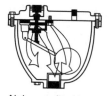

Float

During the filling of the line, air entering the valve body will be exhausted to atmosphere. When the air is expelled and water enters the valve, the float will rise and cause the orifices to be closed.

The large and small orifices of the air-and-vacuum valve are normally held closed by the buoyant force of the float.

While the line is working under pressure, small amounts of trapped or entrained air are exhausted to atmosphere through the small orifice.

Air is permitted to enter the valve and replace the water while the line is being emptied.

FIGURE 6-15 Operation of an air-and-vacuum relief valve
Courtesy of GA Industries, Inc.

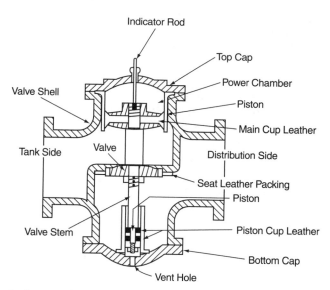

FIGURE 6-16 Principal parts of an altitude valve

Large plug valves are most often used where pressure is relatively high and positive shutoff is required, such as on the discharge of a high-lift pump. Plug valves may also be used for throttling flow without damage.

Ball valves (Figure 6-6) operate similarly to plug valves, except that the operating element is a sphere with a circular hole bored through it. Both plug and ball valves are widely used for water service connections.

Butterfly Valves

Butterfly valves (Figure 6-17) consist of a body in which a disk rotates on a shaft to open or close the valve. In the fully open position, the disk is parallel to the axis of the pipe and the flow of water. In the closed position, the disk seals against rubber or synthetic elastomer bonded either on the valve seat of the body or on the edge of the disk. Because the disk of a butterfly valve stays in the water path in the open position, the valve creates a somewhat higher resistance to flow (i.e., a higher pressure loss) than a gate valve. On the other hand, butterfly valves have the advantage of operating easily and relatively quickly. Water pressure on both halves of the disk is relatively equal as the valve is operated.

The laying length (i.e., the length of the valve body along the pipe axis) of butterfly valves is quite short. In the open position, the disk may extend beyond the valve body into the piping on either side. For instance, a 16-in. (400-mm) diameter flanged-end butterfly valve has an 8-in. (200-mm) laying length. A gate valve for the same pipe diameter is 17 in. (430 mm) long.

Wafer-type butterfly valves require even less laying length because they have no flanges. Instead they are sandwiched between flanges on the adjoining piping. Wafer-type valves are also considerably less expensive because they are easier to manufacture.

The lug-wafer design is basically the same as the wafer design except for the addition of lugs around the body. The lugs are drilled and tapped to allow the valve to be attached directly to one of the flanges for easy pipeline modification or cleaning.

The cost of large butterfly valves is substantially less than the cost of large gate valves. Butterfly valves are also easier and quicker to operate. There is one main disadvantage to using butterfly valves in a distribution system. If it should ever be necessary to clean a main by the use of pigs or swabs, the operation would be blocked by the valve disks. In addition, when butterfly valves are used in a distribution system, they must not be operated too quickly or serious water hammer could result.

The short laying length of butterfly valves makes them particularly useful in piping for pumping stations and treatment plants, where space is often limited. Although butterfly valves are primarily designed for on–off service, they can be used in some situations for throttling under low-flow and low-pressure conditions. They should not be used for high-pressure throttling, however, because the disk will vibrate and eventually be damaged.

Check Valves

Check valves are designed to allow flow in only one direction. They are commonly used at the discharge of a pump to prevent backflow when the power is turned off. Foot valves are a special type of check valve installed at the bottom of the pump suction so that the pump will not lose its prime when power is turned off. If allowed to close unrestrained, a check valve may slam shut, creating water hammer that can damage pipes and valves. A variety of devices are available to minimize the slamming of check valves. These devices include external weights, restraining springs, and automatic slow-closing motorized drives. Various designs of check valves commonly used in water systems are illustrated in Figure 6-18. Backflow-prevention devices are also a type of check valve. They are described in more detail in chapter 11.

FIGURE 6-17 Butterfly valve with electric actuator
Courtesy of Tyco Valves & Controls

VALVE OPERATION

The principal methods of operating water system valves are

- manual,
- electrical,
- hydraulic, and
- pneumatic.

The method used for operating valves depends primarily on how the valve is used and how frequently it must be operated.

Manual Operation

Manual operation of an exposed valve, such as in a treatment plant, is usually with a hand-wheel, chainwheel, or floor stand operator, as shown in Figure 6-19. Buried distribution system valves are usually fitted with a 2-in. (50-mm) square operating nut for manual operation from the surface. Figure 6-20 illustrates a water main valve-operating key. Service-line valves are equipped with tee heads and require a slotted key for operation.

Slanting Disk Check Valve Cushioned Swing Check Valve

Rubber Flapper Swing Check Valve

Double Door Check Valve

Foot Valve

FIGURE 6-18 Five types of check valves
Reprinted with permission of APCO/Valve & Primer Corp.

Power Actuators

Valves may be operated by any of several different types of power actuators. Each has some advantages and disadvantages in terms of speed of operation, reliability, and ease of control.

Electric actuators

Electric actuators use a small electric motor to rotate the valve stem through a gear box. The actuator is turned on by a switch or by remote signal. It continues to operate until turned off by a limit switch when the valve is fully open or closed. Some actuators are fitted with controls that can be adjusted to leave the valve partially open for use in throttling. An electric valve actuator is shown attached to a butterfly valve in Figure 6-17. The handwheel on the unit is for emergency operation of the valve by hand in the event of a power failure.

Hydraulic actuators

Valves in plants and pumping stations are frequently operated by hydraulic cylinders using either plant water pressure or hydraulic fluid. The fluid is admitted to the cylinders through electric solenoid valves to operate the valve in each direction.

FIGURE 6-19 Handwheel operator
Courtesy of M&H Valve and Fire Hydrant

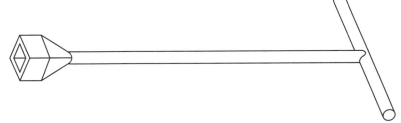

FIGURE 6-20 Valve key for water main valves

Although water-operated valve cylinders are widely used, there are two warnings that should be observed. Under some water quality conditions, deposits may build up within the cylinders. The cylinders will have to be cleaned periodically to maintain smooth operation. The other problem is that if system pressure is used to operate the valves, there may be no means of operating the valves in the event of low system pressure. Some water plants guard against this by having all valve actuators operated by a separate pressure system that has a backup pressure tank.

Pneumatic actuators

Pneumatic actuators operate much like hydraulic actuators, except that compressed air is used as the operating force. The installation requires a source of compressed air with sufficient reserve capacity to provide continued operation after a power failure. Although pneumatic actuators can be used on isolation valves, they are more common on control valves.

Actuator operating speed

It is usually advisable for valve actuators to operate slowly to prevent water hammer in the system. However, if there is a power failure, the pump discharge valves must close fast enough to prevent water from flowing backward through the pumps. The operating speeds of hydraulic and pneumatic operators are usually individually adjustable for each direction by means of needle valves that control the flow of fluid or air to the cylinder.

VALVE STORAGE

Whenever possible, valves should be stored indoors. If outside storage is necessary, gears, power actuators, cylinders, valve ports, and flanges should be protected from the weather. Valves stored outside in freezing climates should be closed and have their disks in a vertical position. If the disks are in a horizontal position, rainwater will collect on top of the disk and may seep into the valve body, freeze, and crack the casting.

VALVE JOINTS

Iron and brass valves 2 in. (50 mm) and smaller are generally furnished with female (inside) iron pipe thread. The most common joint types furnished with larger metallic valves are flanged, mechanical, and push-on. Plastic valves are also commonly furnished with solvent-weld joints. Figure 6-21 illustrates valves with typical joints.

VALVE BOXES AND VAULTS

Buried valves are made accessible for operation either by a valve box placed over them or by a valve vault built around them. Since valves are placed at various depths, valve boxes are made in two or more pieces that telescope to adjustable depths underground. Figure 6-22 shows a typical valve box installation.

When cast-iron valve boxes are used, they should rest above the valve so the weight of traffic passing over them will not be transferred to the valve or the pipe. The bottom flared edge of the box may require extra support. Valve boxes must be installed vertically and centered around the operating nut of the valve to ensure that the valve wrench can engage the nut.

Valve vaults are also often installed around valves to provide access for operation and maintenance. Round vaults are usually constructed in the field. They are formed of either

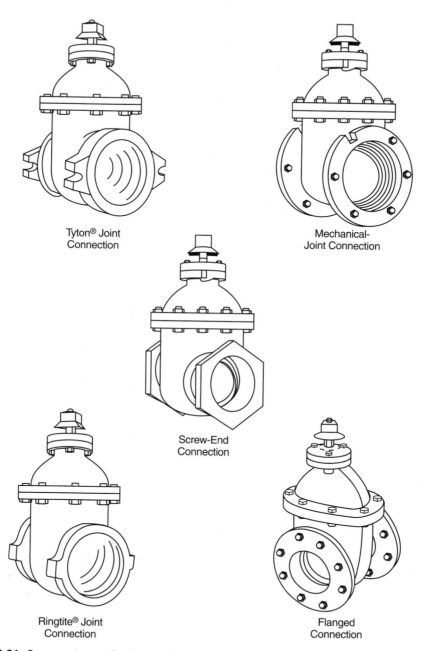

Tyton® Joint
Connection

Mechanical-
Joint Connection

Screw-End
Connection

Ringtite® Joint
Connection

Flanged
Connection

FIGURE 6-21 Common types of valve couplings

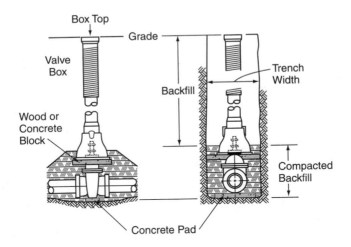

FIGURE 6-22 Valve box installation
Courtesy of the Ductile Iron Pipe Research Association

rounded concrete blocks or precast concrete segments (Figure 6-23). When valves that have exposed gearing or operating mechanisms must be installed below grade (underground), a vault must be installed around them. When a vault is constructed around a valve, the operating nut must be accessible for operation with a valve key from the surface through the access cover or through a special opening in the top slab.

Care must be taken when constructing valve vaults to ensure that the weight of the vault will not rest on the water main. Valve vault drains should not be connected to storm drains or sanitary sewers because the water system could become contaminated if the sewer should back up and flood the vault. Vaults are usually drained to underground absorption pits.

VALVE RECORDS

It is extremely important that accurate records be maintained on each valve in a distribution system. Records of each valve's location must include measurements from at least three different permanent reference objects. This method makes it possible to locate the valve even if some of the reference objects are changed or moved over a period of years. The record should also include the make, size, date of installation, and type of valve to facilitate obtaining the proper parts if the valve must be repaired. Valve records are discussed in more detail in chapter 16.

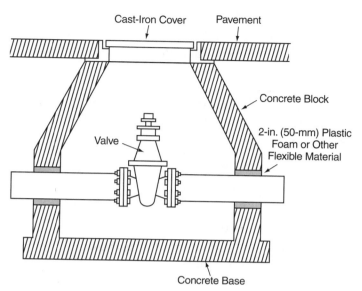

FIGURE 6-23 Valve vault

BIBLIOGRAPHY

AWWA Standard for Air-Release, Air/Vacuum, and Combination Air Valves for Waterworks Service. ANSI/AWWA C512. Denver, Colo.: American Water Works Association (latest edtion).

AWWA Standard for Ball Valves, 6 In. Through 48 In. (150 mm Through 1,200 mm). ANSI/AWWA C507. Denver, Colo.: American Water Works Association (latest edition).

AWWA Standard for Electric Motor Actuators for Valves and Slide Gates. ANSI/AWWA C542-09. Denver, Colo.: American Water Works Association (latest edition).

AWWA Standard for Hydraulic and Pneumatic Cylinder and Vane-Type Actuators for Valves and Slide Gates. ANSI/AWWA C541-08. Denver, Colo.: American Water Works Association (latest edition).

AWWA Standard for Metal-Seated Gate Valves for Water Supply Service. ANSI/AWWA C500. Denver, Colo.: American Water Works Association (latest edition).

AWWA Standard for Protective Interior Coatings for Valves and Hydrants. ANSI/AWWA C550. Denver, Colo.: American Water Works Association (latest edition).

AWWA Standard for Reduced-Wall, Resilient-Seated Gate Valves for Water Supply Service. ANSI/AWWA C515. Denver, Colo.: American Water Works Association (latest edition).

AWWA Standard for Resilient-Seated Gate Valves for Water Supply Service. ANSI/AWWA C509. Denver, Colo.: American Water Works Association (latest edition).

AWWA Standard for Rubber-Seated Butterfly Valves. ANSI/AWWA C504. Denver, Colo.: American Water Works Association.

AWWA Standard for Swing-Check Valves for Waterworks Service, 2-In. Through 24-In. (50-mm Through 600-mm) NPS. ANSI/AWWA C508. Denver, Colo.: American Water Works Association (latest edition).

AWWA Standard for Underground Service Line Valves and Fittings. ANSI/AWWA C800. Denver, Colo.: American Water Works Association (latest edition).

Manual M44, Distribution Valves: Selection, Installation, Field Testing, and Maintenance. 2006. Denver, Colo.: American Water Works Association.

Manual M49, Butterfly Valves: Torque, Head Loss, and Cavitation Analysis. 2001. Denver, Colo.: American Water Works Association.

Manual M51, Air-Release, Air/Vacuum, and Combination Air Valves. 2001. Denver, Colo.: American Water Works Association.

Water Distribution Operator Training Handbook. 1999. Denver, Colo.: American Water Works Association.

CHAPTER 7

Fire Hydrants

Fire hydrants are one of the few parts of a water system visible to the public, so keeping them well maintained can help a water utility project a good public image. In smaller towns, hydrants are so seldom used for fighting fires that it may be easy to forget how important it is to keep them well maintained for quick and reliable service when needed. A hydrant that does not operate when needed can result in a loss of life and property.

FIRE HYDRANT USES

In addition to fighting fires, hydrants have other uses. This section discusses the different uses of fire hydrants.

Fire Fighting

The major purpose of fire hydrants is for public fire protection. The water utility is usually responsible for keeping hydrants in working order, although the fire department assumes this responsibility in some communities. Water system operators should be aware that they can be held liable for damages if property is destroyed by fire because neglected hydrants fail to operate.

Following are guidelines on where hydrants should be located:

- Hydrants should not be located too close to the buildings they are intended to protect. Fire fighters will not position their fire trucks where a building wall could fall on them if the building should collapse during a fire.

- Hydrants should preferably be located near street intersections. This way, hose can be strung to fight a fire in any of several directions.

- Hydrants should be placed back far enough from a roadway to minimize the danger of them being struck by vehicles.

- On the other hand, hydrants should be close enough to pavement to ensure a secure connection between the pumper and hydrant without the risk of the truck getting stuck in mud or snow.

- In areas of the country with heavy snow, hydrants must be located where they are least liable to be covered by plowed snow or struck by snow-removal equipment.

- Hydrants that are very low or in a hole will be difficult for fire fighters to use. They are more quickly covered by snow and may be difficult to find in winter and more difficult to dig out for use. Hydrants should generally be high enough that valve caps can be removed with a standard wrench, without the wrench hitting the ground.

- Some residents like to screen hydrants by planting bushes around them, but this may make it difficult for fire fighters to find the hydrant in an emergency. Residents should

be politely told that the hydrant must be kept exposed for proper protection of their property.
- Some residents are bothered by a hydrant's bright color. They may take it upon themselves to repaint the one in front of their house a darker color to match the vegetation. If the change in color could make it harder to find the hydrant in an emergency, the darker color should not be allowed.

Miscellaneous Fire Hydrant Uses

Other frequent uses for fire hydrants include

- flushing water mains;
- flushing sewers;
- filling tank trucks for street washing, tree spraying, and other uses; and
- providing a temporary water source for construction jobs, such as for mixing mortar and settling dust.

Restricting miscellaneous uses

Authorized uses of hydrants for miscellaneous purposes should be rigidly controlled and generally discouraged. *Unauthorized* use should be absolutely prohibited. One reason for discouraging miscellaneous uses is that the water is often not paid for. A more important reason is that inexperienced persons can unknowingly or carelessly cause a hydrant to be damaged or become inoperable.

Inexperienced persons can damage hydrants by

- failing to note that a hydrant has not properly drained after being shut off;
- not shutting the main valve completely, resulting in slight leakage;
- leaving attachments on the hydrant when it is not in use; and
- using a pipe wrench for operation, which will damage the top nut.

Allowing miscellaneous uses

Several points should be observed in allowing miscellaneous uses of hydrants. First, it is preferable to require permits for anyone outside of the utility or local government staff to use hydrants. When the permit is issued, the water utility can check that the person or firm has the proper equipment and understands the proper operation of hydrants. At the same time, it is possible to charge a fee for the water that will be used, either through a flat permit fee or by metering the water taken from hydrants. Figure 7-1 illustrates a hydrant meter used for metering water use at a construction site.

When the utility issues a permit, if should specify whether certain hydrants are not to be used. Types and locations of hydrants that should be excluded from use if possible include

FIGURE 7-1 Fire hydrant meter
Courtesy of Neptune Technology Group Inc.

- very old hydrants (they may not seat properly when closed),
- hydrants on dead-end mains (their use may stir up sediment),
- hydrants on busy streets (they may disrupt traffic), and,
- hydrants in locations with a high groundwater table (they may not drain properly).

Everyone who is allowed to use hydrants should be instructed in their proper operation. These people must be explicitly told that only appropriate operating wrenches and hose connections with a throttling valve are to be used. They should also be instructed to close hydrants slowly, check for draining, and guard against any possibility of creating a cross-connection.

The importance of restricting hydrant use should be explained to police and fire department personnel and other municipal staff. These groups should be asked to cooperate by immediately reporting any suspected unauthorized use.

SYSTEM PROBLEMS CAUSED BY HYDRANT OPERATION

Problems can arise if the distribution system is not designed to handle the demands placed on it by hydrants during a fire. For example, if the mains are not large enough to provide adequate fire flow, a fire department pumper can create a negative pressure in the main. This in turn can cause any cross-connections on the system to siphon nonpotable water into the distribution system. It can also cause hot water heaters to drain back through service lines and damage customer meters.

Standard practice is to install hydrants only on mains 6 in. (150 mm) or larger. Larger mains are often necessary to ensure that the residual pressure during fire flow remains greater than 20 psi (140 kPa).

Whenever hydrants are operated, the increase in flow causes the water to move faster through the mains. The faster flow scours any sediment that may have accumulated. This material can produce discolored and cloudy water. It will usually result in customer complaints unless the public has been notified of the problem beforehand and instructed on how to deal with it.

Hydrants should always be closed slowly. Stopping flow quickly can cause a water hammer condition that could move the hydrant backward if it is not firmly blocked. It could also lead to other distribution system damage.

TYPES OF FIRE HYDRANTS

The types of hydrants generally available are classified as

* dry-barrel hydrants,
* wet-barrel hydrants,
* warm-climate hydrants, and
* flush hydrants.

Dry-Barrel Hydrants

As illustrated in Figure 7-2, dry-barrel hydrants are equipped with a main valve and a drain in the base. The barrel is filled with water only when the main valve is open. A small drain valve connected to the operating stem opens as the main valve is closed, allowing water to drain from the barrel.

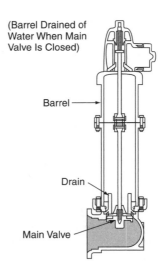

(Barrel Drained of Water When Main Valve Is Closed)

Barrel

Drain

Main Valve

FIGURE 7-2 Dry-barrel hydrant
Courtesy of Waterous Company

Dry-barrel hydrants are primarily designed for use in freezing climates, but they are also used in warmer parts of the country. An advantage of dry-barrel hydrants is that there is no flow of water from a broken hydrant.

Wet-top hydrants

One type of dry-barrel hydrant is the wet-top hydrant, which is constructed so that the threaded end of the main rod and the operating nut are not sealed from water when the main valve is open (Figure 7-3).

Dry-top hydrants

The other type of dry-barrel hydrant is the dry-top hydrant. As shown in Figure 7-4, the threaded end of the operating stem is sealed from water in the barrel when the hydrant is in use. This design reduces the possibility of the threads becoming fouled by sediment or corrosion.

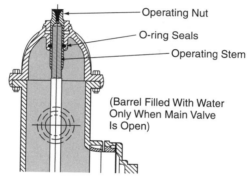

FIGURE 7-3 Wet-top hydrant
Courtesy of Mueller Company, Decatur, Ill.

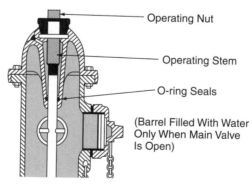

FIGURE 7-4 Dry-top hydrant
Courtesy of Mueller Company, Decatur, Ill.

Valve types

Dry-barrel hydrants are also classified according to the type of main valve. The primary types of valves that have been used, as illustrated in Figure 7-5, are described here.

- In a standard compression hydrant, the valve closes with the water pressure against the seat to aid in providing a good seal.
- In a slide gate hydrant, the valve is a simple gate valve, similar to one side of an ordinary rubber-faced gate valve.
- In a toggle (Corey) hydrant, the valve closes horizontally, and the barrel extends well below the branch line.

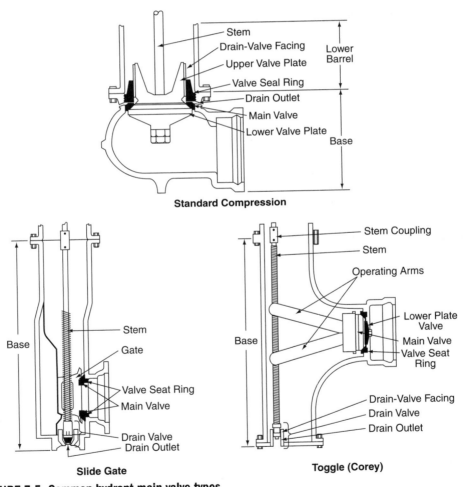

FIGURE 7-5 Common hydrant main valve types

Breakaway hydrants

Early dry-barrel hydrants were generally constructed with a solid barrel extending from the base to the top, as shown in Figure 7-6. Many of these hydrants are still in use today. The principal disadvantage with this design is that when the barrel is broken, usually by being struck by a vehicle, it is necessary to excavate down to the base and replace the entire barrel. The operating stem is usually also bent and must be replaced. The parts and labor for repairing a broken hydrant of this type are quite expensive. If there is deep frost in the ground, the repair will be particularly difficult or may have to be deferred until spring.

The breakaway or "traffic" hydrant design is now in general use. It has a two-part barrel with a flanged coupling just above the ground line (Figure 7-7). The flange is designed to break on impact without further damage to the barrel. The operating stem is also in two pieces, with a coupling designed to break on impact. In most cases, workers can repair a hydrant that has been struck by simply replacing the breakaway flange and stem coupling at nominal cost for parts and labor. No excavation is required. The design of dry-barrel hydrants is covered in American Water Works Association (AWWA) Standard C502, *Dry-Barrel Fire Hydrants* (most recent edition).

Wet-Barrel Hydrants

A wet-barrel hydrant is completely filled with water at all times (Figure 7-8). The hydrant itself has no main valve. Instead, each nozzle is equipped with a valve. Wet-barrel hydrants

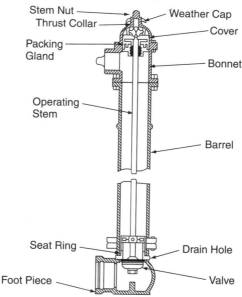

FIGURE 7-6 Dry-barrel hydrant

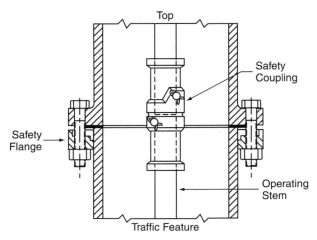

FIGURE 7-7 Detail of one type of breakaway flange and stem coupling
Courtesy of American Cast Iron Pipe Company

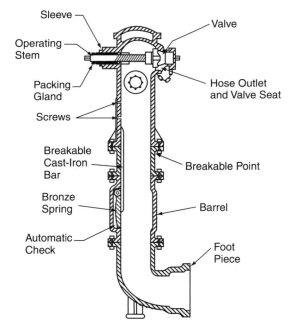

FIGURE 7-8 Wet-barrel hydrant

cannot be used in climates where temperatures fall below freezing. Another disadvantage of this type of hydrant is that large amounts of water will flow from a broken hydrant until a repair crew can shut it off. The design of wet-barrel hydrants is covered in AWWA Standard C503, *Wet-Barrel Fire Hydrants* (most recent edition).

Warm-Climate Hydrants

Warm-climate hydrants have a two-part barrel. The main valve is located at the ground line, and the lower barrel is always full of water and under pressure. The main valve controls flow from all outlet nozzles, and there is no drain mechanism.

Flush Hydrants

The entire standpipe and head of a flush hydrant are below ground. As shown in Figure 7-9, the operating nut and outlet nozzles are encased in a box with a removable cover that is at the ground surface level. Flush hydrants are used on aprons and taxiways at airports,

FIGURE 7-9 Flush hydrant
Courtesy of Mueller Company, Decatur, Ill.

pedestrian malls, and other locations where post-type hydrants are considered unsuitable. Flush hydrants are usually of the dry-barrel type.

HYDRANT PARTS

This section discusses the components of fire hydrants. The upper section and lower section of a hydrant are discussed separately.

Upper Section

The upper barrel section of a hydrant is often called the nozzle section or head. Figure 7-10 shows some of the principal parts. The operating nut is usually a five-sided nut so that it cannot be operated with a regular socket wrench. Special wrenches are used for hydrant operation (Figure 7-11). Opening a hydrant usually involves turning the operating nut counterclockwise. An arrow and the word *open* should be cast in relief on the top of the hydrant to designate the direction of opening.

The top cover or closure on the hydrant upper barrel is usually referred to as the bonnet. The bonnet may or may not be pressurized. The upper barrel is a gray cast-iron or ductile-iron section that carries water from the lower barrel to the outlet nozzles.

The outlet nozzles are threaded bronze outlets on the upper barrel. They are used (1) to connect hose lines for direct use of main pressure or (2) for connecting a suction hose from the hydrant to the pumper truck. Most water systems use hydrants with two 2 ½-in. (64-mm) nozzles and one 4 ½-in. (114-mm) nozzle. The smaller-diameter outlet nozzles are for connecting to fire hoses. The larger nozzle, called a pumper outlet or steamer outlet, is used for connecting to a pumper suction hose. Outlet-nozzle threads are normally National American Standard threads (*Standard for Fire Hose Connections* is available from the National Fire Protection Association—see appendix B). However, some utilities use special threads. Both utility and fire department personnel must be aware of the thread type used so that it will be compatible with the fire-fighting equipment.

The outlet-nozzle caps are cast-iron covers that screw onto outlet nozzles. These caps protect the nozzles from damage and unauthorized access. Caps are usually furnished with a nut the same size as the hydrant operating nut. They should be removed only with a hydrant wrench.

Lower Section

The lower section of a hydrant includes the lower barrel, main valve, and base. The lower barrel is made of static-cast or centrifugally cast gray or ductile iron. It carries water flow between the base and the upper barrel. It should be buried in the ground so that the connection to the upper barrel is approximately 2 in. (50 mm) above ground line.

The main valve assembly includes the operating stem, resilient valve gasket, and other attached parts. The base is also known as the shoe, inlet, elbow, or foot piece. The base is usually cast iron. Many hydrant mechanisms include a travel-stop device to prevent the main rod

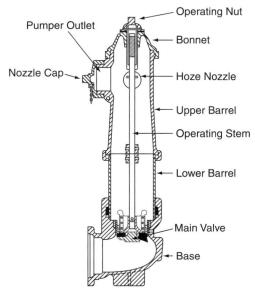

FIGURE 7-10 Main parts of a typical fire hydrant
Courtesy of U.S. Pipe and Foundry Company

Pumper Outlet

Nozzle Cap

Operating Nut

Bonnet

Hoze Nozzle

Upper Barrel

Operating Stem

Lower Barrel

Main Valve

Base

FIGURE 7-11 Adjustable hydrant wrench

and valve from moving down too far. If a hydrant doesn't have such a device, a stop must be built into the middle of the base. The valve assembly will then rest against the stop.

Auxiliary Valves

An auxiliary valve should be installed on every hydrant. This way, each hydrant may be individually turned off for maintenance or repair. Some water systems may have old hydrants that were installed without auxiliary valves. It is advisable to install valves on these hydrants as time becomes available so that repairs can be made without shutting down an entire section of the distribution system.

The type of auxiliary valve most often used is directly connected to the hydrant by a flanged connection, as shown in Figure 7-12. One advantage of this arrangement is that the valve cannot separate from the hydrant. Another is that all valves are at a standard location in relation to the hydrant. The valves are then easily located in an emergency.

INSPECTION AND INSTALLATION

This section discusses the steps required before a hydrant can be put into service. These steps include inspection, installation, and testing.

Inspection of New Hydrants

Hydrants should be inspected at the time of delivery in order to verify that they meet specifications and were not damaged during shipment. Points that should be checked include

- direction to open the hydrant,
- size and shape of the operating nut,
- depth of bury (distance to the main connection below the ground surface),
- size and type of inlet connection,
- main valve size,
- outlet-nozzle sizes and configuration, and
- nozzle thread dimensions.

The hydrant should be cycled to the fully open and fully closed positions to ensure that no internal damage or breakage has occurred during shipment and handling. All external bolts should be checked for proper tightness.

After inspection, the hydrant valves should be closed and the outlet-nozzle caps replaced to prevent foreign matter from entering. Stored hydrants should be protected from the weather whenever possible. If they must be stored outside in cold weather, they should be placed with the inlet down to prevent water from entering and freezing.

Installation Procedures

Some important concerns during the installation of a fire hydrant include

- location,
- footing and blocking,
- drainage, and
- color for painting.

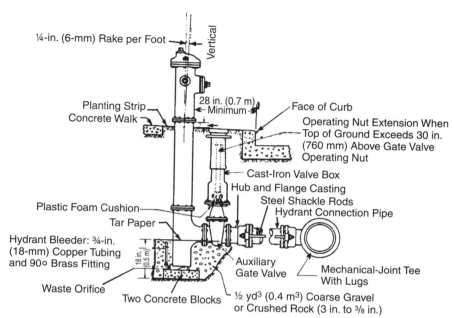

FIGURE 7-12 Hydrant auxiliary valve installation

Location

Hydrants should generally be set back at least 2 ft (0.6 m) from the curb. This will place the auxiliary valve box behind the curb and locate the hydrant beyond normal vehicle overhang. In rural areas or where there is no curb, hydrants must be protected from traffic with a larger setback. The pumper outlet nozzle should always face the street. If the location is particularly vulnerable to traffic, guard posts should be installed.

Footing and blocking

Hydrants must be set on a firm footing that will not rot or settle. A flat stone or concrete slab is ideal. A carpenter's level should be used to ensure that hydrants are set plumb (i.e., exactly vertical).

Hydrants must also be securely blocked or restrained from movement in the direction opposite the main connection (usually the back of the hydrant). If they are not, they may move and open a joint if there is a water hammer in the system. The excavation for a hydrant should, if possible, be made in a manner that will preserve undisturbed earth a short distance behind the hydrant base. After the hydrant is placed, a concrete block can be poured between the base and the undisturbed earth. Installers must carefully place the concrete so as not to block the drain hole in the base.

When it is not possible to obtain good support for a block behind a hydrant, restraint must be provided by rods or fittings. A common practice is to use two shackle rods ¾ in. (19 mm) in diameter to tie a hydrant valve to the tee in the main.

Drainage

Unless the soil surrounding a hydrant is very porous, special provisions must be made to carry off the drainage from dry-barrel hydrants. The accepted method is to excavate an area around the base of the hydrant to be large enough to permit the placement of approximately ⅓ yd³ (0.25 m³) of clean stone. The top of this layer should be slightly above the drain openings. To prevent the drainage stone from being clogged with dirt, it should be covered with heavy polyethylene or tar paper before backfill is placed.

If hydrants must be installed where the barrel will not fully drain because the water table is above the hydrant base, it is advisable to plug the hydrant drains. These hydrants must then be specially marked so that the barrel will be pumped dry with a hand pump after each use.

Hydrant painting

Hydrants should be painted with colors that are easily visible both day and night. Red, orange, and yellow are usually the most visible colors. Some water utilities paint the operating nut or portions of the nozzles with reflective paint to make the hydrants readily visible at night.

Color coding hydrants to indicate their capacity helps the fire department select the most productive hydrants for fighting a fire. Table 7-1 shows a commonly used color scheme for painting hydrant tops and/or caps to indicate the expected flow rate, as determined by hydrant flow tests. Another scheme for color coding might be based on water main size.

Testing

New hydrants installed at the same time as a new main should not be pressure-tested along with the main. All hydrant auxiliary valves should be closed during the main pressure test. After the main test is completed—and preferably before hydrant excavations have been backfilled—the auxiliary valves can be opened and the hydrants tested as follows:

1. Open the hydrant fully and fill it with water. Then close all outlets.
2. Vent air from the hydrant by leaving one of the caps slightly loose. After all air has escaped, tighten the cap before proceeding.
3. Apply a pressure up to a maximum of 150 psig (1,000 kPa [gauge]) by using a pressure pump connected to one of the nozzles. If it is not practical to apply higher pressure, system pressure will suffice.
4. Check for leakage at flanges, outlet nozzles, and the operating stem.
5. If leakage is noted, repair or replace components until the condition has been corrected.

TABLE 7-1 Standard hydrant color scheme to indicate flow capacity

Hydrant Class	Color[†]	Usual Flow Capacity at 20 psig (140 kPa [gauge])*	
		gpm	*(L/mm)*
AA	Light blue	1,500	(5,680)
A	Green	1,000 to 1,499	(3,785 to 5,675)
B	Orange	500 to 999	(1,900 to 3,780)
C	Red	Less than 500	(Less than 1,900)

*Capacities are to be rated by flow measurements of individual hydrants at a period of ordinary demand. See AWWA Standard C502 for *Dry-Barrel Fire Hydrants*, most recent edition, for additional details.
†As designed in Federal Standard 595B, General Services Administration, Specification Section, Washington, D.C.

It is not uncommon for outlet nozzles and connecting bolts to become loose as a result of rough handling in shipment, storage, and installation. Loose flange bolts and nozzles should be tightened.

Utility personnel can test dry-barrel hydrants for proper drainage by placing the palm of one hand over the outlet-nozzle opening immediately after the main valve is shut. The barrel usually drains fast enough that a noticeable suction can be felt. If there is a question of whether the barrel has completely drained, a string with a small weight should be dropped in through one nozzle to see if it comes out wet.

After backfilling is complete, hydrants should be operated so that any foreign material is flushed out. When nozzle caps are replaced, they should be made tight enough to prevent removal by hand, yet not excessively tight. New hydrants should promptly be painted the color preferred by the water system.

OPERATION AND MAINTENANCE

This section discusses common operation and maintenance procedures for hydrants that have been placed in service.

Hydrant Operation

Hydrants are designed to be operated by one person using a 15-in. (380-mm) wrench. The use of a longer wrench or piece of pipe added to a standard wrench (cheater) is not good practice. If one person cannot operate a hydrant with a standard wrench, the hydrant should be repaired or replaced. To prevent damage to the operating nut, wrenches not specifically designed for hydrant operation should not be used.

If the main valve in a dry-barrel hydrant is not fully closed, there may be leakage through the drain valves. This will eventually saturate the surrounding soil and prevent proper draining of the barrel. In soil that can easily wash away, the leakage can also undermine the hydrant footing or blocking.

The main valve of a dry-barrel hydrant should always be completely opened when in use. This will ensure that the drain valve is completely closed. In addition, throttling the main valve may cause damage to the valve seat and rubber. If flow from a hydrant must be throttled, a nozzle cap fitted with a gate valve should be used, with the hydrant valve fully opened.

Hydrant Maintenance

Regularly scheduled inspection of hydrants is necessary to ensure satisfactory operation. All hydrants should be inspected at least annually. In freezing climates, each hydrant should be inspected in the autumn to make sure no standing water is in the barrel. Some water systems do this by "stringing" the hydrant—dropping a weighted string down the barrel to see if it comes out wet.

Water systems that have a problem with sediment accumulating in their mains usually perform an annual system flushing and hydrant checking at the same time. If a hydrant is found to be inoperable and cannot be repaired immediately, a barricade or barrel should be placed over it. The fire department should then be notified that the hydrant is out of service.

Inspection Procedures

Following are some general inspection procedures:

* Check that there is nothing near the hydrant that will interfere with its operation.
* Ensure that there is nothing attached to the outlet nozzles.
* Visually check the hydrant to be sure it is not leaning. Look for other indications that it may have been struck by a vehicle.
* Remove one outlet-nozzle cap from dry-barrel hydrants and check for water or ice standing in the barrel.
* Use a listening device to check for seat leakage on dry-barrel hydrants. Visually inspect each valve on wet-barrel hydrants.
* On dry-barrel hydrants, replace the outlet-nozzle cap and open the hydrant to a fully open position while venting air from the barrel. Check the ease of operation. If the stem action is tight, repeat the action several times until it is smooth and free.
* While the hydrant is under pressure, check for leakage at joints, around outlet nozzles and caps, and at packing or seals. If leakage is observed, it might be necessary to (1) tighten or recaulk outlet nozzles, (2) lubricate and tighten compression packing, or (3) replace O-rings, seals, and gaskets. If leakage cannot be corrected with the tools at hand, record the nature of the problem for prompt attention by the repair crew.

- Wet-barrel hydrants require that a special outlet-nozzle cap be used to test and operate each valve.

- Remove an outlet-nozzle cap and attach a section of hose (or a flow diverter if necessary) to direct flow into the street. Open the hydrant and allow it to flow for a short time to remove foreign material from the interior and lateral piping.

- On dry-barrel hydrants, close the main valve to the position at which the drains open. Allow flow through the drains under pressure for about 10 seconds to flush the drain. Then close the main valve completely.

- Remove all outlet-nozzle caps and inspect them for thread damage from impact or cross threading. Clean and lubricate outlet-nozzle threads. Use caps to check for easy operation of threads. Be sure outlet-nozzle cap gaskets are in good condition.

- Lubricate operating-nut threads in accordance with manufacturer's instructions. Some hydrants require oil in the upper operating-nut assembly. This lubrication is critical and must not be overlooked during routine maintenance.

- Check that the barrel drains properly on each dry-barrel hydrant. You should feel suction rapidly being created when your hand is placed over an outlet nozzle during drainage.

- Use a listening device to check again for seat leakage on dry-barrel hydrants. Visually check wet-barrel hydrants.

- If dry-barrel hydrants do not drain, pump out any residual water. Check with a listening device to make sure that the main valve is not leaking.

- Check outlet-nozzle cap chains for free action on each cap. If you observe binding, open the chain connection around the cap until the action is free. This prevents kinking when the cap is removed under emergency conditions.

- On traffic hydrants, check the breakaway device. Inspect couplings, cast lugs, special bolts, and other parts for damage due to impact or corrosion.

Hydrant Repair

Any condition that cannot be corrected during the regular inspection should be recorded and reported for subsequent action by a repair crew. Before any repair is started, the fire department must be notified that the hydrant will be out of service.

Turning off the hydrant auxiliary valve is usually the first step in repairing the internal parts of a hydrant. The hydrant is then disassembled according to procedures supplied by the manufacturer. In addition to any worn or damaged parts, all gaskets, packing, and seals should be replaced while the hydrant is apart, regardless of whether they appear worn or not.

When the repair is completed, the auxiliary valve should be opened. The hydrant should then be thoroughly tested for proper operation and leaks. The fire department may then be notified that the hydrant is back in service. The date and details of the repair should be promptly recorded on the hydrant record card.

Flow Testing

As a municipality expands, the increased demand on the water distribution system causes higher velocities in mains. This results in increased head loss in the piping. At the same time, because of corrosion, scale, and sediment in the pipes, the carrying capacity of old piping in the system may slowly decrease. It is therefore necessary to check the capacity of the system periodically to determine (1) the need for additional feeder or looping mains and (2) the need to clean and line existing pipes.

Hydrants should periodically undergo flow testing to provide data on distribution system flow and fire flow capabilities. Testing will also identify other problems, such as a valve that has inadvertently been left closed. Flow capability should also be retested after any major changes are made in the distribution system piping.

HYDRANT RECORDS

A meaningful inspection and maintenance program requires that a record card, sheet, or computerized record be kept on each hydrant. Some initial information that should be recorded at the time each hydrant is installed includes the make, model, and location. The record should also list any other pertinent information that may be of use in the future when repairs are required. Most water systems assign a number to each hydrant to assist in record keeping.

Each time a hydrant is inspected, an entry should be made in the record indicating the date of inspection and the condition of the hydrant. This record must be accurately maintained. It can serve as proof of inspection in the event the water utility is ever charged with negligence because of a hydrant failing to operate properly in an emergency. A record should also be kept of all repair work that is performed. Additional details and samples of hydrant records are included in chapter 16.

HYDRANT SAFETY

In addition to the general safety precautions detailed in other chapters for installing piping and protecting the public, special precautions must be taken to prevent injury and damage to private property during hydrant flushing. Following are several safety concerns that should be taken into account:

- Remember, in addition to getting people wet, the force and volume of water from a full hydrant stream are sufficient to seriously injure workers or pedestrians.
- If traffic is not adequately controlled, drivers trying to avoid a hydrant stream might stop quickly or swerve. An accident might result.
- If the temperature is below freezing, water that is allowed to flow onto pavement may freeze and cause accidents.
- If flow is diverted with a hose to a sewer, care must be taken not to create a cross-connection.
- If flow is diverted with a hose, the end of the hose must be securely anchored. A loose hose end can swing unpredictably and could cause serious injury.

FIGURE 7-13 Flow diffuser
Courtesy of Pollardwater.com

During flow tests, the hydrant nozzle must be unobstructed, so the only way of protecting property is to choose the nozzle that will do the least damage. Barricades should be provided to divert traffic. Other precautions may have to be taken to minimize property damage and prevent personal injury.

Steps can be taken to divert flushing flow to prevent property damage. Flow diffusers or a length of fire hose should be used where necessary to direct the flow into a gutter or drainage ditch (Figure 7-13). A rigid pipe connected to a hydrant outlet and turned at an angle to divert flow down a gutter is not considered a good idea. The torque produced by the angular flow could be enough to twist or otherwise damage the hydrant.

BIBLIOGRAPHY

AWWA Standard for Dry-Barrel Fire Hydrants. ANSI/AWWA C502. Denver, Colo.: American Water Works Association (latest edition).

AWWA Standard for Wet-Barrel Fire Hydrants. ANSI/AWWA C503. Denver, Colo.: American Water Works Association (latest edition).

Fire Hydrants. 2006. DVD. Denver, Colo.: American Water Works Association.

Hydrant Flow Tests. 2010. DVD. Denver, Colo.: American Water Works Association.

Maintaining and Replacing Fire Hydrants. 2006. DVD. Denver, Colo.: American Water Works Association.

Manual M17, Installation, Field Testing, and Maintenance of Fire Hydrants. 2006. Denver, Colo.: American Water Works Association.

Manual M31, Distribution System Requirements for Fire Protection. 2008. Denver, Colo.: American Water Works Association.

Mays, L.W. *Water Distribution Systems Handbook.* 1999. New York: McGraw-Hill.

Standard for Fire Hose Connections. 1998. Quincy, Mass.: National Fire Protection Association.

Standard for the Installation of Standpipe, Private Hydrants, and Hose Systems. NFPA 14. 2000. Quincy, MA: National Fire Protection Association.

Water Distribution Operator Training Handbook. 2005. Denver, Colo.: American Water Works Association.

CHAPTER 8

Motors and Engines

Electric motors are used as prime movers to power more than 95 percent of the pumps used in municipal water supply operations. They are available in a wide range of types, speeds, and power capabilities. Their inherent characteristics of smooth power output and high starting torque make them well suited for direct connection to centrifugal pumps.

Most electric motors used by water utilities are powered by alternating current (AC) electricity supplied by power utilities. (Direct current [DC] electric motors are not covered in this chapter because of their limited application.) Alternating current is commonly used for motors because of its ease of generation and lower transmission costs. It is also easy to convert the current from higher to lower voltage by the use of transformers.

Alternating current flows first in one direction and then in the other. The current strength rises from zero to a maximum, returns to zero, and then falls and rises in the opposite direction. This sequence is called a cycle. The frequency of an AC current is the number of cycles that are completed in 1 second. Power systems in the United States operate at 60 cycles per second, usually expressed as 60 Hz. Power systems in some other countries operate at 50 Hz.

Current with one peak per cycle is known as single-phase power. Three-phase power (three peaks per cycle) is also commonly used, especially for powering larger motors.

The terms *volt*, *ampere*, and *ohm* are used to express measures of electrical pressure, flow, and losses, respectively. A volt (V) is a measurement of electrical pressure. It can be compared to the measurement of water pressure in pounds per square inch (or kilopascals). The common voltages are 110/120 V and 220/240 V for lighting and operation of small motors. Larger motors are operated with 440/480 V and higher. An ampere (A), or amp, is the unit used to measure the flow of electric current. If no current is flowing, only electrical pressure (voltage) is present. Once flow starts, amperes can be measured. An ohm is a measure of electrical resistance or impedance. It can be likened to friction loss when water flows through a pipe. As electrical current flows through wires, there is a drop in electrical pressure caused by resistance.

The interrelationship between volts, amps, and ohms is called Ohm's law and is expressed as

$$E \text{ (volts)} = I \text{ (amps)} \times R \text{ (ohms)}$$

These concepts are discussed in more detail in *Basic Science Concepts and Applications*, another book in this series.

PRINCIPLES OF ELECTRIC MOTOR OPERATION

In basic terms, an electric motor operates by magnetism. A magnet has two poles, one on each end, designated north (N) and south (S). When two magnets are placed near each other, opposite poles attract each other and like poles repel each other.

A magnet can be formed if current is passed through a wire wrapped around a wire rod—this is called an electromagnet. The poles of an electromagnet are reversed when the flow of current in the wire is reversed.

A simple motor is formed when a rotating magnet is placed near an electromagnet. As current is passed through the electromagnet (stator), it creates a magnetic pole that attracts the unlike pole of the other magnet (rotor). Reversing the current changes the poles of the electromagnet, and the stator now attracts the other pole of the rotor. As a result, the rotor spins and rotates the shaft. In large motors, both the stator and the rotor are electromagnets.

There must be at least two poles in a motor but there can be more, depending on how the stator is made and how the current-carrying coils (the windings) are wired and connected.

Speed

The speed at which the magnetic field rotates is called the motor's synchronous speed. It is expressed in revolutions per minute (rpm). A motor that operates on an electric power system having a frequency of 60 Hz will have a maximum synchronous speed of 3,600 rpm, or 60 revolutions per second. In other words, because the electric current changes its flow direction 60 times per second, the rotor can rotate 60 times per second. This speed is achieved with a two-pole motor.

Motors may be designed to run at fractions of 3,600 rpm if the number of poles in the stator is increased. For instance, a four-pole motor has a synchronous speed of 1,800 rpm, and a six-pole motor has a synchronous speed of 1,200 rpm. Depending on its type (synchronous or induction), the motor may run at exactly its synchronous speed or somewhat less (for example, 3,450, 1,750, or 1,150 rpm). Electronic motor controllers are now available that can vary the speed of some types of motors over a wide range.

Starting current

It generally takes considerably more electrical current to start motors than it does to keep them running. The current drawn by the motor the instant the motor is connected to the power system (while the motor rotor is still at rest) is called the locked-rotor current. It is sometimes referred to as the motor-starting current or the inrush current. The locked-rotor current is often 5 to 10 times the normal full-load current of the motor. When the motor is started, the locked-rotor current starts out at its maximum value, then decreases to the motor's ordinary current draw as the motor reaches full speed.

The time required for a motor to start from rest and reach full speed varies with the load and type of motor, but it generally will be from 1 to 8 seconds. During this time, the excessive current drawn by the motor causes an abnormal voltage drop in the power system. Excessive voltage drops during startup can cause dimming of lights and other problems in the electrical system, such as the dropout of relays, solenoids, or other motor starters. If this problem occurs, the electrical system should be evaluated and modified to correct the problem.

The size of the motor has a considerable influence on the motor-starting current and its effects. In addition to the usual charge for kilowatt-hours of power used, many electric utilities also have a demand charge based on either maximum demand or average demand during a month or other period of time. This encourages water systems to purchase motors and controls that will minimize the motor-starting current required.

SINGLE-PHASE AND THREE-PHASE MOTORS

The AC electricity used to power electric motors in water supply systems is supplied from power utilities in the form of three-phase current. It may be reduced to single-phase current within the plant to provide lighting and to operate small equipment. Power is also available at 220 V single-phase, but this option is rarely used in water system applications. Most utility pumping operations utilize three-phase motors.

Single-Phase Motors

Single-phase motors are generally only used in fractional-horsepower sizes, but they can be supplied with ratings up to 10 hp (7.5 kW) at 120 V or 240 V. A motor with a straight single-phase winding has no starting torque (power needed to bring the motor up to speed), but if it is started and brought up to speed by an outside device, it will run and continue to carry its load. Thus, some means must be provided to start the motor turning. The starting mechanism (starting winding) is usually built into the motor in the form of specially wired additional windings. The motors therefore start on the high-torque starting winding. Then an internal switch changes to the other (running) winding as soon as the motor is up to speed.

Single-phase motors used as prime movers are usually one of three types:

1. In a split-phase motor, the motor rotor has no windings. This type of motor has a comparatively low starting torque and requires a comparatively high starting current.
2. A repulsion–induction motor is more complex and expensive than a split-phase motor. It has a high starting current.
3. A capacitor-start motor has a high starting torque and a high starting current. It is limited to applications for which the load can be brought up to speed very quickly and infrequent starting is required.

Three-Phase Motors

Most motors ½ hp (4 kW) and larger in water treatment and distribution systems are three-phase, and they may be rated at 230, 460, 2,300, or 4,000 V (Figure 8-1). There are three general classes of three-phase motors:

1. Squirrel-cage induction
2. Synchronous
3. Wound-rotor induction

Squirrel-cage induction motors

The squirrel-cage induction motor is the simplest of all AC motors. The rotor windings consist of a series of bars placed in slots in the rotor and connected together at each end. This gives the rotor the appearance of a squirrel cage. The stator windings, located in the frame, are connected to the power supply. Current flowing through them sets up a rotating magnetic field within the motor. This rotating field induces a current flow in the rotor windings, which generates an opposing magnetic field. The force between the two magnetic fields causes the rotor to turn.

No elaborate controls are required for squirrel-cage motors. Simple starting controls are adequate for most of the normal- and high-starting torque applications for which these motors are used.

Synchronous motors

The power supply leads of a synchronous motor connect to the stationary windings in such a way that a revolving magnetic field is established, rotating at synchronous speed. The rotor is constructed to have poles that match the poles of the stator (i.e., there are the same number of rotor and stator poles). The rotor pole pieces are supplied with direct current, so the rotor's magnetic field is constant. A slip-ring assembly (called a commutator) and brushes (graphite connectors) must be used to connect the direct current to the rotating parts of the motor. Thus, in the synchronous motor, alternating current is supplied to the stator and direct current is supplied to the rotor. Where power consumers pay a penalty for low power-factor conditions, the synchronous motor is often a logical choice because it has a power factor of 1.0. Where speed must be held constant, the synchronous-type motor is necessary.

Wound-rotor induction motors

The wound-rotor induction motor has a stator similar to that of the squirrel-cage induction motor. A number of pairs of poles and their associated stator windings are connected to the three-phase power supply. The rotor has the same number of poles as the stator, and the rotor windings are wired out through slip rings (rotating contacts). The resistance of

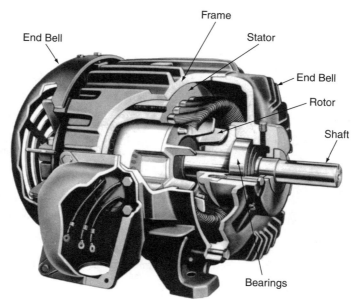

FIGURE 8-1 Motor components

the rotor circuit can be controlled while the motor is running, which varies the motor's speed and torque characteristics.

The wound-rotor motor offers ease of starting in addition to variable-speed operation. The starting current required by a wound-rotor motor is seldom greater than normal full-load current. In contrast, squirrel-cage and synchronous motors may have starting currents 5 to 10 times their normal full-load currents.

Variable-Frequency Drives

The use of variable-frequency drives for pumping applications incorporates a basic wound-rotor induction motor and a controller that modulates the current to the motor. Many units have silicon-controlled rectifier systems to accomplish this modulation.

In operation, the standard three-phase current is converted to DC power in a control unit. The DC power is controlled by a rheostat and is then converted to a modulated-output AC power, which is supplied to the motor. Because the power is not constant, the motor runs at less than full speed, depending on how the power level is selected for the pump speed. If the level is set and the unit starts on a command from a controller, the motor slowly powers up until it reaches its preset speed. This provides a "soft start" to the motor, which reduces the starting current required as compared with a standard motor.

ELECTRIC MOTOR CONSTRUCTION

Industrial electric motors are used for a wide range of loads, environmental conditions, and mounting configurations. To meet these conditions, motor manufacturers have standardized most motors and motor modifications in sizes up to 200 hp (1,500 kW). Larger motors are not necessarily standardized, although they have certain standard construction features.

Temperature

Motors convert electrical energy into mechanical energy and heat. The heat given off by a motor can be readily calculated from the motor's horsepower (watt) rating and its efficiency. For example, if a 100-hp (750-kW) motor is 95 percent efficient, 95 percent of the energy supplied to it generates useful work—the remaining 5 percent is lost as heat. This heat must be removed from the motor as fast as it is produced to prevent the motor temperature from rising too high. The construction of a motor is influenced greatly by its cooling and venting requirements.

Motors are designed to operate with a maximum ambient (surrounding) temperature of 104°F (40°C). However, the internal temperature can be much higher, depending in great part on the type of insulation used for the motor windings. The air supplied to ventilate a motor should be at the lowest temperature possible and must not exceed 104°F (40°C). The useful life of any motor is extended by having plenty of cool ventilating air available.

Mechanical Protection

Various motor designs provide different protection to the windings and internal motor parts from falling, airborne, or wind-driven particles and from various atmospheric and surrounding conditions. Motor designs commonly available include open, drip-proof (Figure 8-2), splash-proof, guarded, totally enclosed (Figure 8-3), totally enclosed with fan cooling, explosion-proof, and dust-proof. In most cases, the name clearly describes the unique features of each type of construction. Note that internal explosions could occur in explosion-proof motors. However, the explosions would be contained within the motor housing, preventing the surrounding area from being damaged or ignited.

MOTOR CONTROL EQUIPMENT

This section discusses the numerous devices used to control and protect motors.

Motor Starters

The first electric motors incorporated a simple manual switch to start and stop the motors. Each starting and speed-control operation had to be performed by hand, with the operator moving a manual switching device from one position to another. Controlling large motors required great physical effort, and the open construction of the equipment also posed a considerable danger of shock or electrocution.

FIGURE 8-2 Open drip-proof motor

FIGURE 8-3 Exploded view of a typical squirrel-cage induction motor
Courtesy of Leeson Electric Corp.

The development of magnetic starters enabled the operator to use push-button stations and control switches that can be operated with practically no effort or danger. Magnetic starters also permit push-button stations and control switches to be in a different location than the starting equipment. This in turn allows centralized automatic control.

As illustrated in Figure 8-4, motor control equipment includes not only the motor starter but also all of the miscellaneous devices that protect the motor and the motor-driven equipment during operation. The functions of motor control fall into two general categories:

1. Those functions that are for the protection of the motor and the motor feeder cables.
2. Those functions that determine when and how the motor is to operate.

Full-Voltage and Reduced-Voltage Controllers

A motor controller will be either a full-voltage or reduced-voltage type. A full-voltage (across-the-line) controller uses the full line voltage to start the motor. The starting current is drawn directly from the power line. A reduced-voltage controller is used when the starting current of the motor is so high that it may damage the electrical system. The controller uses a reduced voltage and current to start the pump motor. When the motor gains full speed, full voltage is applied.

Motor Control Systems

Automatic controls to start and stop pumps eliminate the need for constant station attendance. In some cases, they provide closer control of operations. Control switches are usually operated by changes in discharge line pressure or in the water level in elevated storage tanks.

Figure 8-5 illustrates three common motor control systems frequently used in a water system. One control system has a level (or pressure) transmitter located at the elevated tank. This transmitter switches the high-service pumps on and off to maintain the water level within preset limits in the tank. The transmitter is also set to provide an alarm if the level in the tank should get too high or low.

The high-service pumps, in turn, draw water from a reservoir that is furnished by three wells. Level controls in the reservoir are set to operate increasing numbers of well pumps as the reservoir level drops in order to keep up with the demands of the high-service pumps. High- and low-level alarms are also provided to alert the operator of malfunctions in the system.

The third type of control operates a booster pump in the distribution system. A pressure sensor detects low pressure on a remote or elevated section of the system and automatically activates the local booster pump.

Other control methods include time clocks and computer control based on demand changes.

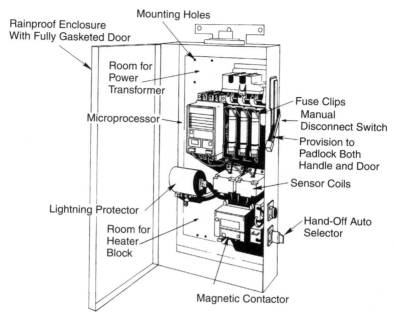

NOTE: Lightning protector is more effective when installed above power, in the area where transformer can be placed.

FIGURE 8-4 Combination motor starter

Location of the control point

If the control switch is operated by pressure changes close to the pumping station, erratic operation (known as racking or seeking) can be a problem. Racking occurs when the pressure surge resulting from pump startup immediately causes the pump control pressure switch to shut the pump off again. For this reason, pump operation is often controlled based on pressure changes measured at an elevated storage tank.

Another reason for locating the control point near the elevated tank is that the discharge pressure measured at the pump station can be affected by system demand as well as by the level of water in the elevated tank. The primary function of the pump may be to maintain a specified level of water in the tank. That level cannot consistently be determined based on the pressure at the pump.

Control methods

The most commonly used pump control system is the pressure sensor. But no matter where in the system a pump control pressure sensor is located, there is some chance that it will be improperly activated or deactivated by sudden pressure surges. These surges do not accurately reflect the actual system pressure and will result in inappropriate commands being sent to start or stop the pumps. If a system is affected by pressure surges at a point where a

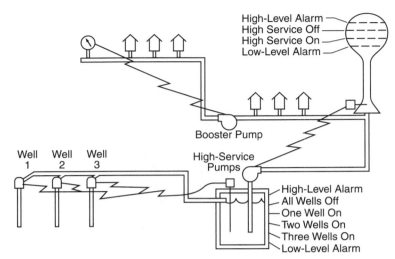

High-Level Alarm
High Service Off
High Service On
Low-Level Alarm

Booster Pump

Well 1   Well 2   Well 3

High-Service Pumps

High-Level Alarm
All Wells Off
One Well On
Two Wells On
Three Wells On
Low-Level Alarm

FIGURE 8-5 Typical pump installation with a number of controls

pressure sensor is installed, the sensor should be provided with a device that will delay the sensor's signal to the pumps. One method is to mount the sensor on an air chamber that has a small orifice at the connection with the pipeline. The orifice restricts flow into the chamber, and the air compresses to absorb the shock of sudden high- and low-pressure surges. This provides a substantial time delay between a pressure change and activation of the pressure sensor. Thus, the pressure change will have to have a duration longer than a surge in order for the sensor to send a signal to the pumps. A similar delay may be obtained if an electrical time delay relay is used in the control circuit.

The utility may also control pumps based on the water level at a storage tank by using (1) a float switch, (2) electrodes suspended within the tank, or (3) a mercury U-tube with electrical contacts at the base of the tank. However, if the pump station is located close to the elevated tank, it is sometimes feasible to lay a pipe from the tank to the station and simply use pressure switches directly connected to the pump controls, rather than installing electrical circuits. To ensure that storage is immediately replenished in case of a sudden system draw due to fire flow, dual control is often provided by an additional pressure switch that starts the pump immediately when system pressure lowers significantly.

Multiple-pump controls

Automatic controls may be provided to operate any combination of pumps simultaneously or in sequence to match pumping capacity with demand. The pumps may be activated by individual pressure switches or a sequence matrix panel (a series of switches that can be programmed to change the pump-starting sequence), based on the level in a storage tank. Switchboard circuits can be set to stop one or more smaller pumps when larger pumps are in operation. For maximum flexibility, a single control device with multiple contact points can be used to control all pumps.

Accessory controls

Pumping stations with automatic controls for starting and stopping the pumps require only occasional attendance to ensure proper functioning. To increase reliability and further reduce the cost of attendance, additional controls are provided to guard against mechanical or electrical defects and abnormal hydraulic conditions. Reliability is further increased if alarms are installed to notify operators at a central location immediately in case of any problems that require human response. Pressure- and flow-recording instruments are used to provide a full record of operation.

Motor Protection Equipment

Some of the automatic relays used to control and protect motors are described in the following paragraphs.

Thermal-overload relays are provided on starters to prevent the pump motor from burning out if abnormal operating conditions increase the load beyond the pump's design capacity. When current is excessive, the relays (sometimes called heaters) placed on each phase of the power supply open the control circuit and stop the motor. These relays are normally set to stop the motor when current exceeds the design load by 25 percent.

Fuses or circuit breakers are placed in the main power wiring to each motor to protect against short circuits. They are normally located in the safety switch just ahead of the starter.

Overcurrent relays are often referred to as overload relays. Their purpose is to sense current surges in the power supply and to disconnect the motor if a surge occurs. They can be fitted with time-delay or preset thermal-overload mechanisms. The overload relays may reset automatically or they may require manual resetting after they are tripped.

Lightning surge arresters are installed at many pump stations to prevent serious damage by high-voltage surges that can be caused when lightning strikes the power line. A surge arrester is usually installed in the service entrance between the power lines and the natural ground. The arrester acts as an insulator to normal voltage; it automatically becomes a low-resistance conductor to ground when the line voltage exceeds a predetermined amount.

Voltage relays are frequently used to detect a loss of power and to initiate a switchover to an alternate power source. Under-voltage relays are also used to shut down motors if the voltage drops too low. Voltage relays generally have a timing mechanism that allows minor functions (ones that won't damage the motor) to continue before power is actually disconnected from the motor.

Frequency relays respond to changes in the frequency (in cycles per second) of an AC power supply. They are most often used where local power generation is involved. Frequency relays are also used on synchronous motor starters to sense when the motor has reached synchronizing speed.

Phase-reversal relays are installed to detect whether any two of the three lines of a three-phase power system are interchanged. If the phase sequence should be reversed, all motors will run backward. A phase-reversal relay senses this change and opens the control circuit to disconnect the motors. A phase reversal can be particularly serious for deep-well pump applications. The pump shafting can become unscrewed, allowing the pump to fail.

Loss-of-phase relays are installed in most wiring systems to detect the loss of one of the three phases. Loss of one phase is not unusual, and if the pump power is not cut off, the motor may burn out within a few minutes while operating on only two phases.

Differential relays are frequently used on large equipment or switchgear. These units check whether all of the current entering a system comes back out of the system. If it does not, the relay closes a contact that shuts down the equipment. An equipment shutdown by a differential relay indicates major trouble, so an electrician should always be called.

Reverse-current relays sense a change in the normal direction of current or power flow and activate an alarm for the operator. They can also open circuits to isolate the faulty portion of the system.

Time-delay relays are used when some condition needs to last for a specified length of time before some other action is begun. For example, if a pump motor must be given sufficient time to reach full speed before the discharge valve opens, a time-delay relay can be energized when the motor is started. The relay can be set to close its contacts several seconds later to activate the valve-opening control circuits.

Bearing-temperature sensors can be placed on any pump or motor to monitor bearing temperature. The sensors incorporate a contact that opens or closes at a predetermined temperature. This action can be used to sound an alarm and initiate an emergency electrical control shutdown.

Speed sensors are used to trigger an alarm or emergency shutdown if a motor should for any reason exceed its normal speed. The sensors are also used on variable-speed motors.

IMPROVING THE EFFICIENCY OF ELECTRICALLY DRIVEN PUMPS

The power consumed by electrically driven pumps is one of the most costly items in many water utility budgets. Power costs can be reduced in three major ways:

1. Total power use can be reduced by an increase in system efficiency.
2. Peak power use can be reduced if the pumping load is spread more evenly throughout the day.
3. Power-factor charges can be reduced.

Although most major reductions in power costs require replacing equipment, some pumping schedule changes can reduce costs with little or no investment.

Reducing Total Electric Power Usage

In addition to the obvious steps of maintaining motors and pumps in peak condition, running the pumps at the peak of their efficiency curve will reduce power usage. This can readily be accomplished in systems that have multiple pumps.

Where system pressure is higher than necessary for fire and other demands, pressure reduction can reduce the load on pumping facilities. Utilities can often reduce pressure by using high-service pumps in areas where fire protection is critical or by having two-speed motors that are pressure-activated. A reduction in system pressure from 85 to 75 psi (590 to 520 kPa) could result in energy savings of 5 to 25 percent, depending on the system flow and pump arrangement.

Reducing Peak-Demand Charges

Unfortunately, water systems and electric systems experience peak demands during the same periods—early mornings, late afternoons, and the hottest afternoons and evenings of the summer season. The extra capacity installed by electric utilities to provide for peak-demand periods is paid for by demand charges, which high-demand customers pay in addition to the usual rate per kilowatt-hour of electricity.

The demand charge for a billing period is generally calculated based on the average or peak kilowatt demand during the billing period. Demand charges can be minimized if electricity is used at a minimal, constant rate—i.e., if the peak demand is kept as low as possible. Demand charges will also be lower if the water utility's peak demand occurs during a period that is off-peak for the electric utility.

Methods that can be used to reduce demand charges include

- using gravity-feed storage,
- using engine-driven pumps,
- reducing peak water usage, and
- changing the time of demand.

Gravity-feed storage

Gravity-feed storage is an effective way to smooth demand, as well as minimize friction head and reduce overall energy use. For a water utility, the energy conservation that gravity storage provides in meeting peak demand can be substantial. By providing sufficient storage to minimize variations in the pumping rate, the water utility can minimize peak electrical demand and the corresponding demand charges.

Engine-driven pumps

Using engine-driven pumping units instead of electric-motor–driven units during periods of peak demand may substantially reduce electricity usage and demand charges. This approach may be more energy efficient than installing electrical generators to power pump motors during periods of peak demand. In some areas, natural gas could be used economically to power engine-driven pumps, because summer is the off season for natural gas. Engine-driven units should be operated periodically under load anyway, so a good time to run them is during peak flow conditions.

Reducing peak water usage
Utilities can reduce peak residential water usage to some degree by lowering service pressure, using flow restrictors, or instituting pricing policies that discourage the use of lawn sprinklers. However, the reduction that can actually be accomplished is usually quite small.

Changing the time of demand
The methods discussed for reducing peak electrical demand also apply to changing the time of demand to the time of day when off-peak electrical rates are available. Additional gravity-feed storage may be used to permit increased off-peak pumping.

Power-Factor Improvement
The power factor of a motor indicates how effective the motor is at using the power available to it. A value of 1.0 is the highest possible. A power factor of less than 1.0 does not indicate that the motor is inefficient. Instead, it means that the motor requires larger power lines and transformers than are needed for a motor with a power factor of 1.0. For example, a 50-kW motor with a power factor of 1.0 draws 50 kW·h of energy during an hour. The electric utility needs to supply 50 kW of power to the motor. A 50-kW motor with a power factor of 0.8 must be supplied with 50/0.8 = 62.5 kW of power to operate properly (even though the motor will actually use the same amount of energy [50 kW·h] in an hour).

The electric utility imposes a charge for the lower power factor, because its facilities must be able to produce and supply the full 62.5 kW. Power factors vary with the type of motor and the load imposed on the motor. They can also be affected by other components in the motor circuit. Methods of improving the power factor include

- changing the motor type,
- changing the motor loading, and
- using capacitors.

Changing the motor type
Premium-efficiency motors have higher power factors than standard motors. When purchasing a new motor, a utility should carefully consider the cost savings that might be achieved over a period of time with a motor that is initially more expensive.

Changing the motor loading

Although induction-motor efficiencies remain relatively high between 50 and 100 percent of rated horsepower, the power factor decreases substantially and continuously as the rated load falls below 100 percent. For example, at 100 percent of rated load, the power factor of an induction motor may be 80 percent. At 50 percent of rated load, the power factor may be 65 percent, and at 25 percent of rated load, it may drop to 50 percent. Therefore, although reducing the load on motors may decrease power usage somewhat, this advantage is offset to some extent by the reduction in power factor.

Using capacitors

A utility may improve the power factor of an induction-motor circuit by installing a capacitor in parallel with the motor. The power-factor improvement resulting from a certain capacitance depends on the motor characteristics. Power-factor improvement beyond a certain point may cause problems to an electrical system, so the motor manufacturer should be consulted to determine the capacitance that may be used with a particular motor.

MAINTENANCE OF ELECTRIC MOTORS

A preventive maintenance schedule should be developed and followed for all motors in the system. Motors should be inspected at regular intervals, usually once a month. Under severe conditions, more frequent inspections are advisable. Log cards should be maintained for each motor. These cards should list details of all inspections carried out, maintenance required, and general conditions of operation. These records allow the operator to check the motor's condition at a glance. Any maintenance required should immediately be apparent. Regular annual or semiannual in-depth inspections are better than frequent casual checks. Items to be checked during an inspection are

- housekeeping,
- alignment and balance,
- lubrication,
- brushes,
- slip rings,
- insulation,
- connections, switches, and circuitry, and
- phase imbalance.

Housekeeping

Routine housekeeping includes keeping the motor free from dirt or moisture, keeping the operating space free from articles that may obstruct air circulation, checking for oil or grease leakage, and routine cleaning. If the correct motor enclosure has been chosen to suit the conditions under which the motor operates, cleaning should be necessary only at infrequent intervals. Dust, dirt, oil, or grease in the motor will choke ventilation ducts and deteriorate insulation. Dirt or moisture will also shorten bearing life.

Alignment and Balance

The alignment and balance on new motors or motors that have been removed and replaced must be carefully checked. They should be rechecked semiannually. Operators should perform the check according to the manufacturer's instructions, using thickness gauges, a straightedge, and/or dial indicators.

The most likely causes of vibration in existing installations are imbalanced rotating elements, bad bearings, and misalignment resulting from shifts in the underlying foundation. Therefore, vibration should always be viewed as an indication that other problems may be present. Both the magnitude and the frequency of vibration should be measured and compared with the measurements made when the unit was first installed.

Lubrication

The oil level in sleeve bearings should be checked and replenished as needed (every 6 months or so). The type and grade of oil specified by the manufacturer must be used. The oil ring, if fitted, should be checked to ensure that it operates freely. Oil wells should be filled until the level is approximately ⅛ in. (3 mm) below the top of the overflow. Overfilling should be avoided.

Grease in ball or roller bearings should be checked and replenished when necessary (usually about every 3 months) with grease recommended by the manufacturer. Bearings should be prepared for grease according to manufacturer's instructions.

New motors fitted with ball or ball-and-roller bearings should be supplied with the bearing housing correctly filled with grease. The bearings should be flushed and regreased every 12 to 24 months or so. The type of grease recommended by the motor manufacturer should be used. If the motor operates in an environment with a high ambient temperature, a grease with a high melting point should be selected. Where the ambient temperature varies considerably throughout the day, or where there are climatic changes, the manufacturer should be asked for specific advice on the type of grease to use.

It is usually convenient to regrease the bearings after removing the old grease and cleaning the bearings. The operator can perform these tasks by taking off the outside bearing covers when the motor is dismantled for periodic cleaning. The bearing housing should not be overfilled, and the operator should take care that no grit, moisture, or other foreign matter enters the dismantled bearings or the housing. A grease gun can be used for motors fitted with grease plugs.

Oil or grease should not be allowed to come into contact with motor windings. This could cause deterioration of the insulation. In all motors fitted with ball bearings, roller bearings, or both, the bearing housings are fitted with glands to keep the grease in and the dirt out. However, over lubrication of the bearings will result in grease leakage. An over-greased bearing tends to run hot, causing the grease to melt and creep through the gland and along the shaft to the windings. Over-lubricated sleeve bearings can also contaminate motor windings.

Brushes

Brushes should be checked quarterly for wear. They should be replaced before they wear beyond the manufacturer's recommended specification. For the three-phase motors commonly used by water utilities, brushes generally should not be allowed to wear below ½ in. (12 mm).

New brushes should be bedded in to fit the curvature of the slip rings. Operators can accomplish this by placing a strip of sandpaper between the slip rings and brushes, with the rough surface toward the brushes. Emery cloth should not be used because the material is electrically conductive and may contaminate components. Normal brush-spring pressure will hold the brushes against the sandpaper. The sandpaper should be moved under the surface of the brushes until the full width of each brush contacts the surface of the slip ring. Where the motor rotates in one direction only, it is best to move the sandpaper in this direction only and then release the brush-spring pressure when the sandpaper is moved back for another stroke.

All carbon dust should be removed after the bedding-in process is complete, including dust that may have worked between the brushes and brush holders. Brush-spring tension should be set as specified by the manufacturer. A small spring balance is useful for checking spring tension. The brush must slide freely in the brush holder, and the brush spring should bear squarely on the top of the brush.

Slip Rings

When properly maintained, slip rings acquire a dark glossy surface with complete freedom from sparking between slip rings and brushes. Sparking will quickly destroy a slip ring's surface, and rapid brush and slip-ring wear will result. Sparking may result from excessive load (starting or running), vibration (for example, from the driven machine), worn brushes, sticking brushes, worn slip rings, or an incorrect grade of brushes. Periodic inspection and checking will ensure that trouble is reduced to a minimum. If inspection reveals slip rings with a rough surface, the rotor must be inspected for wear and horizontal movement, then repaired as necessary.

Insulation

Motor-winding insulation should be checked periodically to make sure it remains uncontaminated with oil, grease, or moisture. In general, insulation should exhibit at least 1 megohm of resistance.

Connections, Switches, and Circuitry

All electrical circuits *must* be de-energized and "locked out" before any inspection or maintenance is performed.

Wiring should be checked semiannually or annually. Wires should be examined, and all connections should be checked to ensure that they have not worked loose. Continuous vibration and frequent starting will gradually loosen screws and nuts. Vibration can also break a wire at the point where it is secured to a terminal.

Circuit-breaker and contactor contacts should be examined and cleaned regularly. Slightly pitted and roughened contact surfaces that fit together when closed should not be filed. The burned-in surfaces provide a far better contact than surfaces that have been filed down, no matter how carefully. Any beads of metal that prevent the contacts from completely closing should be removed. Contactor retracting springs should be examined and tested to ensure that they retain sufficient tension to provide satisfactory service.

Moving parts should be operated by hand to check the mating of main, auxiliary, and interlocking contacts. Faces of the contactor armature and holding electromagnet should be clean and should fit together snugly. Anything on the faces that prevents an overall contact when the contactor is closed is likely to lead to chatter and excessive vibration of the mechanism, which increases wear and the possibility of a failure.

Moving parts must be kept clean to avoid any tendency to stick. Check the settings to make sure that the adjustments have not been moved by vibration or altered by unauthorized persons. Oil dashpots used with magnetic current relays may need replenishing with oil of the correct grade in order to maintain the required triggering time delay.

Phase Imbalance

Phase imbalance is caused by defective circuitry in the electrical service. It is the principal cause of motor failures in three-phase pumps. Phase imbalance produces a large current imbalance between each phase or leg of a three-phase service. This imbalance, in turn, will produce a reduction in motor starting torque, excessive and uneven heating, and vibration. The heat and vibration will eventually result in the failure of the motor windings and bearings.

Some phase imbalance will occur in any system but it generally should not exceed 5 percent. Each phase value should be checked monthly with a volt-ammeter for balance, and the values should be logged. If excessive imbalance is found, an electrician or the power utility should be called to isolate and repair the problem.

When a fuse blows on one leg of a three-phase circuit, a phase-imbalance situation known as single-phasing occurs. The fuse may have blown because of an insulation failure in the motor or because of problems with the power line. Unless the motor is manually or automatically turned off, it will continue to try to run on the two remaining phases. However, the windings that are still under power will be forced to carry a greater load than they were designed for, which will cause them to overheat and eventually fail.

TYPES OF COMBUSTION ENGINES

Internal-combustion engines, and occasionally external-combustion (steam) engines, are used by water utilities to power pumps for emergency backup and for portable dewatering applications. They are also used in some remote installations where electric power is not available, and they can be used to reduce peak-demand electrical charges.

Internal-combustion engines commonly used by utilities include gasoline engines and oil (diesel) engines. Gasoline engines can also run on propane, natural gas, or methane with minor modifications. External-combustion engines include reciprocating and turbine-type steam engines.

Gasoline Engines

Gasoline engines are generally used as standby or emergency units. Their initial cost is low, but fuel and maintenance costs are relatively high. They also present somewhat more of a hazard than diesel or natural gas engines because of the danger of a gasoline leak. Small gasoline engines may be used to power portable dewatering pumps. Gasoline engines, available in sizes from 1 hp to several hundred horsepower, generally operate at variable speeds between 600 and 1,800 rpm. They can be connected directly to centrifugal pumps. A common practice is to install an electric motor at one end of a pump shaft and a gasoline engine at the other or through a right-angle drive (Figure 8-6). When a pump is connected to both an electric motor and an engine, the engine must be furnished with a clutch.

FIGURE 8-6 Combination right-angle gear drive
Courtesy of Amarillo Gear Company

Diesel Engines

Diesel engines are best operated at constant, low speed. Though higher in initial cost, they are reliable and generally economical to operate. Diesels are often used to power pumps where electricity is unavailable, expensive, or unreliable. They also frequently function as emergency generator drives, furnishing power to an entire station that is normally supplied with power from an electric utility. Diesel engines are available in sizes from 50 to 10,000 hp, with 1 to 16 cylinders. The operation and maintenance of modern diesels require special training and skill.

Steam Engines

Reciprocating steam engines were widely used in early water systems but are rarely used for water pumping today. A number of these units have been preserved as historical civil engineering works. The more modern steam turbine is occasionally used to operate pumps in very large installations.

OPERATION AND MAINTENANCE OF INTERNAL-COMBUSTION ENGINES

Proper engine starting and operating procedures should be posted near each internal-combustion engine. The following sections describe general service procedures to be performed before an engine is started, during initial and continued operation, and after shutdown. A more detailed general checklist is provided in Table 8-1.

Service Prior to Operation

Before putting an engine into operation, operators should perform a general inspection of the engine and all attached components. Items to check include all fluid levels (fuel, oil, coolant), all rubber components (belts, hoses), and any regularly adjusted items, such as V-belts. Any leaks or loose components should be repaired or, if minor, noted for future repair.

    If the engine has a separate cooling-water source, make sure all valves are turned on before starting the engine. Also make sure that the clutch is disengaged.

Initial-Operation Service

As soon as the engine starts, check idle speed, oil pressure, ammeter, water temperature, and any other gauges that are attached. If any unusual gauge readings are observed, shut the engine down and correct the problem.

    If idle-speed checks show no problems, allow the engine to warm up according to manufacturer's recommendations. Then engage the clutch or other transmission device and put the engine under load. With the engine under load, keep a close watch on the engine and its gauges to make sure it is running properly. Watch for leaks and loose components and listen for unusual noises. If the engine will not come up to speed or will not hold speed under the load, shut it down and correct the problem immediately. A fouled fuel filter or a sticking governor may be reducing engine performance or the engine may require major maintenance.

TABLE 8-1 Operation of internal-combustion engines

Service Before Operation	Check fire extinguishers for ease of removal and for tight mounting, full charge, and closed valves. See that valves and nozzles are not corroded or damaged.
	Check the engine for signs of tampering, damage, or injury, such as loosened or removed accessories or drive belts.
	Check the amount of fuel in the tanks, and note signs of leaks or tampering. Add fuel if necessary. Fuel should be protected against freezing by use of appropriate additives.
	Check the oil level and add oil if necessary. Check the level and condition of the coolant. During the season when antifreeze is used, test the coolant with a hydrometer. Add antifreeze and/or water if necessary. Do not use alkali water. Check for any appreciable change in fuel or water level since the last after-operation service.
	Check all attachments such as carburetor, turbocharger, compressor unit, generator, regulator, starter, fan, fan shroud, and water pump for loose connections or mountings.
	Look for signs of fuel, oil, water, or gear-oil leaks. Check the cooling system for leaks, especially the radiator core and connecting hose. Check for leaks from the engine crankcase, oil filters, oil tanks, oil coolers, and lines. Check the fuel system for indications of leaks. Trace all leaks to their source and correct them.
	If the engine is cooled by a heat exchanger that uses an external source of water, make sure the water source is turned on.
	Check the electrolyte level in the batteries. Make sure that the battery is not leaking and the battery, cables, and vent caps are clean and secure. Check the voltmeter to see that it registers at least a nominal battery rating.
	Set the choke or operate the primer if necessary.
	Activate the starter. Note whether the starter has adequate cranking speed and engages and disengages properly without unusual noise.

Table continued on next page

TABLE 8-1   Operation of internal-combustion engines   (Continued)

Initial-Operation Service	When the engine starts, adjust the throttle to normal (fast idle) warm-up speed and continue the servicing procedure. *Caution:* The engine may be damaged or its service life appreciably reduced if it is placed under load before it reaches normal operating temperature.
	Observe the operation of the oil-pressure gauge or light indicator. If these instruments do not operate properly within 30 seconds, stop the engine immediately and determine the cause. Do not operate the unit if the oil pressure drops below indicated normal range at normal operating speed.
	If the engine has a manual choke, reset the choke as the engine warms up to prevent overchoking and dilution of engine oil.
	The ammeter may show a high charging rate for the first few minutes after starting, until current used in starting is restored to the battery. After this period, the ammeter should register zero or a slight positive charge with accessories turned off and engine operating at fast idle. Investigate any unusual drop or rise in reading. An extended high reading may indicate a dangerously low battery or faulty generator regulator.
	Note whether the tachometer indicates the engine rpm and varies with engine speed through the entire speed range.
	Check voltmeter for proper operation. It should register at least nominal battery voltage and will register slightly higher if the generator or alternator is working properly.
	Check the temperature gauge to be sure engine temperature increases gradually during the warm-up period. If engine temperature remains extremely low after a reasonable warm-up period, the engine cooling system or temperature control device needs attention. Engines should not be operated for an extended time at temperatures outside the normal operating range. All stationary gasoline engines will be equipped with thermostatically controlled coolant regulators recommended by the manufacturer.
Service During Operation	The clutch or engine drive train should not grab, chatter, or squeal during engagement or slip when fully engaged. If the clutch lever does not have sufficient free travel before the clutch begins to disengage, the clutch may slip under load. Too much free travel may keep the clutch from disengaging fully, causing clashing gears and damage during shifting.

Table continued on next page

TABLE 8-1 Operation of internal-combustion engines (Continued)

Service During Operation (continued)	The transmission gears should shift smoothly, operate quietly, and not creep out of mesh during operation. Gears jumping out of mesh indicate wear in the shifting mechanism or gear teeth, or incorrect alignment of transmission or clutch housing. Note poor engine performance such as a lack of usual power, misfiring, unusual noise, stalling, overheating, or excessive exhaust smoke. See that the engine responds to controls satisfactorily and that the controls are correctly adjusted and are not too tight or too loose. Check the instruments regularly during operation. As a general rule, do not let the oil pressure drop below the indicated normal range at normal speed.
Short-Stop Service	Short-stop service is performed whenever the engine is stopped for a brief period. It consists of correcting any defects noted during operation and making the following inspections: • If fuel tank is engine mounted, check fuel supply and, if engine is operated continuously, fill tank. When refueling, use safety precautions for grounding static electricity. Allow space in filler neck for expansion. See that filler-vent caps are open and pressure-cap valves are free. Replace caps securely. • Check crankcase oil level and add oil if necessary. Make sure oil drain is closed tightly. If there is evidence of loss or seepage, determine the cause. • Remove radiator filler cap, being careful of steam, especially if a pressure cap is used. See that coolant is at proper level and replenish as necessary. Do not fill to overflowing, but leave enough space for expansion. If engine is hot, add coolant slowly while engine is running at fast idle speed. • Check for leaks. • Check accessories and belts. Make sure that fan, water pump, and generator are secure and that their drive belts are properly adjusted and undamaged. • Check air cleaners. Inspect air cleaners and breather caps to see that they are delivering clean air. Clean and service as necessary.

Table continued on next page

TABLE 8-1 Operation of internal-combustion engines (Continued)

Service After Operation	Check fuel, oil, and water. Add if necessary so that the engine is prepared for the next use.
	Check engine operation.
	Check instruments.
	Check battery and voltmeter.
	Check accessories and belts. Inspect the carburetor, generator, regulator, starter, fan, fan shroud, water pump, and other accessories for loose connections or mountings. Inspect the fan and accessory drive belts and adjust if necessary. Belts should deflect. Replace any damaged or unserviceable belts.
	Check electrical wiring. Make sure that all ignition wiring is securely mounted and connected, clean, and not damaged. Repair or replace if necessary.
	Check air cleaner and breather caps. Oil in the air cleaner must be kept at the correct level. Rub a drop of oil between fingers. If it feels dirty, drain and refill with fresh oil. If operating in a dusty area, remove and clean the air cleaners and breather caps whenever necessary. Check all fuel filters for leaks.
	Check engine controls. Check linkage for worn or damaged joints and connections. Correct or report any defective linkage.
	Check for leaks. Inspect for any fuel, oil, or water leaks, and correct or report any found.
	Check gear-oil levels. Check the lubricant level in drives and transmissions after they have cooled enough to be touched by hand. If the lubricant is hot and foamy, check to determine the level of the liquid below the foam.

Service During Operation

Once the engine is placed into normal service, gauges and overall operation should be checked periodically to ensure no problems have occurred. Fuel level should be monitored, especially for diesels, because running out of fuel can cause major damage to diesel fuel-injection systems. Short-stop service (i.e., correcting defects noted during operation and generally inspecting the engine) should be performed when the engine is stopped for a short period.

Service After Operation

After the engine is shut down and disengaged from the pump, any minor problems that were noticed during operation should be corrected. This is also a good time to again check the lubricant levels in the engine and gear cases. Lubricant should be checked for foaming, which may indicate contamination by water.

Regular Maintenance

Traditionally, internal-combustion engines have been used to provide backup and emergency power for pumping during periods of peak demand or when there are electric power outages. In these applications, it is essential that the engine function perfectly on very short notice. A program of regular inspection and maintenance is necessary to ensure this reliability. As part of this program, all engines should be started and operated continuously for at least 15 minutes each week. Running the engine under load is advisable because it will be more likely to reveal any problems that have developed.

Regular maintenance checks should be scheduled according to manufacturer's recommendations. The necessary tasks include draining and replacing oil and other fluids, changing filters and belts, inspecting and replacing hoses and other rubber parts, and performing more major operations as necessary. Where antifreeze is used, its strength should be checked every 6 months, more often if leakage has required frequent coolant replenishment. Battery electrolyte levels should be checked at least quarterly and more often in hot climates. A schedule of battery testing should be established, especially for standby units that may require maximum battery efficiency to start during winter months.

One item of particular importance for diesels is fuel filtering. The injection-type fuel system used in diesels is a close-tolerance mechanical system that can easily be fouled or damaged by improperly filtered fuel. Only qualified personnel should repair a diesel fuel-injection system. Diesel engines require large volumes of air, so the air cleaners must be well maintained.

MOTOR AND ENGINE RECORDS

An equipment and maintenance record system should be maintained to assist in scheduling inspections and needed service work, evaluating pump equipment, and assigning personnel. Most water systems maintain a card or notebook sheet with a listing of the make, model, capacity, type, date and location installed, and other information. The remarks section should include the serial or part numbers of special components (such as bearings) that are likely to require eventual replacement.

A separate operating log should be kept, listing all units along with a record of the operating hours. This record is an essential feature of any reasonable periodic service or maintenance schedule. In addition, a daily work record should be kept on each piece of equipment. Many water utilities keep computer records of motor and engine parts and maintenance as part of a comprehensive maintenance management program.

MOTOR AND ENGINE SAFETY

Operators must follow special safety precautions when dealing with motors and engines. In addition to all the other safety concerns associated with water distribution (as discussed in other chapters), operators must be cautious around electrical devices and be aware of fire safety guidelines.

Electrical Devices

There is no safety tool that will protect absolutely against electrical shock. Operators should use plastic hard hats, rubber gloves, rubber floor mats, and insulated tools when working around electrical equipment. However, these insulating devices do not guarantee protection, and the operator using them should not be lulled into a false sense of security.

Electrical shocks from sensors are possible in many facilities, such as pumping stations, because many instruments do not have a power switch disconnect. It is important to tag such an instrument with the number of its circuit breaker so that the breaker can quickly be identified. After the circuit breaker has been shut off, an operator should tag or lock the breaker so other employees will not reenergize the circuit while repairs are being performed. Even after a circuit is disconnected, it is good practice to check the circuit with a voltmeter to be certain that all electrical power has been removed.

It is very easy to damage an electrical or electronic instrument by inadvertently shorting a circuit while making adjustments. Insulated screwdrivers should be used for electronic adjustments to reduce the chance for damage.

Blown fuses are usually an indication of something more than a temporary or transient condition. The cause of the overload should be identified. Never replace a blown fuse with a fuse of higher amperage than the circuit's designed rating.

Broken wires should be replaced instead of repaired. Secure terminal connectors should be used on all wires at all points of connection. Charred insulation on a wire is a warning of a serious problem.

Extreme care should be taken in working around transformer installations. Maintenance activities should be performed only by authorized employees of the power company. Power to a transformer cannot be locked out except by the power company, so operators should always assume that all transformers are energized and observe full safety precautions.

Electric switchboards should be located and constructed in a manner that will reduce the fire hazard to a minimum. They should be located where they will not be exposed to moisture or corrosive gases, and their location should allow a clear working space on all sides.

Adequate illumination should be provided for the front (and back, if necessary) of all switchboards that have parts or equipment requiring operation, adjustment, replacement, or repair. All electric equipment, including switchboard frames, should be well grounded. Insulating mats should be placed on the floor at all switchboards.

Open switchboards should be accessible only to qualified and authorized personnel and should be properly guarded or screened. Permanent and conspicuous warning signs should be installed for panels carrying more than 600 V. Areas screened off because of high voltage should be provided with locks to prevent unauthorized persons from entering. However, the lock must be operable by anyone from the inside so that there is no chance of someone being locked inside the enclosure.

Switches should be locked open and properly tagged when personnel are working on equipment. Fully enclosed, shockproof panels should be used when possible. Such equipment should be provided with interlocks so that it cannot be opened while the power is on.

Fire Safety

Suitable fire extinguishers should be kept near at hand and ready for use. The locations and operation of fire extinguishers should be familiar to all employees. Water or soda-acid extinguishers should never be used on electrical fires or in the vicinity of live conductors. Carbon dioxide or dry-powder extinguishers are recommended for these situations.

Detailed fire safety requirements vary considerably from one installation to another and from one locality to another. Therefore, it is recommended that the advice (and where necessary, the approval) of one or more of the following organizations be secured: local fire department, state fire marshal's office, fire insurance carrier, or local and state building and fire prevention bureaus. Generally, the fire insurance carrier and one of the fire protection agencies will provide all of the necessary advice.

BIBLIOGRAPHY

AWWA Standard for Horizontal and Vertical Line-Shaft Pumps. ANSI/AWWA E103. Denver, Colo.: American Water Works Association (latest edition).

AWWA Standard for Submersible Vertical Turbine Pumps. ANSI/AWWA E102. Denver, Colo.: American Water Works Association (latest edition).

Centrifugal Pumps. 2007. DVD. Denver, Colo.: American Water Works Association.

Jones, G.M., R.L. Sanks, , G. Tchobanoglous, and B.E. Bosserman II. 2008. *Pumping Station Design*, 3rd ed. New York: Butterworth-Heineman.

Water Distribution Operator Training Handbook. 2005. Denver, Colo.: American Water Works Association.

CHAPTER 9

Instrumentation and Control

Instruments in a water utility are usually used for measurement or control. Instruments allow an operator to monitor and control flow rates, pressures, levels, and other important parameters from all parts of the distribution network.

An operator's primary responsibilities are supervision and control. *Supervision* means examining system performance information and deciding if it is acceptable. If, in the operator's opinion, performance is unacceptable, an element of system operation must be changed to bring performance back to an acceptable condition. A setup for which a human operator evaluates the performance continually is said to have open-loop control. Control equipment allows the operator to change valve settings, turn pumps on and off, and otherwise adjust the system for efficient operation.

When instruments are provided to make the necessary change or correction automatically without the intervention of the operator, the system is said to have closed-loop control. However, regardless of the extent to which automatic control is used, the operator still may need to intervene manually during abnormal or emergency situations.

Most utilities feed instrument readings to one or more computers both to gather information and to assist in providing control. However, some smaller systems continue to maintain simpler, onsite instrumentation and controls. Computers and electronic control equipment continue to become less expensive, less complicated, and more reliable. This is leading to even more widespread use of computerized systems.

The main categories of instrumentation and control discussed in this chapter are as follows:

- *Primary instrumentation:* Sensors that measure process variables such as flow, pressure, level, and temperature
- *Secondary instrumentation:* Instruments that respond to and display information from primary instrumentation
- *Control systems:* Manual, automatic, and digital systems that operate final control elements such as pumps and valves

This chapter will also discuss maintenance and operation safety for this instrumentation.

PRIMARY INSTRUMENTATION

Sensors that measure flow, pressure, and other parameters are essential to water system operations. The measurements allow the operator to efficiently maintain the quality and quantity of drinking water.

Basic Instrument Components

Various instruments are used to measure, display, and record the conditions and changes in a distribution system. Some instruments indicate what is happening at a given instant. Others guard against equipment overload and failure. Still others provide permanent records that are used to determine operating efficiency and the need for regularly scheduled maintenance.

The simplest instruments have only two parts: a sensor and an indicator. The sensor responds to the physical condition (parameter) being measured, converting it to a signal that can activate the indicator. The signal may be a simple physical motion or it may be an electrical current or a change in pneumatic (air) pressure.

The indicator may display the result immediately or it may be replaced or supplemented by recorders or totalizers to monitor conditions over a period of time. The main categories of indicators are direct, remote, and distant. Indicators will be discussed in more detail later in this chapter.

The following descriptions introduce the most common types of sensors.

Flow Sensors

The most significant measurement in distribution facilities is the flow of water. Day-to-day operational decisions and long-term planning are based on the measurements from flowmeters. Knowledge of flow is needed to bill customers, check the efficiency of pumps, monitor for leaks, and help control or limit the delivery of water. Analyses of water system operations are also derived from distribution system flow information.

The flow passing through distribution system meters might simply be totalized on the meter register. It can also be recorded locally on a chart recorder or transmitted to a central location for recording. It is often necessary to know the flow rate at various times of the day, as well as the flow in various parts of the distribution system.

The various types of meters used in water systems are discussed in chapter 10. The following paragraphs review various meter technologies for mainline measurements.

Differential-pressure flowmeters

The measurement devices most commonly used for measuring large quantities of water are differential-pressure flowmeters (also called head meters). Their popularity is largely due to a combination of flexibility, simplicity, ease of installation, and reliability. The flow reading of a differential-pressure meter is calculated from the difference between two pressures measured in the meter. Differential-pressure flowmeters are available in many forms, such as the Venturi, modified Venturi, flow tube, or orifice plate. The measurement of flow is a function of detecting two pressure heads, usually one in the normal pipe size and one in a constricted region, called the throat, within the meter. The flow is proportional to the square root of the difference between the two pressure readings.

Velocity-type flowmeters

Velocity-type, or current-type, flowmeters include magnetic, turbine, propeller, multijet, proportional, and sonic flowmeters. In each case, the flow velocity is measured, and the quantity is calculated from the product of velocity times the cross-sectional area of the pipe. The Pitot meter is also a velocity-type measuring meter but it determines the velocity based on the difference between the flow's dynamic pressure and the static pressure.

Other flowmeters

Other meters are available for mainline flow sensing, such as the vortex-shedding flowmeter and the variable-area flowmeter. The open-channel flow detectors (weirs and flumes) represent another type of differential-pressure flowmeter. In both cases, the flow is calculated from the fluid depth or the head that drives the flow.

Pressure Sensors

Pressure sensors are used to determine suction and discharge pressures at pumps, pressure regulators, and selected points in a distribution grid. They also determine pressures of plant waters, eductors, storage tanks, and air compressors. Pressure instrumentation may range from simple, direct-reading pressure gauges to complex pressure-sensing equipment that transmits readings to remote locations. The four most common types of pressure sensors are:

1. Strain gauges
2. Bellows (low pressure)
3. Helical elements (medium pressure)
4. Bourdon tubes (high pressure)

Strain gauges

The diaphragm strain gauge sensor is the type used most widely in modern instrumentation. As shown in Figure 9-1A, it consists of a section of wire fastened to a diaphragm. As the variable being measured (such as pressure) changes, the diaphragm moves, changing the length of the wire, thus increasing or decreasing its resistance. This changing resistance can be measured and transmitted by electrical circuits. Similar electronic sensors include the variable-capacitance pressure cell, variable-reluctance pressure cell, and vibrating wires.

Direct-reading pressure gauges

The remaining three types of sensors are direct-reading pressure gauges, which were once widely used but are seldom installed today. The bellows sensor (Figure 9-1B) is a flexible copper can. The sensor expands and contracts with changes in pressure. The helical sensor (Figure 9-1C) is a spiral-wound tubular element that coils and uncoils with changes in

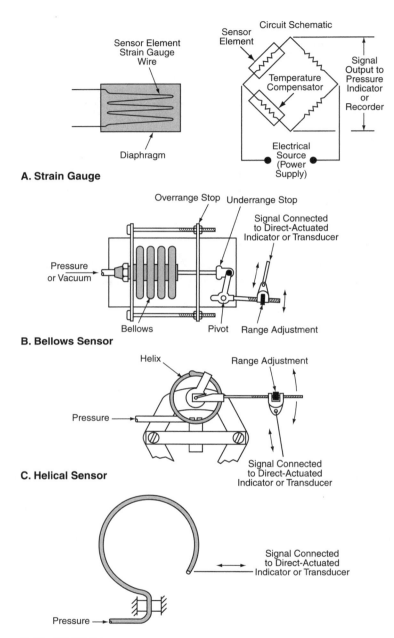

FIGURE 9-1 Types of pressure sensors

pressure. The Bourdon tube (Figure 9-1D) is a semicircular tube with an elliptical cross section that tends to assume a circular cross-sectional shape with changes in pressure, thereby causing the C-shape to open up.

Level Sensors

Level sensors are commonly used to measure the elevation of water in wells, as well as the depth of water in storage reservoirs and tanks. They are also used to measure the levels of stored chemicals in tanks. A pressure sensor can readily be adapted to level measurement if it is installed at the base of a tank. As level increases in the tank, the pressure reading increases. The reading can be calibrated in feet (or meters) of liquid. In elevated tanks, the level measurement needed is the level in the elevated portion of the tank, rather than in the tank and riser. Transmitting mechanisms can be calibrated so that "zero" represents the bottom or minimum level in the elevated portion of the storage tank.

The common types of level sensors are

- float mechanisms,
- diaphragm elements,
- bubbler tubes, and
- direct electronic sensors.

Float mechanisms

Float mechanisms (Figure 9-2A) have a float that rides on the water surface and drives the transducer through an arm or cable. Floats are very inexpensive and simple to operate but they cannot be used where the liquid surface may be rough or may freeze.

Where extreme accuracy in level-sensing systems is required, a special adaptation of a float-type level transmitter called a stage recorder has been used. It uses larger floats and a more precise pulley system to increase positioning accuracy. These devices may also be equipped with a digital sensing system.

Diaphragm elements

Diaphragm elements (Figure 9-2B) have a flexible bulb or diaphragm connected by a tube to a pressure sensor. They operate on the principle that the confined air in the tube will compress in relation to the head of water above the diaphragm. The change of pressure that is sensed is then related to a change in the head of water.

Bubbler tubes

Bubbler tubes (Figure 9-2C) maintain a constant low flow of air discharging from the end of a tube suspended in the liquid. They operate on the principle that the pressure required to discharge air from the tube is proportional to the head of water above the tip of the tube.

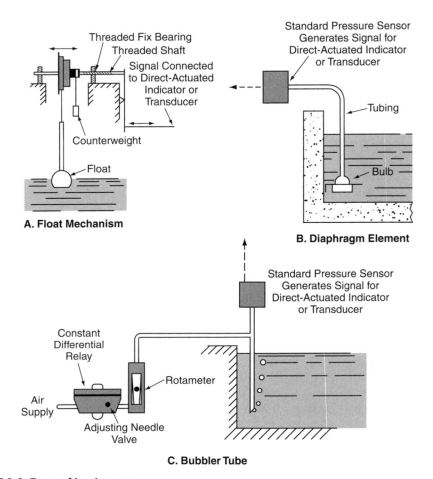

FIGURE 9-2 Types of level sensors

Both the diaphragm and the bubbler-tube sensors commonly use a strain gauge sensor to convert the changes in air pressure into an electrical signal.

Bubbler tubes were once widely used because they are almost maintenance-free and can be used in almost any type of liquid. They do require an air compressor to supply a continuous stream of air. Bubbler tubes have now generally been replaced with newer electronic equipment.

Direct electronic sensors

Several direct electronic devices have been developed for level measurement. Among these are various types of probes, variable-resistance devices, magnetic systems, radio frequency devices, and ultrasonic systems.

Probes. In a probe system, an insulated metallic probe is installed in the reservoir. As the water level rises and falls in the reservoir, the capacitance changes between the probe and the water. This capacitance signal can then be converted into a signal that indicates the level. When the probe is installed in a nonmetallic tank, a second electrode is required.

Variable-resistance devices. A variable-resistance level sensor consists of a wound resistor inside a semiflexible envelope. As the liquid level rises, the flexible outer portion of the sensor presses against the resistor and a portion of the resistor element is temporarily shorted out, which changes the resistance of the sensor. The resistance is then converted into a liquid-level output signal.

Magnetic systems. Magnetic level gauges consist of a float chamber, a float, and an external indication device (Figure 9-3A, B). The float chamber is basically a column of 2 ½-in. (63.5-mm) pipe with process connections to match those of the storage tank, reactor, drum, column, or other equipment where level is to be measured. These connections may be side couplings or flanges, or top and bottom flanges. The float moves up and down inside the chamber as the process level changes.

The float type is determined by the process fluid specific gravity, pressure, and temperature. Also, the materials of construction must be compatible with the process fluid. The float must be light enough to maintain buoyancy and have properties that allow it to withstand the pressure and temperature combination. Each float is engineered for proper buoyancy for each particular application for level or interface measurement. Contained inside the sealed float is a very strong magnet assembly.

The indicator tube and scale assembly is attached to the outside of the float chamber. Inside the transparent, hermetically sealed indicator tube is a lightweight magnetized indicator. This indicator is magnetically coupled to the float and moves up and down inside the indicator tube as the float moves up and down with the rising and falling of the liquid level in the vessel. The indicator is brightly colored and allows the operator to read the level from 100 ft (30.48 m) away. The only moving parts are the float and the indicator.

A transmitter providing a 4–20-mA output signal can be added. This signal can be sent to a remote location for indication or control functions. Additionally, level switches can be added, external to the float chamber, for alarm or control. These switches can easily be moved up or down the chamber while not disturbing the process.

A magnetic level transmitter is a stand-alone, loop-powered device, usually mounted in the top of a tank or other container. It may also be used as a level transmitter when mounted externally. The sensor consists of a stainless-steel tube with a heat-treated wire stretched through the center of it. A magnetic float is placed around the tube to track the fluid level.

A low-power microprocessor-controlled circuit produces a high-current pulse, which is transmitted down the heat-treated wire. The magnetic field in the float interferes with the high-current pulse, creating a torsional wave that is "bounced" back to the electronics. The transit time of the wave is converted to a 4–20-mA signal, proportional to the fluid level.

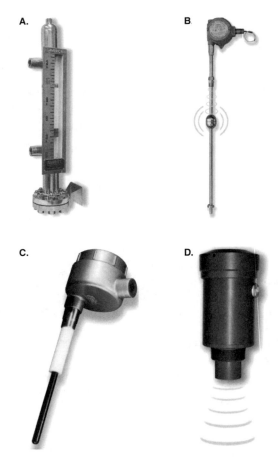

FIGURE 9-3 A. Magnetic liquid level gauge, B. magnetostrictive liquid level transmitter, C. radio frequency level, D. ultrasonic liquid level

Courtesy of Babbitt Level Controls, www.babbittlevel.com

Radio frequency level switches. A radio frequency balanced impedance bridge circuit (Figure 9-3C) is used to detect if the probe is in contact with the material that is to be sensed. A high-frequency/low-power radio signal is emitted by the electronics in the head of the level switch. The probe serves as an antenna to detect the frequency. Calibration is done when material is not in contact with the probe. The bridge is balanced by turning the adjustment pot to find the bridge balance threshold. When material comes in contact with the probe, the bridge becomes unbalanced and the comparing circuit realizes the change. This causes a relay output to change state. This relay output may be used for high- or low-level alarming or to control level by turning pumps on or off or by opening and closing valves.

Ultrasonic systems. In ultrasonic level-sensing systems (Figure 9-3D), an ultrasonic generator is installed above the water level. This generator sends ultrasonic signals toward the water surface, and the signals bounce back and are detected by a receiver. The time required for this signal to echo is calibrated to produce a water-level output signal. Air temperature variations must be compensated for because the speed of sound in air is a function of temperature. Excessive humidity in the air above the liquid may also significantly interfere with proper operation.

Temperature Sensors

Temperature sensors are commonly used in a water system to monitor the condition of pump bearings, switchgears, motors, building heat, and water. The temperature sensors commonly used in distribution systems are thermocouples and thermistors.

Thermocouples use two wires of different materials, represented as metal A and metal B in Figure 9-4A. These wires are joined together at two points: the sensing point and the reference junction. Temperature changes between the two points cause a voltage to be generated, which can be read out directly or amplified through a transducer.

Thermistors (Figure 9-4B), also called *resistance temperature devices*, use a semiconductive material, such as cobalt oxide, that is compressed into a desired shape from the powder form. The material is then heat-treated to form crystals to which wires are attached. Temperature changes are reflected by a corresponding change in resistance of the thermistor, as measured through the attached wires.

Electrical and Equipment Sensors

Measurements of electric power and equipment status are important for maintaining operational efficiency.

Electrical sensors

Four important electrical parameters are monitored for various pieces of equipment in the distribution system:

1. voltage, measured in volts
2. current, measured in amperes
3. resistance, measured in ohms
4. power, measured in watts

Because the equipment used to measure these parameters is itself electrical in nature, it is difficult to make clear distinctions as to what part of the equipment is the sensor and what part is the indicator. In fact, the measurements of volts, amperes, and ohms are all made with the same unit. Electric current passing through the meter's coil creates a magnetic field.

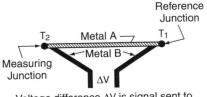 

Voltage difference ΔV is signal sent to
direct-acting indicator (voltmeter)
or transducer.

A. Thermocouple

Change in electrical resistance ΔΩ is the
signal sent to direct-acting indicator
(ohmmeter) or transducer.

B. Thermister

FIGURE 9-4 Types of temperature sensors

The field reacts with the field of the permanent magnet surrounding it, causing the coil and the attached indicator needle to move.

Digital multimeters are used to measure volts, ohms, and amperes (Figure 9-5). These meters are highly accurate and durable. Many meters are resistant to transient voltages and have features to graph and display the output in a variety of ways.

The electric energy used by a utility, which determines the utility's electric bill, is measured in units of kilowatt-hours (kW•h). One thousand watts (that is, 1 kW) drawn by a circuit for 1 hour results in an energy consumption of 1 kW•h. The meters used to determine kilowatt-hour usage are essentially totalizing wattmeters (Figure 9-6).

On most kilowatt-hour meters, a rotating disk can be seen through the glass front or through a small window in the front of the case. An operator can count the revolutions of the disk by watching for the black mark on the disk. The speed of the disk indicates the kilowatt load. Checking the disk speed with a stopwatch is a quick and accurate way of determining kilowatts used over a short period of time. The following formula may be used:

$$\text{kilowatts} = \frac{\text{disk watt} - \text{hours constant} \times \text{revolutions} \times 3{,}600}{(\text{seconds} \times 1{,}000)}$$

Equipment status monitors

A number of operating conditions associated with major equipment should be monitored. Vibration should be monitored in a plant process control system, particularly for large, expensive equipment such as pumps and blowers. Excessive vibration can quickly cause significant damage to this equipment, particularly when the equipment is operated at speeds greater than 1,800 rpm. Vibration sensors are available that, when mechanically attached to the particular piece of equipment, will activate a contact closure when vibration levels exceed a specific "*g*" value. To minimize problems associated with adapting equipment to accommodate these sensors in the field, vibration sensors should, if possible, be listed in the specifications as accessories to be included with equipment.

Position and speed are two other equipment status conditions that can be monitored. Position transmitters normally work with variable-resistance devices mechanically linked

FIGURE 9-5 Digital multimeter
Reproduced with permission of Fluke Corporation

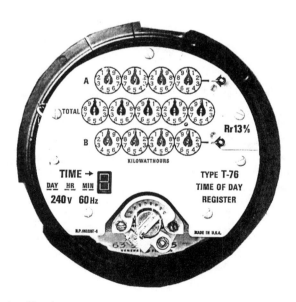

FIGURE 9-6 Totalizing wattmeter

to a piece of equipment. The output of the variable-resistance device is then converted to a signal for monitoring. Speed transmitters are usually tachometers driven by the equipment being monitored. The tachometers produce a voltage that is converted into standard current values for monitoring purposes. The instrument uses an electrical coil that produces a voltage when excited by magnets mounted on a spinning shaft. This voltage is used to operate a milliammeter that provides an indication that is proportional to speed.

With some equipment, particularly clarifier drives, torque is another variable that is sometimes monitored. Torque sensing is normally used to shut down circuits to prevent damage to the equipment. Torque-sensing equipment can be supplied to produce a contact closure when torque rises above a preset value. It can also be supplied with converters that produce a 4–20-mA direct current (DC) signal proportional to the magnitude of the torque so that the reading can power an indicator at a remote location.

Larger pumps and motors incorporate discrete sensors that measure parameters such as temperature, pressure, and position. These sensors are installed for safety monitoring and sequential control operations.

Process Analyzers

Because of increasingly stringent requirements for water quality, analytical equipment that measures process variables is used more frequently in water systems. Water quality analyzers are becoming more common in distribution systems. Increasing demands by customers and regulatory requirements are making the use of these sensors cost effective. Process-analyzing equipment is discussed in greater detail in *Water Treatment* and *Water Quality*, other books in this series.

Turbidity monitors

Turbidity monitoring is a must for the effluent from granular-media filters. US Environmental Protection Agency regulations define the requirements for turbidity monitoring in treatment plants. The turbidity of raw water, as well as that of settled water and finished water leaving the treatment plant, is a critical parameter when optimizing performance. Turbidity measurement in the distribution system can provide valuable information on changes in quality as water travels throughout the system.

Early turbidity-monitoring equipment used a sensor to detect changes in the amount of light transmitted directly through a sample. This concept of sensing was found to be nonresponsive when very low turbidity levels were being measured. Now, turbidity-sensing devices monitor the intensity of light scattered from particles when a beam of light shines on the sample. The most common method measures the amount of light scattered in a direction at 90° to the path of the light beam.

At higher turbidity levels, this concept is not practical because the scattered light at high levels of turbidity cannot be measured accurately. One successful method of measuring the higher turbidity levels normally encountered in raw water is the surface scatter concept.

Light scattered from turbidity particles at the surface of a sample compartment is measured to provide an indication of the turbidity level.

Laser nephelometers (Figure 9-7) employ laser technology and fiber optics to improve sensitivity and speed response. These are used primarily in the 0–1 ntu range that is of most interest for filtered water. The instruments require little maintenance and can be easily calibrated as compared to particle counters.

pH monitors

Another analytical instrument that has been used for some time in water treatment plant applications is the pH monitor. Readings of the pH level can be extremely useful in monitoring and controlling a treatment process. Chemical reactions affecting coagulation, corrosion, and softening can be predicted, monitored, and changed through proper pH control. This measurement can provide a sensitive measure of water quality in the distribution system.

The pH monitors detect the level of hydrogen ion activity in a sample and convert it to a signal. A pH-sensing system consists of a glass electrode and a reference electrode. The glass electrode develops an electrical potential that varies with the pH of the process fluid. The potential developed between the two electrodes is amplified and converted to a signal representative of the pH.

Residual chlorine monitors

Residual chlorine monitoring is often required as water enters the distribution system and is desirable at key locations throughout the system. Types of residual-chlorine monitoring systems are (1) a chlorine-permeable membrane probe that allows chlorine to diffuse through a membrane system on the end of the probe. The chlorine concentration passing through the membrane generates a current in an electrode system that corresponds to the chlorine level; (2) the amperometric type where two dissimilar metals are placed in a measurement cell containing an electrolyte. A voltage is applied to these two metal electrodes, and the amount of current flowing between the electrodes is proportional to the amount of chlorine present in the solution; and (3) a colorimetric measurement that uses the DPD test (common for field measurements) to develop a color. The intensity is directly related to the concentration.

Chlorine residual may be the most important measurement of water quality in the distribution system. This measurement may reflect the microbiological safety of the water. Currently, continuous monitoring of pathogens in drinking water is not possible, so the presence of an adequate chlorine residual may provide a "real-time" measurement that can be used to ensure water quality.

Particle counters

Particle counters are somewhat similar to turbidity monitors. However, they provide a more detailed analysis of particulate matter in the sample. Particle counters use different types of sensor technologies. One type transmits a laser beam across a sample stream, and the beam is detected by a sensor on the opposite side of the sample. When particles pass

FIGURE 9-7 Laser nephelometer
Courtesy of Hach Company, USA

through the beam, they interrupt the beam. The sensor also senses particle size based on the amount of the beam that is interrupted. Particle counters can be programmed to provide information on particle count, total particle count below a specified particle size, and particle distribution. The primary applications for these devices has been to monitor filter effluent in water treatment plants. However, particle counters may also monitor the effectiveness of coagulation and flocculation. The usefulness of particle counters in distribution systems is still under development.

Other water quality sensors
Some other sensors that may be useful in distribution systems are

- conductivity (an indirect measurement of dissolved solids),
- temperature,
- color (may give an indication of organic content),
- ultraviolet absorbance (dissolved organic substances),
- dissolved organic carbon, and
- online gas chromatography or gas chromatography mass spectrometry (capable of detecting specific organic compounds).

SECONDARY INSTRUMENTATION AND TELEMETERING

Secondary instrumentation displays the signals from the sensors of primary instrumentation. It also allows distant control. Secondary instruments are usually panel mounted. They can be mounted in local control panels, filter control consoles, area control panels, or main control panels.

Before World War II, most measuring and control instruments were mounted adjacent to the process being controlled, with direct connections to the process. These primary measuring devices required operators to move throughout the plant to take readings and make control adjustments. With a small process plant and an abundant workforce, this was sufficient.

Then, as plant monitoring, maintenance, and compliance requirements increased, monitoring and control functions needed to be more centralized. Direct connection was no longer possible, so secondary instrumentation was developed. Secondary instrumentation measures the parameters and transmits signals that correspond to the measurements. Secondary instrumentation required the development of a signal transmission method, as well as field and panel-mounted hardware to perform the monitoring and control functions.

Pneumatic and Electronic Signal Transmission

The first signal transmission methods used air pressure (i.e., pneumatic) transmission. Electronic signal transmission, developed later, uses electrical current or voltage signals. Both methods perform the same functions, and the instrument exteriors look more or less identical, both in the field and in the control room. As development and usage have progressed, both systems have also standardized signal levels. The pneumatic standard range is 3–15 psig (20–100 kPa [gauge]), whereas the electronic standard is 4–20 mA DC. The practical advantage of this standardization is that control room instruments' operating mechanisms are based on common signal units in all locations. That is, the panel may contain indicators, recorders, and controllers handling pressure, temperatures, flows, and other process variables, but the input and output signals operate over the same standard range. The only differences among them are the display scales or charts employed.

Pneumatic and electronic signal transmission systems can also be used within the same process control system, allowing a great deal of flexibility in providing instrumentation for a plant. Equipment can be chosen that most effectively suits the application and environment. Signal converters are readily available to convert pneumatic pressure to electric current (P/I converters) and electric current to pneumatic pressure (I/P converters).

Although pneumatic systems are still in use, most new systems are electronic. The costs of maintenance and installation of pneumatic systems have made them virtually obsolete. The remainder of this chapter describes the components of electrical systems.

Receivers and Indicators

Receivers convert the signal sent by the transmitter to an indicator reading for the operator to monitor. The indicators may be

- a direct-reading display that shows the current value of the parameter being monitored,
- a recorder that preserves the information for later examination,
- a totalizer that gives the total accumulated value since the instrument was last reset, or
- some combination of these units.

Indicator displays and recorders are of two types: analog and digital. These terms are commonly used in describing various components and functions of instrumentation. Analog values range smoothly from the minimum to the maximum value of a given range. An analog signal is either a variable voltage or current. The dial indicators shown in Figures 9-8 and 9-9 are examples of analog displays. Analog indicators include dial gauges and strip or circle charts (Figures 9-10 and 9-11). The indicated values on an analog display range smoothly from the display's zero to its maximum.

FIGURE 9-8 Analog indicator

FIGURE 9-9 Analog and digital indicator

FIGURE 9-10 Circular recorder

FIGURE 9-11 Strip chart recorder

Digital values, on the other hand, take on only a fixed number of values within a range. Digital indicators, like digital watches, display decimal numbers. The number of possible readings within a given range is limited by the number of digits displayed.

Most parameters measured in water distribution are continuous in nature, like an analog signal or display. However, analog-to-digital converters allow continuous values to be displayed on digital indicators or transmitted over digital transmission channels. And digital-to-analog converters allow the reverse conversion.

Digital indicators are usually more accurate than analog indicators. They are not subject to the errors associated with electromechanical or mechanical systems and they are easier to read correctly. However, analog indicators may be preferable for at-a-glance monitoring to ensure a value remains within a given range or to observe its rate of change.

Telemetry

To monitor conditions at very distant locations, such as a remote pump station or reservoir, a telemetry system may be used. With this system, a sensor is connected to a transmitter, which sends a signal over a transmission channel to a combination receiver and indicator.

The type of signal used must be designed to maintain its accuracy over a long distance. Older equipment used audio tones or electrical pulses, but most equipment now transmits the information by a digital signal. The signal is transmitted either through direct wiring, through a leased telephone line, or by radio or microwave transmission (Figure 9-12). The receiver converts the signal to operate the indicator.

Telemetry systems allow flow rate, pressure, and other distribution system parameters to be sensed at one or more remote sites and indicated at a central location. Every telemetry system has the following three basic components:

1. Transmitter
2. Transmission channel
3. Receiver

The transmitter takes in data from one or more sensors at the remote site. It converts the data to a signal that is sent to the receiver over the transmission channel. The receiver changes the signal into standard electric values that are used to drive indicators and displays, recorders, or automatic control systems.

Telemetry transmission channels

The transmission channel in a telemetry system may be cable owned by a water utility that extends for short distances, such as between two buildings on a common site. In most cases, however, the channel is either a leased telephone line, a radio channel, or a microwave system. A system using space satellites is also available, though it is expensive. The leased telephone line may be a dedicated metallic pair, which is relatively expensive but

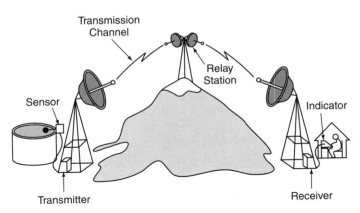

FIGURE 9-12 Relay station for radio or microwave telemetry system

highly reliable and interference-free or it may be a standard voice-grade phone line. Most modern transmitters generate signals that are designed to be sent over voice-grade lines and fiber-optic cables.

Radio channels can be in the VHF (very high frequency) or UHF (ultra high frequency) band. Both radio and microwave systems generally require a line-of-sight path, which is unobstructed by buildings or hills between the transmitter and the receiver. To bypass obstructions or to ensure signal strength over very long distances, relay stations may be required.

Analog signal systems

Analog signals that are commonly used include:

- Current—the DC current generated by the transmitter is proportional to the measured parameter.
- Voltage—the DC voltage generated by the transmitter is proportional to the measured parameter.
- Pulse-duration modulation (PDM)—the time period that a signal pulse is on is proportional to the value of the measured parameter.
- Variable frequency—the frequency of the signal varies with the measured parameter.

Current and voltage signals can be used only for short-distance systems with utility-owned cable or a leased metallic pair. The signals can be damaged by line loss and other factors over telephone lines. PDM and variable-frequency signals can be used for any distance over any type of channel.

Digital signal systems

Digital systems generally send binary code, in which the transmitter generates a series of on–off pulses that represent the exact numerical value of the measured parameter (for example, off–on–off–on represents 5). These signals can be used over long or short distances with any transmission channel. The binary code signal is well adapted for connection to computerized systems. Figure 9-13 illustrates analog and digital telemetry signals.

In digital systems, the remote or transmit device is normally referred to as a remote terminal unit (RTU) and the receiver is known as the control terminal unit (CTU). A rather significant difference between analog and digital telemetry systems is that in digital systems, the RTU does not itself directly measure the variables in the system. Instead, the variables are measured by a transducer device, which normally converts the physical value of the process variable into a current output signal of 4–20 mA DC. These signals in turn serve as input to the RTU. The RTU converts this data into a message consisting of digital words that are transmitted to the CTU.

Multiplexing

Multiplexing systems allow a single physical channel, such as a single phone line or radio frequency, to carry several signals simultaneously. Tone-frequency multiplexing accomplishes this by having tone-frequency generators in the transmitter and tone-frequency filters in the receiver. The signal representing each measured parameter is assigned a separate audio frequency. The transmitter sends data representing each signal only over its assigned frequency, and the filters in the receiver allow it to respond to each frequency separately. Up to 21 distinct frequencies can be sent over a single voice-grade line. Tone-frequency multiplexing can be used with PDM and digital systems.

Scanning

Scanning is a second method of sending multiple signals over a single line or transmission channel. A scanner at the transmitter end checks and transmits the value of each of several parameters one at a time, in a set order. The receiver decodes the signal and displays each value in turn. Scanning can be used with all types of signals and with all types of transmission channels. Scanning and tone-frequency multiplexing can be combined to allow even more signals over a single line. A 4-signal scanner combined with a 21-channel tone frequency multiplexer would yield 84 distinct signal channels on a single line.

Polling

Another method of using a single line or channel to send several different signals is known as polling. In a polling system, each instrument has a unique address, or identifying number. A system controller unit sends messages over the line telling the instrument at a given address to transmit its data. The process of asking an instrument to send data is called polling.

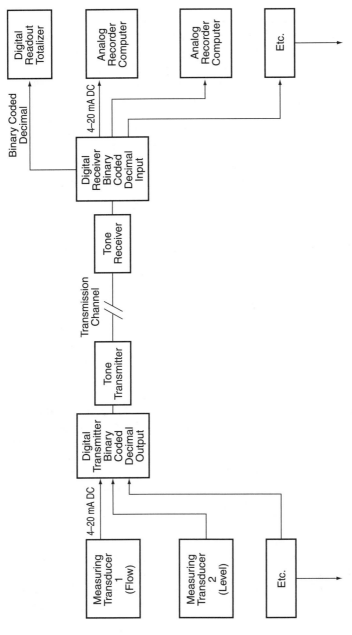

FIGURE 9-13 Typical digital telemetering system

In many systems, the controller is programmed to poll instruments as often as necessary to monitor the system—some instruments may need checking more often than others. In more sophisticated systems, the controller regularly scans the status of each instrument to see whether there is new information to be transmitted. If the status indicates that new information exists, the controller instructs the instrument to send its data. In some systems, critical instruments also have the capability to interrupt a long transmission by another instrument in order to call the controller's attention to urgent new data. Since data is transmitted only when needed under the polling system, a single line or channel can handle more instruments than with a simple scanning system.

Duplexing

In many telemetry installations, the instrument signals received at the operator's central location may require the operator to send control signals back to the remote site. Duplexing allows this to be accomplished with a single line:

1. Full duplex allows signals to pass in both directions simultaneously.
2. Half duplex allows signals to pass in both directions but only in one direction at a time.

Full-duplex systems usually make use of tone-frequency generators to divide the line into transmission and receiving channels. Half-duplex systems may use tone-frequency generators or they may simply rely on timing signals (similar to scanners) or on signals indicating status, such as end-of-transmission or ready-to-receive.

CONTROL SYSTEMS

Control systems consist of three distinct components:

1. *Signal conditioners* receive either pneumatic, electric, or electronic signals from a controller. These signals are then conditioned or amplified and used to initiate the actuator. Signal conditioners include solenoids, starters, and positioners.
2. *Actuators* produce either rotary or linear movement of the final element. Actuators are usually motors or hydraulic cylinders and their related gearing (e.g., valve operators and motor controllers). These controls were discussed in the chapters on valves and motors.
3. *Control elements* are equipment such as pumps and valves that change the process fluid.

Understanding the relationship among these three elements is important. In many cases, the elements are supplied by different manufacturers, and each component must meet the system requirements. The combination of these three will produce a final control element with its own distinct characteristics.

Two different types of controls are required in a process control system: two-state or continuous. Two-state control requires the final control element to be either on (open) or off (closed). Continuous control (also called modulation control) requires the element to vary its operation between the minimum and maximum points. An example is a valve operator, which may be designed to (1) operate a valve either fully opened or closed or (2) throttle the valve at intermediate positions.

In addition to the two types of control and the three components of final control elements, a variety of control media (air, electric, hydraulic) and several actuator types are available.

Control Classifications

Control equipment can be completely independent of instrumentation or it may operate in direct response to instrument signals. The principal classifications of control are

- direct manual,
- remote manual,
- semiautomatic, and
- automatic.

Direct manual control

In a direct manual control system, the operator directly operates switches or levers to turn the equipment on or off or otherwise change its operating condition. A valve operated by a handwheel is a common manually controlled piece of equipment. Operating electrical equipment requires throwing levers on the motor starter. Manual control has the advantages of low initial cost and no auxiliary equipment that must be maintained, but equipment operation may be time-consuming and laborious for the operator.

Remote manual control

With remote manual control, the operator is also required to turn a switch or push a button to operate equipment. However, the operator's controls may be located some distance from the equipment itself. When the operator activates the control switch, an electric relay, solenoid, or motor is energized, which in turn activates the equipment. Power valve operators and magnetic motor starters are common examples of remote manual control devices.

The solenoids and relays used for remote control are common components of all types of control systems. A solenoid (Figure 9-14) is an electric coil with a movable magnetic core. When an electric current is passed through the coil, a magnetic field is generated that pulls the core into the coil. The core can be attached to any piece of mechanical equipment that needs to be moved by remote control. Solenoid valves (Figure 9-15) are very common in water system controls.

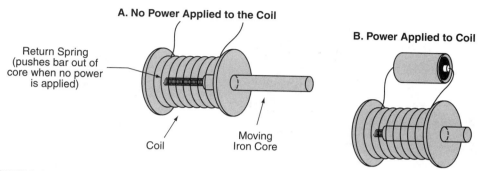

FIGURE 9-14 Operation of a solenoid

FIGURE 9-15 Solenoid valves

Photo courtesy of ASCO®

A relay is constructed of a solenoid that operates an electrical switch or a bank of switches. The most common use of a relay is to allow a relatively low-voltage, low-current control circuit to activate a high-voltage, high-current power circuit. A typical power relay and a control relay are illustrated in Figure 9-16.

Semiautomatic control

Semiautomatic control combines manual or remote manual control with automatic control functions for a single piece of equipment. A circuit breaker, for example, may disconnect automatically in response to an overload, then require manual reset.

Power Relay   **Control Relay**

FIGURE 9-16 Typical relays
Courtesy of Danaher Controls

Automatic control

An automatic control system turns equipment on and off or adjusts its operating status in response to signals from instruments and sensors. The operator does not have to touch the controls under normal conditions. Automatic control systems are quite common. Simple examples are a thermostat used to control a heating system and automatic activation of lighting systems at night.

A number of modes (logic patterns) of operation (Figure 9-17) are available under automatic control. Two common modes are on–off differential control and proportional control.

1. On–off differential control is used to turn equipment full on when a sensor indicates a preset value, then turn it full off when the sensor indicates a second preset value.
2. Proportional control is used to open a valve or increase a motor's speed when the sensor shows a variation from a preset intended value. A common application is control of a chemical feeder in response to a flowmeter or residual analyzer signal.

Automatic controllers attempt to imitate human decision making, but they cannot achieve the level of complexity in decision making that humans can. Therefore, automatic control is limited to the more simple process situations. However, many water utility processes are in this category. The various pieces of equipment used to control each process parameter form a control loop of information processing. The direction of the information flow loop can be in either the same direction as the process (feedforward control) or the opposite direction of the process (feedback control).

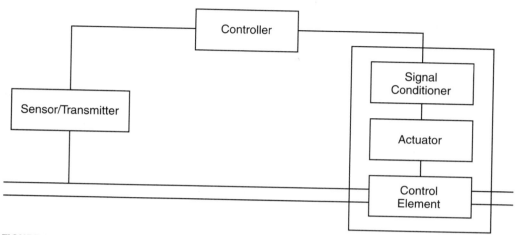

FIGURE 9-17 Components of control

A feedforward control loop measures one or more inputs of a process, calculates the required value of the other inputs, and then adjusts the other inputs to make the correction. Figure 9-18 illustrates this method of control, for which a chlorinator feed rate is automatically controlled (paced) in response to a signal from a flow transmitter.

Because feedforward control requires the ability to predict the output, this type of control is sometimes called predictive control. Furthermore, since feedforward control does not measure or verify that the result of the adjustment is correct, it is also referred to as open-loop control. A consequence of open-loop control is that if the measurements, calculations, or adjustments are wrong, the control loop cannot correct itself. In the example in Figure 9-18, there is no check built into the control system that the treated flow of chlorine residual is actually at the desired concentration.

A feedback control loop measures the output of the process, reacts to an error in the process, and then adjusts an input to make the correction. Thus, the information loop goes backward. Figure 9-19 illustrates the control of a chlorinator in response to a residual analyzer. The analyzer continuously adjusts the feed rate of the chlorinator to provide the desired residual in the treated flow. It automatically adjusts for changes in flow rate or chlorine demand.

Because the process only reacts to an error, it is also called reactive control. Furthermore, since feedback control checks the results of the adjustment, it is said to be closed-loop control. Thus, feedback control, unlike feed-forward control, is self-correcting. If the initial adjustment in response to changed conditions does not produce the correct output, the closed-loop system can detect the problem and make another adjustment. This process can be repeated as often as necessary until the output is correct.

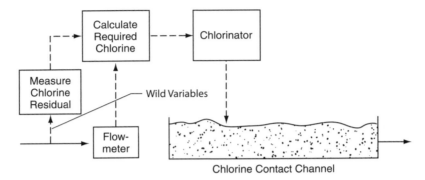

FIGURE 9-18 Feedforward control of chlorine contact channel

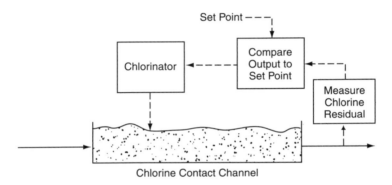

FIGURE 9-19 Feedback control of chlorine contact channel

Direct Wire and Supervisory Control

Within a plant or an attended pump station, equipment is usually connected to the central control panel through electric wiring. This approach is known as direct-wire control. When a remote station is unattended, the station equipment is controlled from an operator's control location by supervisory control equipment, which transmits control signals over telemetry channels. The unattended site is sometimes called the outlying station, and the operator's station may be called the dispatch station. Many large utilities have a large, central facility to monitor and control the entire water distribution system.

SUPERVISORY CONTROL AND DATA ACQUISITION

Supervisory control and data acquisition (SCADA) combines many of the elements already discussed in this chapter, such as RTUs and control elements.

SCADA System Components

In a SCADA system, the control can be remote or automatic. SCADA subsystems consist of

- remote terminal units (RTUs),
- communications (telemetry transmission),
- a master station, and
- human–machine interface (through graphical format at a central console).

The first two subsystems have already been discussed in this chapter.

Master station

The main function of a master station includes scanning the RTUs, processing the data, transmitting operator commands, and maintaining a database of historical data (such as valve positions, flow, and pressure). A master station consisting of a single computer is called a centralized station. A distributed system consists of several computer control devices.

Centralized computer control. The earliest SCADA systems consisted of a single master station controlling several RTUs. A problem with this system is its dependency on a single computer installation and the communications links to the equipment. For instance, if a remote pumping station is completely operated by a central computer, loss of the computer or the telephone line to the station could mean complete loss of use of the station.

Distributed computer control. Smaller, more powerful, and less expensive computers have made it possible to integrate computer control for subsystems of RTUs as well as for individual pieces of equipment. In the example, each remote pumping station can now be operated by its own computer, so that it continues to function properly in spite of the loss of the communications link with the central computer. Smart equipment (computer controlled) can adjust themselves and monitor their own operation. This reduces, or eliminates, the load on the central computer and decreases the system's dependency on that computer for operation.

Human–machine interface

This interface is the connection between the operator and the computer or SCADA system, usually through a central console, keyboard, and mouse. SCADA systems allow personnel with no programming knowledge to set up the display. Operators can design graphics and tabular displays that precisely meet their needs. These displays may be interactive. For example, the symbol for a pump may change color or shape depending on status or a reservoir icon may "fill" as the reservoir level increases. Figure 9-20 shows a typical computer control center.

The future of supervisory control

Few water systems exactly fit into any one category of control system. Some very small systems are still primarily operated manually, but there are more that are almost completely automated. Most water systems are somewhere in between. It is clear that the future will bring increasing automation.

State regulatory agencies have mixed emotions about allowing complete automation of water systems. In some ways, automation may be better than manual control. It is usually more precise and it eliminates the possibility of human error that is always possible with manual control. On the other hand, although computers and computer programs are becoming increasingly reliable, a simple "bug" in the computer program could seriously damage equipment or cause contamination of the water supply. State regulators will always want to make sure that every automated system is supplied with all possible auxiliary monitoring. The system must be positively designed to shut down or take other appropriate action if anything goes wrong.

One of the newer innovations in computer control is the expert system, which is an interactive computer program that incorporates judgment, experience, rule of thumb, intuition, and other expertise to provide advice about a variety of tasks. In short, the computer is programmed to attempt to copy the decision-making process that would be made by a human expert.

FIGURE 9-20 Central computer command center
Courtesy of ABB Inc.

Such systems are still in the early stages of development and are extremely complex. However, they are seen as eventually having a place in water utility operation by providing sophisticated control over certain operations, resulting in safer and more economical operation.

OPERATION AND MAINTENANCE

Instrumentation represents about 8 percent of the capital investment in mechanical equipment for the average distribution system. Although this may not be a large sum, the proper operation of controls can make a substantial contribution to cost savings by improving overall system performance.

Most new equipment is much more reliable and requires less maintenance than instrumentation of just a few years ago. There are, however, numerous routine maintenance tasks that an operator can perform to keep the instrumentation functioning properly. An instrument's useful life can be increased significantly if it is routinely checked to be sure it is not exposed to moisture, chemical gas or dust, excessive heat, vibration, or other damaging environmental factors.

Maintenance of Sensors and Transmitters

Sensors often require routine maintenance that can be performed by an operator. Special skills may be required to work on transmitters.

Pressure sensors

Every sensor that in any way responds to liquid pressure will perform poorly if air enters the sensor. Air or gas may enter into the sensing element when released from the fluid line. A vent should be provided on the mechanism and used on a regular basis to allow any accumulated air to escape. A maintenance schedule can eventually be developed as operators gain experience working with the particular sensors. Wherever possible, the sensor should be installed so that trapped gasses flow into the upper part of the monitored line, rather than into the sensor.

In cold climates, freezing is a possibility. Most pressure sensors either have provision for a heater and thermostat or can be equipped with special protective cabinets. Heater tape can also be used for protecting sensing lines.

The small pipe connections used to connect pressure sensors, switches, and gauges to the pipeline can be a continual source of trouble where dissimilar metals (for instance, cast iron and brass or bronze) are used together. A valve cock should be installed between the pipeline tap and the device to allow easy disconnection and service. At least twice a year, the sensor should be removed and the valve cock blown and rodded out if necessary. Corrosion of nipples should be checked, and if any weakness is apparent, the failed part should be replaced.

Flowmeters

Several types of flowmeters are used in distribution systems, and each has its own particular maintenance requirements. A Venturi tube, orifice, or flow nozzle type of meter has small ports that connect the process fluid to the transmitter mechanisms. These parts should be blown out periodically.

Propeller meters are all-mechanical devices. Over a period of time, wear produces lower readings. The manufacturer's recommendations for lubrication should be carried out regularly, and only the recommended lubricant should be used. A magnetic meter is an electrical unit, so regular checks should be made for corrosion or insulation breakdown around conduit connections or grounding straps.

Transmitters

The operator with no special training in transmitter repair can generally do little more in the way of maintenance than to ensure a favorable environment for the equipment. In hot climates, it is important to protect electronic transmitters from exposure to high temperatures. Most electronic transmitters are rated for operation up to 130°F (55°C). Beyond this temperature, many electronic components may break down. Direct exposure to sunlight may contribute to serious heat buildup within units, and it is often necessary to provide a shield or fan if components are confined in a cabinet. Most transmitters are splash-proof and will function with occasional exposure to wetting. However, unless otherwise specified, most units will not withstand being submerged in water. Therefore, metering pits should have pumps and level alarms to warn of flooding.

Maintenance of Receivers and Indicators

Receiving and indicating devices usually require special skills for service and maintenance. However, the operator should attempt to maintain a favorable environment for the equipment. A dirty or damp atmosphere may damage receivers. Vibration, chemical dust (such as fluoride), and high chlorine content in the atmosphere should all be avoided. The high temperatures that affect transmitters also cause problems for receivers, and inking systems on recording indicators may perform poorly in the cold.

Troubleshooting Guidelines

Some general guidelines for troubleshooting instrumentation systems are as follows:

- Never enter the facility with tools of any type—most of us have a tendency to start taking things apart without a thorough diagnosis.
- Make a complete diagnosis and attempt to confirm it by example.
- Consider ways to deal with the safety risks.

- Consider what will happen to the operating system while maintenance is being performed. In particular, where control valves are involved, consider the possibility of water hammer.
- Always inform the proper authorities of planned actions beforehand.

Maintenance Records

The development of a maintenance file on instrumentation has many long-term benefits for any distribution system, including the following:

- Accurate accounting of the cost
- Indication of areas where there may be a need for a change in the type of equipment
- Projection of the life expectancy of each type of equipment
- Knowledge required to maintain parts inventories at minimum cost (based on operating experience)
- Proper parts inventories that can reduce emergency downtime substantially (another form of cost savings)
- Guidance for preventive maintenance steps

Many facilities have computers that print out maintenance schedules (maintenance management software) for each type of equipment as a guide to the preventive measures that need to be performed (Figure 9-21). Where computers are not used, reminders can be placed in an instrument chart so that the reminder comes up as the charts are rotated.

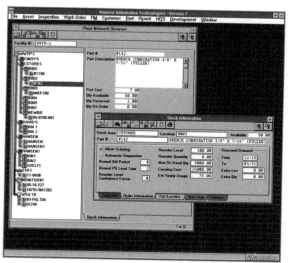

FIGURE 9-21 Computerized maintenance management system
Courtesy of Hansen Information Technologies

Maintenance records should be kept for each instrument. A common problem for maintenance program continuity is the loss of instruction manuals and calibration sheets. A file should be provided for these items, and care should be taken to avoid losing them, especially when responsible personnel are transferred or terminated.

Records should also be kept on services related to instruments not owned by the utility, such as telephone lines and power supplies to pump stations. Such information should be carefully documented so that there is concrete evidence in case it becomes necessary to register a complaint. System diagrams showing the routings for communication links are important. For telemetry systems, it is useful to have a map showing channels or frequencies in each tone spectrum and the facilities to which they are connected.

Safety

Operators who work around instrumentation and control units must follow safety precautions for

- hand tools,
- portable power tools,
- electrical devices, and
- hydraulic and pneumatic devices.

Safety precautions regarding this equipment are detailed in previous chapters.

BIBLIOGRAPHY

Manual M2, Instrumentation & Control. Denver, Colo.: American Water Works Association (latest edition).

Mays, L.W. 2000. *Water Distribution Systems Handbook.* New York: McGraw-Hill.

Pollack, A.J., A.S.C. Chen, R.C. Haught, and J.A. Goodrich. 1999. *Options for Remote Monitoring and Control of Small Drinking Water Facilities.* New York: Battelle Press.

Water Distribution Operator Training Handbook. 2005. Denver, Colo.: American Water Works Association.

CHAPTER 10

Water Meters

Water meters are used to measure and record the volume of water flowing through a line. The primary functions of metering are to help a water utility account for water pumped to the system and to equitably charge customers for the water they use.

The meter types most commonly used on water systems include

- positive-displacement,
- compound,
- current,
- detector-check,
- proportional,
- Venturi,
- orifice,
- pitometers,
- magnetic, and
- sonic.

Various types of water meters are covered by various industry standards (e.g., AWWA Standards C700–C704, C706–C708, C710, C712, and ANSI/NSF Standard 61).

Each type of meter has certain advantages and disadvantages and has been found to work best in specific applications. Desirable meter characteristics include the following:

- Accuracy within the range of anticipated flows
- Minimal head loss
- Durability
- Ease of repair
- Availability of spare parts
- Quiet operation
- Reasonable cost

CUSTOMER WATER METERS

Most public water systems meter the water used by each service connection or customer. The principal reason is to determine billing charges. A secondary reason is to track water use to ensure that there is no undue waste or leakage in the distribution system. In addition, when customers are billed for the exact amount of water used, they have an incentive to use water wisely.

Positive-Displacement Meters

The most common type of meter for measuring water use through customer services is the positive-displacement meter. This type of meter consists of a measuring chamber of known size that measures the volume of water flowing through it by means of a moving piston or disk. The movement of each oscillation of the piston or disk is then transmitted to the register to record the amount of water. There are two types of positive-displacement meters: the piston type and the nutating-disk type.

Piston meters (Figure 10-1) utilize a piston that moves back and forth as water flows through the meter. A known volume is measured for each rotation, and the motion is transmitted to a register through a magnetic drive connection and series of gears.

Nutating-disk meters (Figure 10-2) use a measuring chamber containing a flat disk. When water flows through the chamber, the disk nutates (i.e., wobbles and rotates) and "sweeps out" a specific volume of water on each cycle. The rotary motion of the disk is then transmitted to a register that records the volume of water flowing through the meter.

Positive-displacement meters are generally used for residences and small commercial services in sizes from ⅝ in. to 2 in. (16 mm to 51 mm) because of their excellent sensitivity to low flow rates and their high accuracy over a wide range of flow rates.

Positive-displacement meters underregister when they are excessively worn. To avoid excessive wear, they should not be operated in excess of the flow rates listed in Table 10-1. Continuous operation of a meter at maximum flow will quickly destroy it. A meter is generally sized so that its expected maximum rate will be one half of its safe maximum operating capacity.

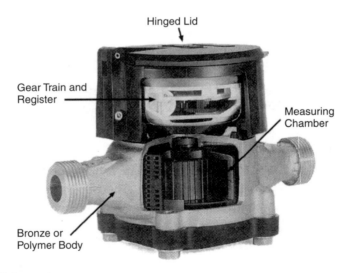

FIGURE 10-1 Piston meter
Courtesy of AMCO Water Metering Systems Inc.

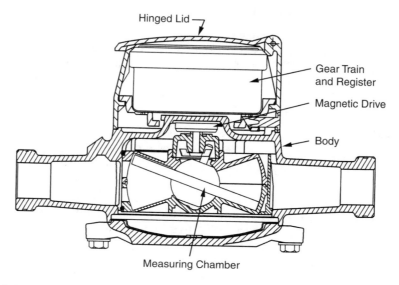

FIGURE 10-2 Nutating-disk meter with a plastic housing
Courtesy of Neptune Technology Group Inc.

TABLE 10-1 Maximum flow rates for positive-displacement meters

Meter Size,		Safe Maximum Operating Capacity,		Recommended Maximum Rate for Continuous Operations,	
in.	*(mm)*	*gpm*	*(L/sec)*	*gpm*	*(L/sec)*
½	(15)	15	(0.95)	7.5	(0.47)
½ × ¾	(15 × 20)	15	(0.95)	7.5	(0.47)
⅝	(17)	20	(1.3)	10	(0.63)
⅝ × ¾	(17 × 20)	20	(1.3)	10	(0.63)
¾	(19)	30	(1.9)	15	(0.95)
1	(25)	50	(3.2)	25	(1.6)
1 ½	(40)	100	(6.3)	50	(3.2)
2	(50)	160	(10)	80	(5.0)

Large-Customer Meters

Examples of customers that use large quantities of water are hospitals, golf courses, large public buildings, apartment houses, and industries. Industries that always use a large amount of water are those that must do a great deal of cleaning and those that incorporate water into their manufactured products. The types of meters most often used for these customers are compound meters, current meters, and detector-check meters.

Compound meters

Compound meters are usually used for customers that have wide variations in water use. There may be some times of the day when their water demand is very high and other times when there is little or no use. The meters furnished for these customers must be relatively accurate at both low and high flow rates.

A standard compound meter consists of three parts: a turbine meter, a positive-displacement meter, and an automatic valve arrangement, all incorporated into one body (Figure 10-3). The automatic valve opens when high flows are sensed, enabling the water to flow with little restriction through the turbine side of the meter. Under low flows, the valve shuts and directs water through a small displacement meter for measurement. The unit therefore combines the favorable characteristics of both turbine and displacement meters into one unit. Compound meters may have separate registers for each meter or their output can be combined to indicate total use on a single register.

Another type of compound meter utilizes two standard meters connected together, as shown in Figure 10-4.

Current meters

Current meters are sometimes used to meter water to large industrial customers. However, they are appropriate only when the minimum use by the customer is within the lower limit of the meter's accuracy. It is recommended that a strainer be installed ahead of a current meter to protect the meter from damage by sediment or other objects in the water.

Detector-check meters

Detector-check meters are designed for service where daily use is relatively low but where very high flow rates may be required in an emergency. The prime example is a building with a fire sprinkler system. The meter consists of a weight-loaded check valve in the main line that remains closed under normal usage. A bypass around the check valve has a displacement-type meter to measure domestic use. When the sprinkler system calls for water, the loaded valve detects the decrease in line pressure in the building and swings completely open to allow full flow through the line.

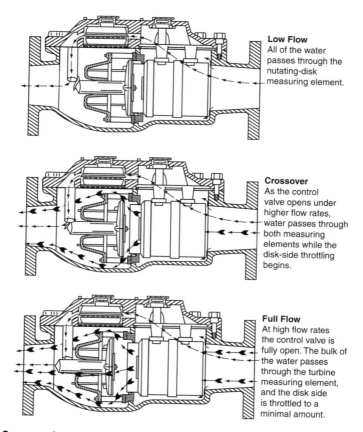

Low Flow
All of the water passes through the nutating-disk measuring element.

Crossover
As the control valve opens under higher flow rates, water passes through both measuring elements while the disk-side throttling begins.

Full Flow
At high flow rates the control valve is fully open. The bulk of the water passes through the turbine measuring element, and the disk side is throttled to a minimal amount.

FIGURE 10-3 Compound meter
Courtesy of Neptune Technology Group Inc.

FIGURE 10-4 Compound meter arrangement that uses two standard meters
Courtesy of AMCO Water Metering Systems Inc.

Meter Selection

There is no hard and fast rule for selecting the correct meter size for a particular service. Proper meter selection involves both size and type considerations. The size of the meter usually depends on several factors, including

- the expected maximum customer demand for water;
- the normal pressure in the system at the point of connection;
- friction losses in the service line, meter, and customer plumbing; and
- the range of flow rates expected on the customer's service.

The meter size does not necessarily have to match the service pipe size. In most cases, the meter is installed one size smaller than the service size. This size meter will still pass as much water as the service can produce once friction losses are considered. If an oversized service is installed to allow for future increases in use, piping should be installed to allow for meter upsizing. However, the meter initially installed should be no larger than needed for immediate use.

Residential meters

Most residential water services are furnished with either a ⅝-in. (17-mm) or ¾-in. (19-mm) size meter. Single-family residences should be furnished with a larger meter only when there are a large number of plumbing fixtures. Usual practice is to start with a small meter. If the customer finds that flow is not adequate, a larger meter can be substituted.

Commercial meters

Positive-displacement meters in sizes 1, 1½, and 2 in. (25, 38, and 51 mm) are primarily used for apartment buildings, businesses, public buildings, and small industries. Buildings having flushometer toilets will usually require an above-average size meter in order to supply the extremely high rate of flow required if several toilets are flushed simultaneously. Following are some general rules for commercial customer metering:

- If the meter will usually operate at 5 to 35 percent of the maximum rated capacity and if accuracy at extremely low rates is not too important, a displacement-type meter should be used.
- If close accuracy at very low flows is important, but if a large capacity is also needed, the compound type will be the best choice. It will have lower pressure loss at high flow rates. However, a compound meter is considerably more expensive than a comparable displacement or turbine meter.

• If large capacity is of primary importance, normal flow is greater than 10 or 15 percent of the maximum rating, and low flow accuracy is secondary in importance, the turbine type should be selected. Newer models, in particular, have very low pressure loss and maintain good accuracy over a wide range.

Additional information on meters is available in AWWA Manual M6, *Water Meters—Selection, Installation, Testing, and Maintenance*, and AWWA Manual M22, *Sizing Water Service Lines and Meters*.

CUSTOMER METER INSTALLATION

Customer meters can be installed either in outdoor meter pits or in the building being served. Installation of meters in basements is quite common. When there is no basement, some utilities allow installation in a crawl space or a utility closet. Meter pits (or meter boxes) are usually located in the parkway, between the curb and sidewalk.

Large meters are often installed in precast concrete or concrete-block vaults if there is no appropriate location in a basement. Large-meter installations are expensive and require considerable planning.

General Considerations

Indoor installations are more common in northern states, where harsh winter weather can cause frost damage to the meter and where houses are more likely to have basements. Outdoor installations are more common in warmer, more temperate climates. Whether the meter is installed indoors or out, there are several general requirements for acceptable installation, including the following:

• The meter should not be subject to flooding with nonpotable water.
• The installation should provide an upstream and a downstream shutoff valve to isolate the meter for repairs.
• The installation should position the meter in a horizontal plane for optimal performance.
• The meter should be reasonably accessible for service and inspection.
• The location should provide for easy reading either directly or via a remote reading device.
• The meter should be reasonably well protected against frost and mechanical damage.
• The meter installation should not be an obstacle or hazard to customer or public safety.
• The meter should have seals attached to the register to prevent tampering.
• There should be sufficient support for large meters to avoid placing stress on the pipe.
• A large installation should have a bypass or multiple meters so that water service does not have to be discontinued during meter replacement or repair.

Manifold Installations

Some water systems commonly install two or three meters in a manifold, or battery, for customers that require a high flow rate. As illustrated in Figure 10-5, a 4-in. (100-mm) service line can have three 2-in. (51-mm) displacement-type meters installed. Advantages of a battery installation include the following:

- Meters can be removed one at a time for servicing without disrupting customer service. This differs from a single large-meter installation, for which the customer is usually given free water through a bypass while the meter is being repaired.
- Meters can be added and the system easily expanded if required.
- There is no need to buy or stock parts for several different-size meters. All meters and valves are the same size.
- The battery can be mounted with the meters stacked along a basement sidewall to conserve floor space.

In a manifold unit with two or three meters (Figure 10-6A and Figure 10-6B), all but one of the meters must have a lightly loaded backpressure valve on their outlets. This way, when the flow is small, only one meter will operate. As flow increases, the backpressure valves open to permit flow through the other meters.

FIGURE 10-5 Manifold connection of three meters

The price difference between a multiple-meter installation and a compound meter is another consideration. Older-model compound meters were very expensive, but newer models are now available that are more competitively priced.

In the past, commercial and industrial services were quite often metered with compound meters. These meters register well over a wide flow range and have a relatively low pressure loss at high flows. Today, however, new types of horizontal turbine meters should be considered for these customers. They are quite sensitive to low flows, may have a lower initial cost, and may require less maintenance.

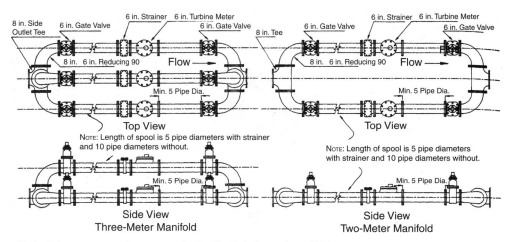

Typical three-meter and two-meter hydraulically balanced manifold.

FIGURE 10-6A Manifold of large meters

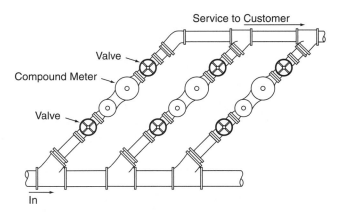

FIGURE 10-6B Diagram of manifold of three large meters

Meter Connections

The types of connections used on water meters vary with the size of the meter. Meter sizes up to 1 in. (25 mm) usually have screw-type connections (Figure 10-7), whereas larger meters usually have flanged connections (Figure 10-8).

Many water utilities use a special device called a meter yoke or horn to simplify the installation of small meters. Figure 10-9 illustrates several types that are commonly used for both interior and meter-pit installations. The purpose of a yoke is to hold the stub ends of the pipe in proper alignment and to maintain spacing to support the meter. Yokes also cushion the meter against stress and strain in the pipe and provide electrical continuity when metal pipe is used.

FIGURE 10-7 A ¾-in. (20-mm) meter with screw connections
Courtesy of Neptune Technology Group Inc.

FIGURE 10-8 A 1½-in. (40-mm) meter with flanged couplings
Courtesy of Neptune Technology Group Inc.

Indoor Installations

Many water systems have developed a diagram to illustrate how a meter should be installed inside a building under normal circumstances. Ideally, the meter should be located immediately after the point where the service pipe enters through the floor or wall. If meters are to be read directly, the location must be kept relatively clear to allow convenient reading. If meters are to be furnished with remote reading devices, the location should still allow for reasonable access. This will allow periodic direct readings to check on the remote device. It will also allow the meter to be changed when repair is required.

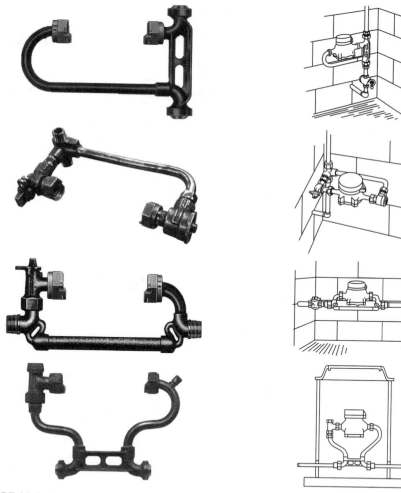

FIGURE 10-9 Various styles of meter yokes
Courtesy of Mueller Company, Decatur, Ill.

Some installation details should be standardized, including

- minimum and maximum heights for the meter above the floor,
- the types of meter connections to be used,
- the required type of valve before and after the meter, and
- the minimum access space required for reading and servicing.

Where the building electrical system uses the water service as a ground, current can flow through the piping and create an electrical hazard to employees who remove the meter or repair the service line. This current also increases the possibility of corrosion of the service line and connections. If the water service pipe is used as a ground and a meter yoke is not used, an electrical ground connection must be installed across the meter. If the water service is plastic pipe, grounding the electrical system to the plumbing system obviously serves no purpose.

Having a diagram of the required meter location and plumbing ensures proper installation for the mutual benefit of the water utility and the customer. A copy of the diagram should be furnished to those applying for water service. A water utility should require compliance with the specifics of the diagram before providing service.

Outdoor Installations

Outdoor meter installation varies depending on the size of the meter involved.

Small-meter installation

A meter box or pit is usually required to protect small outdoor meters. Many factors influence the design, materials, installation details, and overall performance of an outdoor setting. Factors include soil conditions, groundwater level, maximum frost penetration, and accessibility for reading and servicing.

The wide variations in ground frost penetration throughout the country make it impossible to detail a universally practical outdoor setting. As with the indoor setting, a diagram for use by water system employees, building inspectors, contractors, and homeowners should be prepared. This diagram should illustrate a standard for meter pit location, construction, and plumbing. The meter pit is usually constructed and furnished by the owner's contractor. It is particularly important that the contractor be furnished with explicit instructions. Because each community has different requirements on how the installation should be made, it is often hard for contractors to remember what each town requires. The following guidelines for the standard diagram should be observed:

- If at all possible, the meter pit should be located on public property or at least relatively close to the property line.

- The location should be relatively safe from possible damage from vehicles and snow-removal equipment.
- The pit or box lid should be tight fitting, tamper resistant, and placed flush with the ground surface so that it does not create a tripping hazard or interfere with lawn mowing.
- Where ground frost is expected, the riser pipes should be 1 to 2 in. (25 to 50 mm) away from any portion of the meter box walls to avoid freezing.
- The distance below ground surface at which the meter spuds or couplings are to be located must be specified. In general, small meters are raised to facilitate reading and meter replacement.
- The dimensions of the meter box to be used for each size meter must be specified.
- The location and type of curb stop or service control valve must be specified.
- A meter yoke is recommended for pit installation to make it easier to change the meter and to prevent distortion of the meter body due to pipe misalignment caused by shifting over time.

Large-meter installation

There is no uniform standard for large-meter installations. A prime consideration in their design is that large meters are very heavy and require adequate support so that no stress is put on the service pipe. The cover on the meter pit must be made large enough for worker entry and meter removal.

Adequate work space must be allowed around large meters installed in a vault. At least 20 in. (510 mm) of clearance to the vertical walls and at least 24 in. (610 mm) of head space from the highest point on the meter should be allowed. More space is desirable. Test valves should be installed to permit volumetric tests. Provisions should be made for discharging the test water if meters are to be tested in place. The meter and valves should be supported, and thrust blocking should be provided when necessary. A typical large-meter installation is illustrated in Figure 10-10.

Valuable aid and installation recommendations can be obtained from meter manufacturers. These manufacturers can make recommendations on meter vault designs commonly used in the area, as well as on plumbing materials that will provide the most economical and satisfactory installation.

METER READING

Meters used in the United States are generally furnished with registers that record the flow of water in gallons or cubic feet. Registers that read in imperial gallons or cubic meters are also available. Water meter registers are usually of two types: circular or straight, as illustrated in Figure 10-11. Circular registers, which are somewhat difficult to read, have gradually been replaced by straight registers on new meters. Straight registers are read like the odometer on a car. The meter reader simply reads the number indicated on the counting wheels, including any fixed zeros to the right of the counting wheel window.

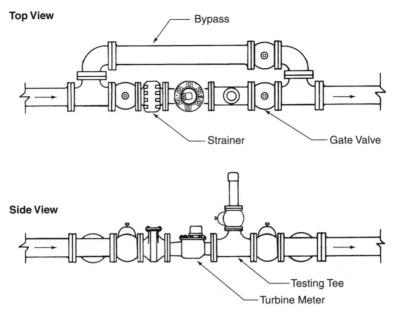

FIGURE 10-10 Large-meter installation diagram
Courtesy of Neptune Technology Group Inc.

The registers on large meters occasionally have a multiplier on them. If so, the multiplier such as "10×" or "1,000×," will be noted on the meter or the register face. This marking indicates that the reader must multiply the reading by the multiplier in order to obtain the correct reading.

Direct Readout

The most common method of meter reading is the direct readout, which involves an individual going from one meter to another and directly reading the registers.

If the meter is located in a home, the following problems may be encountered:

- Some residents are reluctant to admit the meter reader into their home for a variety of reasons including fear of being attacked, damage to property, and inconvenience.
- Some meters are in very inconvenient locations, either because the initial location was poorly planned or because the resident has subsequently built something around the meter to hide it.
- Often no one is home during working hours on weekdays.

Water utilities can use alternative methods to obtain readings from homes where the occupants are not home during the week.

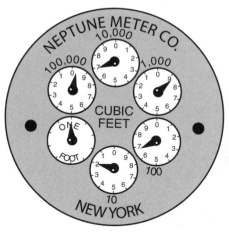

Circular Meter Register
Reading in Cubic Feet

Straight Meter Register
Reading in Gallons

FIGURE 10-11 Meter registers
Courtesy of Neptune Technology Group Inc.

- A meter reader can work on Saturday to make special readings from homes missed during the week.
- The utility can make special arrangements by phone to meet residents at a specific time when they will be home.
- A special card can be hung on the doorknob, asking the resident to read the meter and return the card. Some residents who dislike being bothered by the meter reader will ask that a card be left for every reading.

If the meter is located outside, the following problems may come up:

- The meter pit must be located each time a reading is taken. If grass, dirt, snow, or stones have covered it, the reader must spend time finding and exposing the cover.
- With deeper pits, the pit may fill with water at certain times of the year and must be pumped before the meter can be read.
- With large meters or small meters that haven't been raised, the reader may have to crawl into the pit to make the reading.

Remote Reading Devices

Remote reading devices were developed to eliminate most of the problems with direct meter reading inside buildings. The most common type of remote setup transmits the signal electrically (Figure 10-12). The meter register contains a pulse generator that stores

FIGURE 10-12 Remote meter-reading device
Courtesy of Badger Meter, Inc., Milwaukee, Wis.

energy in a spring-release mechanism. This energy is accumulated very slowly as a specific amount of water—either 10 ft^3 (0.3 m^3) or 100 gal (380 L)—passes the meter. When this point is reached, the energy is released by a mechanism that creates rapid motion between a permanent magnet and a copper coil. This motion generates an electric impulse that is transmitted through a two-conductor electric cable to the remote register mounted at a convenient location on the outside of the building. The remote register contains a counter that advances one digit for each pulse received.

Remote reading devices are quite popular in northern states where most meters are located inside. The use of remote registration has been increasing in areas where pit settings were more common, both in residential and in commercial and industrial installations. In areas where large meters are installed under the street, it is often necessary to send a truck and a crew to direct traffic, manipulate pit entries, and then read the meters. However, conveniently located remote registers make it possible for one person to obtain these meter readings without disrupting traffic or risking personal injury.

When considering the installation of remote registration systems, the water supplier should thoroughly study all related factors. Much of this information is available on request from meter manufacturers, including

• comparisons of new meter setting costs with and without remote reading,

• desirability and economics of retrofitting existing installations,

• interchangeability within meter models and brands used by the water supplier,

- compatibility of the reading data obtained if both direct and remote systems are inter-mixed in the same route,
- original equipment purchase price,
- installation costs,
- probable maintenance costs, and
- ultimate cost of obtaining and processing meter-reading data.

Plug-in type readers

Several manufacturers offer remote meter-reading units that have a plug-in receptacle. With this system, the reading is converted into an encoded electrical message within the meter head. This meter head is connected by a multiconductor cable to a receptacle located on an outside wall of the building. A meter reader can then plug a battery-operated reading device into the remote receptacle to take a reading. The meter reading may be visually displayed and also recorded for later direct entry into a computer.

Remote register or plug-in units are not as advantageous for meters in pits. They are sometimes mounted under the pit cover. Although the cover must be raised to make the reading, this approach is still easier than having a worker crawl into the pit. The alternative is to find a protected spot near the pit where the register can be mounted aboveground, but this is rarely convenient.

Electronic meter reading

A "scanning" unit for electronic meter reading (Figure 10-13A) works well for either indoor or pit-mounted meters. With this type of unit, an inductive coil is mounted either on the wall of the building or in the cover of the meter pit. It is connected by wires to a special register on the meter. The meter reader carries a unit that includes a probe on an extension arm that has to be held only in the proximity of the inductive coil to obtain a reading. Power from the interrogator is transmitted through the inductive coil to the register's microprocessor. The meter identification number and reading are transmitted back to the unit carried by the meter reader.

Automatic Meter Reading

Automatic meter reading (AMR) enables a utility to obtain readings without actually going near the meters. Methods include transmitting a meter signal through the telephone system, through electric power distribution networks, via sound transmission through water lines, through cable TV wiring, by radio, or by satellite (Figure 10-13B) and cellular networks. If radio transmission is used, different options are available, such as operating the entire system from a central radio tower, collecting readings from radio-equipped trucks that drive down the street, or having the readings obtained by a meter reader who is walking or driving by on the street (Figure 10-14A).

FIGURE 10-13A Proximity meter-reading system
Courtesy of Badger Meter, Inc., Milwaukee, Wis.

FIGURE 10-13B Remote meter-reading system
Courtesy of Badger Meter, Inc., Milwaukee, Wis.

In addition to the obvious advantages of avoiding the costs and problems of requiring meter readers to visit each meter, AMR makes it practical to read meters and bill monthly, rather than quarterly as practiced by many water systems. It also makes it easy to perform studies such as determining how much water is being used by various classes of customers on a peak water use day.

FIGURE 10-14A Handheld computer
Courtesy of Neptune Technology Group Inc.

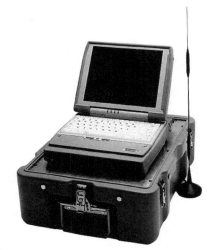

FIGURE 10-14B Mobile data collector
Courtesy of Neptune Technology Group Inc.

METER TESTING, MAINTENANCE, AND REPAIR

Water meters should be tested before use, when removed from service, when a customer makes a request or complaint, and after any repair or maintenance.

Testing New Meters

New meters are tested by water suppliers to identify damaged meters and limit metering errors. Because of the time involved in testing new meters, some state regulatory commissions have adopted regulations that allow a certain number of random meters in a shipment to be tested to determine the accuracy of the entire shipment. For example, if meters are shipped in boxes of 10, a water supplier might test only 1 random meter in that box. If that meter tests out accurately, all of the meters in the box can be assumed to be accurate. If the meter does not test out accurately, the water supplier would have to test all the other meters in the box or return all of them to the supplier for replacement. The state public utility regulatory commission should be consulted concerning any requirements it may have for testing new meters before use.

Frequency Requirements for Testing Meters

Water meters are subject to wear and deterioration. The rate of wear and deterioration is principally a function of how much the meter is used and the quality of the water. Over time, meter efficiency decreases and meters fail to start operating under low flow. Although some customers may sometimes think otherwise, meters very rarely overregister. Occasionally a customer insists on having an old meter replaced because he or she thinks it is running fast. The usual result is that the new meter will be more accurate and the bill will be higher.

To avoid a loss of revenue for the utility, meters must either be field tested or brought in periodically for testing and repair. To help control meter deterioration and limit registration problems, many state regulatory commissions have adopted requirements for the frequency of meter testing. Because of the variability of water characteristics, the cost of testing, and water costs, a nationwide test frequency cannot be established. Most states with frequency requirements for meter testing base the test intervals on the size of the meter—the larger the meter, the more often it should be tested. AWWA recommends that every utility have a scheduled meter-testing program. The determination of the most appropriate test interval should be based on utility experience, local conditions, and cost/benefit considerations. A statistical sampling approach may be the best choice for some systems while a timed approach may be the most suitable for others. Considerations for determining the testing interval are discussed in AWWA Manual M6, *Water Meters—Selection, Installation, Testing, and Maintenance.*

Whether a water utility tests its own meters or has someone else do it depends on the staff size, time available, and facilities. Although some utilities have the facilities to test all their meters, the majority test only smaller meters. Larger meters may be field tested or sent to the manufacturer for testing. Smaller water systems do not usually have the facilities or the time to test meters, so they send the meters to the manufacturer or to a contractor for testing and repair. The economics of simply disposing of old meters and purchasing new ones must also be considered.

Testing Procedures

There are three basic elements to a meter test:

1. Running a number of different rates of flow over the operating range of the meter to determine overall meter efficiency.
2. Passing known quantities of water through the meter at various test rates to provide a reasonable determination of meter registration.
3. Meeting accuracy limits on different rates for acceptable use.

The rates of flow generally used in testing positive-displacement meters are maximum, intermediate, and minimum. For current and compound meters, four or five flow rates are usually run. State regulations should be consulted to determine if specific rates are required.

Equipment for meter testing does not have to be elaborate. A setup like the one shown in Figure 10-15 is suitable for small meters. It is generally more economical to field-test large meters, especially current and compound types. Field testing is essentially the same as shop testing. The difference is that, instead of using a tank to measure the test water, operators compare the meter to be tested and one that has previously been calibrated. The two meters are connected in series, and the test water is discharged to waste. Since the calibrated meter is not 100 percent accurate at all flows, it is necessary to use a special calibration curve to adjust for different rates of flow in order to compute the accuracy of the meter being tested.

Some water suppliers do not field-test large meters over their entire operating range. Instead they check only the lower 5 to 10 percent of the rated or operating range. Utilities may take this approach because the large quantities of water discharged in high-flow tests often present a problem. The underlying assumption for this approach is that the test curve will flatten out after reaching a peak registration that is approximately 10 percent of rated capacity. Presumably the meter will then stay within the required limits of registration.

For field testing to be accurate, both meters must be full of water and under positive pressure. The control valve for regulating flow should always be on the discharge side of the calibrated meter. A control valve on the inlet side of the meter configuration or one located between the two meters should not be used.

Maintenance and Repair

Maintenance and repair for positive-displacement meters involves the following general steps:

1. Dismantle the meter.
2. Thoroughly clean the parts.
3. Inspect all parts for wear, pitting, and distortion.
4. Replace or repair parts as necessary.
5. Reassemble the meter.
6. Retest the meter.

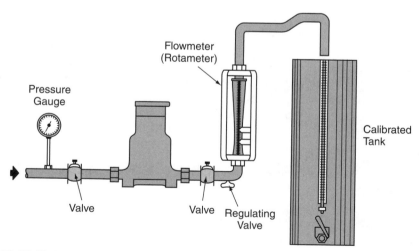

FIGURE 10-15 Meter-testing equipment

Maintenance and repair for other meters generally follow the same procedure, with some variations due to design differences.

Before undertaking a maintenance and repair program, water utilities should evaluate the cost-effectiveness of a meter replacement program versus a meter repair program. It is sometimes more economical simply to replace old parts with new parts or old meters with new meters, rather than to check tolerances, shim disks, or repair registers. Meter manufacturers can furnish detailed repair instructions for each style of meter.

Record Keeping for Meters

Meter records are an essential part of any water distribution system that meters its supply. Meter records should provide information on the installation, repair, and testing of each meter. The records enable field crews to locate meters easily, help repair personnel with meter testing, and aid managers in assessing the value of the system.

One method for maintaining meter records involves having a meter history card for each meter. Basic data recorded at the top of the card include the size, make, type, date of purchase, and location. The remaining portion of the card is divided up to record the location of the meter and, in chronological order, all tests and repair work. Each line of the test and repair section is usually divided into two segments. The upper part is used to record test results when the meter is removed, and the lower part is used to record test results before the meter is reinstalled after maintenance or repair.

In a small- to medium-size shop, test and repair information should be entered on the meter history card by personnel in the meter repair shop. After all information is recorded, meter history cards should be stored in a safe, permanent file in the shop area. They are usually filed in sequence, either by the manufacturer's serial number or by the water system's number.

Many water suppliers use computerized meter maintenance systems in their operation. Information commonly kept on a service or meter history card is entered into the computer system to establish a permanent record. A control number is associated with each service or meter. Any future information concerning work on a customer service line or meter testing and repairs can be entered for the appropriate control number. These systems can be integrated with geographical information systems (GISs) to provide the utility with comprehensive information regarding the meter history, location, service, land ownership, billing, distribution system main and value information, etc.

MAINLINE METERING

Larger water meters are used at various points in water treatment and distribution systems, including at

- well discharges (to record the amount of water being supplied to the distribution system),
- the intake of a surface source (for determining chemical feed and control treatment plant operation),

- intermediate points in the treatment process (for process control), and
- the treatment plant discharge (for pump control and for comparison with customer metering to determine unaccounted-for water).

In addition, if water is purchased from another system, meters at the purchase point determine the payment that must be made. Meters installed on the various zones in the distribution system are used to provide pumpage and pressure control. If water is blended from multiple sources, multiple meters are used to help maintain a uniform blend.

System metering also helps in administering of water rights and in checking the capacity of pumps and pipelines. Meter records are also often used in the engineering design of water system improvements.

Types of Mainline Meters

Following are brief details on the more commonly used mainline meters.

Current meters

Current meters are also commonly called velocity meters. They are principally used to measure flow in lines that are 3 in. (76 mm) and larger. The principal types include turbine, multijet, and propeller meters.

Turbine meters. A turbine meter (Figure 10-16) has a measuring chamber with a rotor that is turned by the flow of water. The volume of water recorded on the meter's register is almost in direct proportion to the number of revolutions made by the rotor. Turbine meters have little friction loss, but water must be moving at a sufficient speed before the rotor will start to rotate.

FIGURE 10-16 Turbine meter
Courtesy of Badger Meter, Inc., Milwaukee, Wis.

Turbine meters can underregister if the blades of the wheel become partially clogged or coated with sediment. This condition can generally be kept under control if a strainer is installed ahead of the meter. Periodic inspection and testing are also necessary. Current models of horizontal turbine meters have improved low-flow accuracy and higher maximum flows than previous instruments.

Multijet meters. A multijet meter has a multiblade rotor mounted on a vertical spindle within a cylindrical measuring chamber. Water enters the measuring chamber through several tangential orifices around the circumference and leaves the measuring chamber through another set of tangential orifices placed at a different level in the measuring chamber.

If the jets in a multijet meter become clogged, the meter will overregister. If the jet orifices become worn, the meter will underregister.

Propeller meters. A propeller meter (Figure 10-17) has a propeller that is turned by the flow of the water. This movement is transmitted to a register. The propeller may be small in diameter in relation to the internal diameter of the pipe, especially in larger sizes. Propeller meters are primarily used for mainline measurement where flow rates do not change abruptly, since the propeller has a slight lag in starting and stopping. Propeller meters can be built within a section of pipe or they can be saddle mounted.

Proportional meters

A proportional meter has a restriction in the line to divert a portion of water into a loop that holds a turbine or displacement meter (Figure 10-18). The diverted flow in the loop is

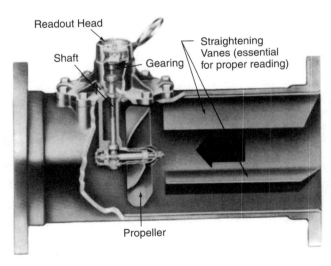

FIGURE 10-17 Propeller flowmeter

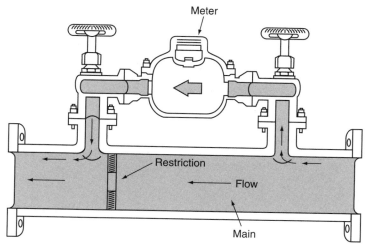

FIGURE 10-18 Proportional flowmeter

proportional to the flow in the main line. A multiplying factor can be applied to the measurement of the diverted flow to record the flow in the pipeline. Proportional meters are relatively accurate but are difficult to maintain. However, they have little friction loss and offer little obstruction to the flow of water.

Venturi meters

Venturi meters consist of a carefully sized constriction in the pipeline, called a Venturi tube (Figure 10-19). The increased velocity of the water through the throat section causes the pressure at that point to be higher than before the throat. The change in pressure is proportional to the square of the velocity. The amount of water passing through the meter is therefore determined by a comparison of the pressure at the throat and at a point upstream from the throat.

Electronic or mechanical instruments are used to compare the pressures, determine the flow rate, and maintain a total of the flow (Figure 10-20). Venturi meters are accurate for a certain range of flows, have little friction loss, require almost no maintenance, and have long been used for measuring flows in larger pipelines.

Orifice meters

As illustrated in Figure 10-21, orifice meters consist of a thin plate with a circular hole in it. This plate is installed in a pipeline between a set of flanges. As is the case with the Venturi meter, the flow is determined by comparing the upstream line pressure with the reduced pressure at the restriction of the orifice. Although an orifice meter is not as expensive as most other large meters and it occupies very little space, it has the disadvantage of creating considerably more head loss than other meters.

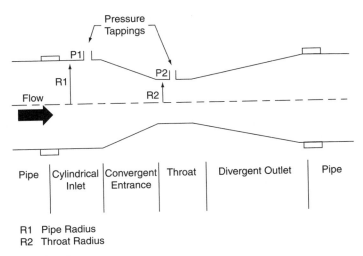

R1   Pipe Radius
R2   Throat Radius

FIGURE 10-19 Venturi meter

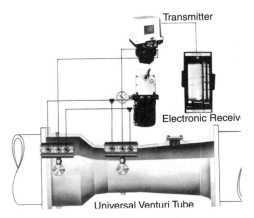

FIGURE 10-20 Venturi metering system
Courtesy of Honeywell, Industrial Measurement and Control

Magnetic meters

Magnetic meters, commonly called "mag meters," measure flow by means of a magnetic field generated around an insulated section of pipe (Figure 10-22). Water passing through the magnetic field induces a small electric-current flow, proportional to the water flow, between electrical contacts set into the pipe section. The electric current is measured and converted into a measure of water flow. Mag meters are particularly useful for measuring the flow of dirty or corrosive liquids that would damage a meter with moving parts or would plug the pressure taps of a Venturi tube.

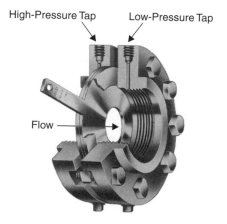

High-Pressure Tap   Low-Pressure Tap

Flow

FIGURE 10-21 Orifice meter
Courtesy of Bristol Babcock Inc., Div. FKI Energy Technology

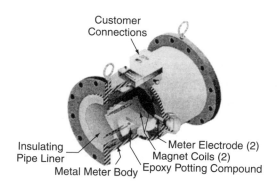

Customer
Connections

Insulating
Pipe Liner

Metal Meter Body

Meter Electrode (2)
Magnet Coils (2)
Epoxy Potting Compound

FIGURE 10-22 Magnetic flowmeters
Courtesy of ABB Inc.

Ultrasonic meters

Ultrasonic meters utilize sound-generating and -receiving sensors (transducers) attached to the sides of the pipe. Sound pulses are alternately sent in opposite diagonal directions across the pipe. Because of a phenomenon called the Doppler effect, the frequency of the sound changes with the velocity of the water. The difference between the frequency of the sound signal traveling with the flow of water and the signal traveling against the flow is an accurate indication of the quantity of water passing the meter.

METERING SAFETY

Safety precautions detailed in chapters 12 and 13 generally apply to working on meters. Personnel must pay particular attention when working with hand tools and working in confined spaces. Workers must also be especially careful to prevent the electric shocks that may be caused by improper grounding of home electrical systems onto the plumbing system. If home meters are not mounted on a yoke or if a permanent jumper wire is not provided across the meter connections, a separate wire with large alligator clips should be used as a temporary bridge between the pipes when meters are removed or installed.

BIBLIOGRAPHY

ANSI/NSF Standard 61: Drinking Water System Components—Health Effects. Ann Arbor, MI: NSF International.

AWWA Standard for Cold-Water Meters—Compound Type. ANSI/AWWA C702. Denver, Colo.: American Water Works Association (latest edition).

AWWA Standard for Cold-Water Meters—Displacement Type, Bronze Main Case. ANSI/AWWA C700. Denver, Colo.: American Water Works Association (latest edition).

AWWA Standard for Cold-Water Meters—Displacement Type, Plastic Main Case. ANSI/AWWA C710. Denver, Colo.: American Water Works Association (latest edition).

AWWA Standard for Cold-Water Meters—Fire Service Type. ANSI/AWWA C703. Denver, Colo.: American Water Works Association (latest edition).

AWWA Standard for Cold-Water Meters—Multijet Type. ANSI/AWWA C708. Denver, Colo.: American Water Works Association (latest edition).

AWWA Standard for Cold-Water Meters—Single-jet Type. ANSI/AWWA C712. Denver, Colo.: American Water Works Association (latest edition).

AWWA Standard for Cold-Water Meters—Turbine Type, for Customer Service. ANSI/AWWA C701. Denver, Colo.: American Water Works Association (latest edition).

AWWA Standard for Direct-Reading, Remote-Registration Systems for Cold-Water Meters. ANSI/AWWA C706. Denver, Colo.: American Water Works Association (latest edition).

AWWA Standard for Encoder-Type Remote-Registration Systems for Cold-Water Meters. ANSI/AWWA C707. Denver, Colo.: American Water Works Association (latest edition).

AWWA Standard for Propeller-Type Meters for Waterworks Applications. ANSI/AWWA C704. Denver, Colo.: American Water Works Association (latest edition).

Bowen, P.T., J.F. Harp, J.M. Entwhistle Jr., J.E. Hendricks, and M. Shoeleh. 1991. *Evaluating Residential Water Meter Performance.* Denver, Colo.: American Water Works Association and American Water Works Association Research Foundation.

Cesario, L. 1995. *Modeling, Analysis, and Design of Water Distribution Systems.* Denver, Colo.: American Water Works Association.

Edgar, T. 1995. *The Large Water Meter Handbook.* New York: Flow Measurement Publishing (available from American Water Works Association).

Manual M6, Water Meters—Selection, Installation, Testing, and Maintenance. 1999. Denver, Colo.: American Water Works Association.

Manual M22, Sizing Water Service Lines and Meters. 2003. Denver, Colo.: American Water Works Association.

Manual M33, Flowmeters in Water Supply. 2006. Denver, Colo.: American Water Works Association.

Manual M36, Water Audits and Loss Control Programs. 2009. Denver, Colo.: American Water Works Association.

US Department of the Interior, Bureau of Reclamation. 2001. *Water Measurement Manual.* Denver, Colo.: US Government Printing Office.

Water Distribution Operator Training Handbook. 2005. Denver, Colo.: American Water Works Association.

CHAPTER 11

Backflow Prevention and Cross-Connection Control

As potable water is transported from the treatment facility to the user, opportunities exist for unwanted substances to cause contamination. Water in the distribution system can become contaminated by backflow of nonpotable substances through cross-connections. The utility operator is responsible for protecting the public water system from hazards that originate on the customers' premises and from temporary connections that may impair or alter the water in the public water system.

All utility operators need to be concerned with this problem, because cross-connections and backflow can and do occur in systems of all sizes. The consequences can be serious.

TERMINOLOGY

The following definitions are important because the conditions they describe aid in understanding and eliminating the potential hazards posed by cross-connections. State or local statutes may define these conditions more specifically.

Backflow

Backflow is the flow of any water, foreign liquids, gases, or other substances back into a potable water system. Two conditions that can cause backflow are backpressure and backsiphonage.

Backpressure is a condition in which a substance is forced into a water system because that substance is under a higher pressure than system pressure.

Backsiphonage is a condition in which the pressure in the distribution system is less than atmospheric pressure. In other words, something is "sucked" into the system because the main is under a vacuum.

Cross-Connections

A cross-connection is any connection or structural arrangement between a potable water system and any other water source or system through which backflow can occur. The existence of a cross-connection does not always result in backflow. But where a cross-connection exists, the potential for backflow is always present if either backpressure or backsiphonage should occur.

CROSS-CONNECTIONS AND LOCATIONS

Cross-connections can be found in almost any type of facility where water is used, including houses, factories, restaurants, hospitals, laboratories, and water and wastewater treatment plants. Many pieces of equipment within these facilities draw water from the potable water distribution system for use in cooling, lubricating, washing, or as an additive to a process. Table 11-1 provides a partial list of the types of fixtures that are particularly likely to have cross-connections that could contaminate a potable water supply.

Cross-connections are frequently created by individuals who are not familiar with the hazards, even though they may be otherwise well trained and experienced in plumbing, steam fitting, pipe fitting, or water distribution work. Many such connections are made simply as a matter of convenience, with no regard given for the problems that may result.

TYPES OF CROSS-CONNECTIONS

A cross-connection will appear to be simply a pipe, hose connection, or any water outlet with no specific or outstanding features. The real distinction is where the connection leads. A cross-connection exists if a connection leads from a potable line to anything other than potable service. Cross-connections can be found in many different places, such as homes, farms, laboratories, and factories. Each connection leads to a vessel containing materials that should not be allowed to enter the potable water system. Cross-connections can be categorized as actual or potential.

Actual Cross-Connections

An actual cross-connection is one for which the connection exists at all times. Examples would be solid piping to an auxiliary supply (Figure 11-1) or into a boiler. An auxiliary supply could be a nonpotable source used for emergencies or for reducing the cost of cooling or washing. The situation could also apply to an older home that has retained its private well (which could be contaminated) but has also been connected to a municipal system, with only a valve separating the two water sources.

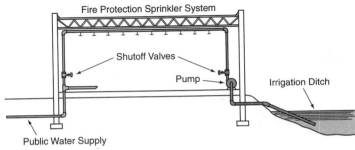

FIGURE 11-1 Cross-connection between a potable water supply and an auxiliary fire sprinkler system

TABLE 11-1 Some cross-connections and potential hazards

Connected System	Hazard Level
Access hole flush	High
Agricultural pesticide mixing tanks	High
Aspirators	High
Boilers	High
Chlorinators	High
Cooling towers	High
Flush valve toilets	High
Laboratory glassware or washing equipment	High
Plating vats	High
Sewage pumps	High
Sinks	High
Sprinkler systems	High
Sterilizers	High
Car wash	Moderate to high
Photographic developers	Moderate to high
Pump primers	Moderate to high
Baptismal founts	Moderate
Dishwashers	Moderate
Swimming pools	Moderate
Watering troughs	Moderate
Auxiliary water supply	Low to high
Garden hose (sill cocks)	Low to high
Irrigation systems	Low to high
Solar energy systems	Low to high
Water systems	Low to high
Commercial food processors	Low to moderate

Potential Cross-Connections

A potential cross-connection is one for which something must be done to complete the connection. In Figures 11-2 and 11-3, the slop sink and water tank are examples of potential cross-connections. In both cases, water has to be added to the vessel before the connection is completed. Although the cross-connection shown in Figure 11-3 appears to pose a very low hazard, the tank could be used for mixing a pesticide, weed killer, or fertilizer, which would greatly increase the danger.

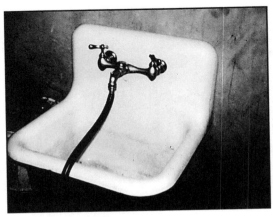

FIGURE 11-2 Slop-sink hose cross-connection

FIGURE 11-3 Water tank cross-connection
Courtesy of USEPA, Region VII, Water Supply Division

Cross-Connection Examples

Figure 11-4 depicts a very common cross-connection, formed when a chemical dispenser is connected to a garden hose and the hose is attached to a sill cock. If a vacuum should occur in the water system while the dispenser is in use, the chemical solution could be sucked back into the house plumbing

Figure 11-5 shows an unprotected hose from a dock used to fill the water tanks on a ship. Ships in port are possible hazards to the potable water supply because their on-board high-pressure fire systems use seawater.

FIGURE 11-4 Garden-hose cross-connection

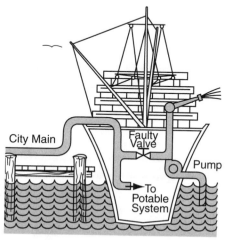

FIGURE 11-5 Unprotected potable water supply from dock to ship in port

Cooking vessels in hospitals, restaurants, and canneries use potable water (Figure 11-6). So do chemical reaction tanks. If the water inlet to one of these vessels is below the overflow rim, it can become a cross-connection.

In hard-water areas, many buildings, particularly homes, have individual water softeners. The softeners form a cross-connection that is not normally hazardous. However, when the drain is connected directly to the sanitary sewer, as shown in Figure 11-7, it becomes potentially very hazardous to both the residents and the community water supply. At least 36 persons were struck by a hepatitis epidemic in California as a result of an arrangement similar to that shown in Figure 11-7.

Figure 11-8 illustrates a cross-connection where the water supply has been protected. An atmospheric vacuum breaker is attached between the soap dispenser and the water faucet.

FIGURE 11-6 Cooking vessel with water inlet beneath overflow rim

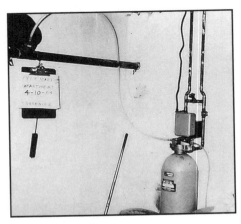

FIGURE 11-7 Water softener cross-connection

Backflow due to backpressure

A typical example of backflow due to backpressure is illustrated in Figure 11-9. When pressure inside the boiler exceeds that of the water supply, backflow can result. A similar situation occurs in Figure 11-10, where a chemical storage tank pressurized by an air compressor is connected to a potable water supply line. A third common type of cross-connection involves the recirculation of potable or nonpotable water on a premises for the purpose of meeting fire demand or processing requirements (Figure 11-11). When the auxiliary system pressure exceeds the pressure in the potable supply, any feeder connections become potential sources of backflow.

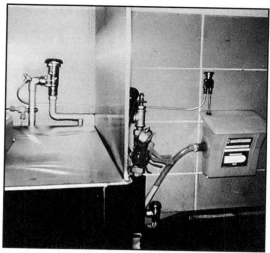

FIGURE 11-8 Atmospheric vacuum breaker installed between soap dispenser and water faucet

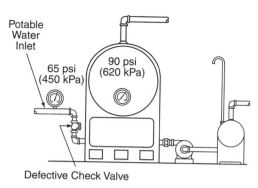

FIGURE 11-9 Backflow from high-pressure boiler due to backpressure

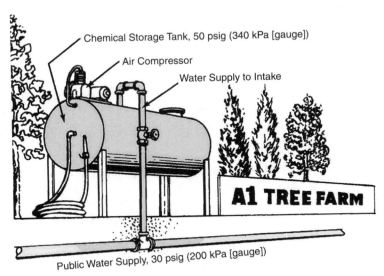

FIGURE 11-10 Cross-connection between pressurized chemical storage tank and lower-pressure potable system

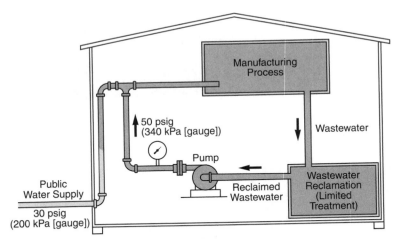

FIGURE 11-11 Backflow from recirculated system

Backflow due to backsiphonage

Because backsiphonage is actually caused by atmospheric pressure, the height to which siphoned water can be lifted is limited to 33.9 ft (10.3 m) at sea level. This is the point at which the downward pressure caused by the weight of a column of water equals the pressure of the atmosphere forcing it upward (Figure 11-12). A typical example of backflow

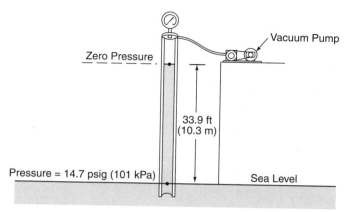

FIGURE 11-12 Effect of evacuating air from a column

due to backsiphonage is shown in Figure 11-13. In this situation, the partial vacuum in the potable system (created by high water flow out of the hydrant) sucks nonpotable materials into the system.

Another example, illustrated in Figure 11-14, can occur if a hose is being used to fill a sink. If someone opens a large valve on a lower floor, the water from the sink can be backsiphoned into the building piping.

A common cause of the less-than-atmospheric pressure (negative gauge pressure) needed to create a backsiphonage condition is overpumping by a fire or booster pump (Figure 11-15). Undersized distribution piping can also create negative pressures. Undersized piping creates a high velocity of water, which in turn causes severe pressure drops. A water service in an undersized area may have negative gauge pressure when water is flowing through the main supplying the service.

A broken main or fire hydrant, particularly at low elevations, can cause backsiphonage, as illustrated in Figure 11-16. When a break occurs in a main, the entire distribution system could become contaminated between the break and the cross-connections. Subsequently reestablishing the pressure could, in turn, contaminate all of the system downstream of the break.

PUBLIC HEALTH SIGNIFICANCE OF CROSS-CONNECTIONS

A water utility goes to great lengths to ensure that the water leaving a treatment plant is free of disease-causing microorganisms, odor, color, and other unwanted materials. Careful installation and constant maintenance of the distribution system components is needed to prevent contaminants from entering the water supply. In spite of these efforts, disease outbreaks, poisonings, and other undesirable water quality changes do sometimes occur. Many of these incidents are attributable to backflow from a cross-connection.

Various sources including The Centers for Disease Control and Prevention, USEPA, American Water Works Association (AWWA), Craun and Calderon (2001), University of

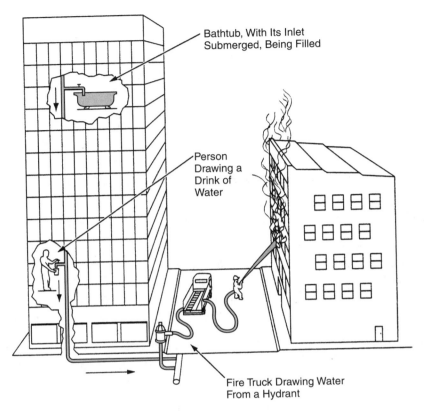

FIGURE 11-13 Backsiphonage due to pressure loss

Southern California Foundation for Cross-Connection Control and Hydraulic Research, American Backflow Prevention Association, and the Cross-Connection Control Committee of the Pacific Northwest Section of AWWA, have summarized outbreaks and linked illness to distribution system deficiencies due to cross-connection and backflow. It is recognized that many incidents go unreported and, as a result, the data likely underestimate the actual number of backflow occurrences.

The USEPA estimated that 459 documented backflow incidents resulted in 12,093 illnesses from 1970 to 2001. Craun and Calderon (2001) found that 30.3 percent of the waterborne disease outbreaks in community water systems from 1971 to 1998 were caused by contamination in distribution systems. They estimated that 50.6 percent of these were attributable to backflow from cross-connections. Other reports estimate that the number of disease outbreaks associated with backflow and cross-connections are much higher (more than 70 percent). Most of the health consequences from backflow incidents involve gastrointestinal disorders.

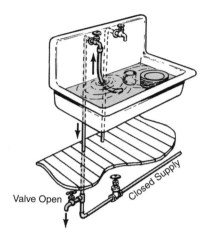

FIGURE 11-14 Backsiphonage (hose forms cross-connection)

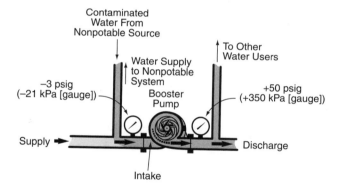

FIGURE 11-15 Backsiphonage from a booster pump

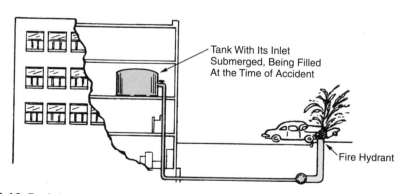

FIGURE 11-16 Backsiphonage due to a broken fire hydrant

Contaminants Associated With Cross-Connections

A large number of contaminants have been introduced into distribution systems by cross-connections and backflow. The severity and scope of the incident depend on several factors including the quantity of the contaminant, toxicity of the contaminant, dilution of the contaminant, and the health status of the exposed individuals. Contaminants generally fall into two broad categories: (1) chemical and (2) biological. Examples of specific contaminants in each category and their potential health effects are described below.

Chemical contaminants

The numerous possibilities under this category are divided into four classifications. Some examples of chemicals identified in backflow incidents and health consequences are given for each.

Pesticides. Most incidents involving pesticides have occurred at residences where pesticides were being used. Common chemicals mentioned include chlordane, malathion, heptachlor, diazinon, 2, 4-D, and Dicamba. Nervous system damage and kidney and liver toxicity are among the health effects.

Metals. Copper and hexavalent chromium are the most commonly reported metals. Backflow of carbon dioxide from carbonated soft drink dispensers dissolving copper from pipes is the most reported situation. Chromium is used as a corrosion inhibitor in heating and cooling systems. Cross-connections between the potable system and cooling/heating system are typical sources of this contaminant. Health effects are vomiting, nausea, and kidney and liver damage.

Organic compounds. Ethylene glycol, propylene glycol, freon, propane, and detergents are the most frequently reported contaminants. The glycols are used in antifreeze. A cross-connection to an air conditioning system, which may contain freon, is the source of many of these incidents. Health effects of these compounds can range from excessive fatigue and dizziness to vomiting and intestinal upsets.

Propane has been introduced into water systems through cross-connections and backflow from propane tank cars or storage tanks. In rare instances this has resulted in explosions and fires.

Soap dispensers and car washes have been implicated as sources for detergents entering the water system. Reported symptoms are burning in the mouth and flu-like reactions.

Nitrates and nitrites. These compounds are used in boilers and cooler systems. Backflow incidents have resulted in vomiting, methemoglobinemia (caused by nitrites), and damage to the spleen (nitrates).

Biological contaminants

Public water supplies have been contaminated by pathogenic microorganisms through cross-connections with untreated water sources, sewer lines, nonpotable reclaimed water supplies, medical facilities, mortuaries, utility sinks, and pools. Other sources are private wells, cisterns, laboratories, and drain lines.

Sewage is responsible for the majority of microbial incidents. In many cases the specific microbe has not been identified. Health effects for most of the pathogens are fever, nausea, and diarrhea. Some incidents have led to life-threatening effects and death.

Contamination by human bodily fluids from funeral homes and hospitals can cause serious disease. These facilities require the highest level of cross-connection and backflow protection.

The most commonly reported microbes that have caused disease attributed to cross-connections and backflow are

- *Shigella,*
- *Escherichia coli,*
- *Salmonella,*
- *Campylobacter jejuni,*
- Cyanobacteria,
- Norwalk and Norwalk-like viruses, and
- *Giardia.*

The consequences from cross-connections and backflow can be severe. Disease and possible death can result from exposure to contaminants. Prevention of backflow from cross-connections is a critical public health responsibility of distribution system operators.

BACKFLOW CONTROL METHODS AND DEVICES

When a cross-connection is found, one of two actions must be taken. Either the cross-connection must be removed or some method to protect the potable water supply from possible contamination must be devised. Where removal is impractical, a protective device should be installed.

The preventive measure chosen depends on the degree of hazard involved, how accessible the premises are, and the type of water distribution system within the cross-connection location. Some protective methods and backflow-prevention devices, in order of decreasing effectiveness, are

- air gaps,
- reduced pressure zone backflow preventers,
- double check valve assemblies,
- vacuum breakers (atmospheric and pressure), and
- barometric loops.

Air Gaps

When correctly installed and maintained, an air gap (Figure 11-17) is the most positive method available for protecting against backflow. It is acceptable in all cross-connection situations and for all degrees of risk. Another advantage is that there are no moving parts to break or wear out. Only surveillance is needed to ensure that no bypasses are added.

The only requirement for installation is that the gap between the supply outlet and the overflow level of the downstream receptacle measure at least two times the inside diameter (ID) of the outlet's tip but no less than 1 in. (25 mm). In situations where an isolated water system is needed to supply nonpotable uses in a factory, a surge tank and booster pump may be installed, as illustrated in Figure 11-18. In this case, the air gap provides positive isolation from the potable water system.

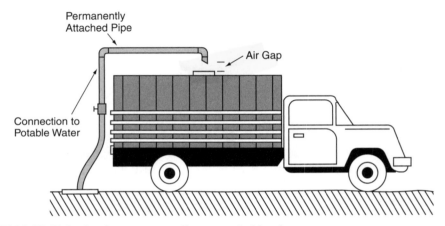

FIGURE 11-17 Water truck cross-connection prevented by air gap

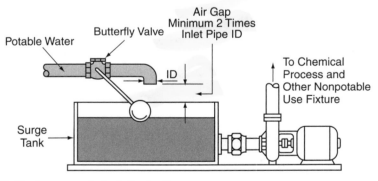

FIGURE 11-18 Air gap on surge-tank feeder line

Reduced Pressure Zone Backflow Preventers

The second type of device that can be used in every cross-connection situation and with every degree of risk is the reduced pressure zone backflow preventer, often abbreviated RPZ or RPBP. This device consists of two spring-loaded check valves with a pressure-regulated relief valve located between them, as shown in Figures 11-19 and 11-20. Flow from the supply at left enters the central chamber against the pressure exerted by check valve 1 (Figure 11-21A). Water loses pressure passing through this valve, so the central chamber is known as the reduced pressure zone. The amount of pressure loss through the check varies with the valve size, flow rate, and valve manufacturer. The second check valve is loaded considerably less (causing about 1 psi [7 kPa] further pressure drop) to keep the total pressure loss within reason.

Two standard check valves connected together is not considered sufficient protection for most hazardous locations, because all valves can leak from wear or obstruction. For this reason, a relief valve is positioned between the two checks. When the unit is operating correctly and the supply pressure exceeds the downstream pressure, the supply pressure opposes the relief valve's spring tension and keeps the valve closed. If a vacuum should occur in the supply line, as shown in Figure 11-21B, both check valves will close and the relief valve will open to drain water from the reduced pressure zone. Backsiphonage is therefore positively prohibited. If the second check valve should leak at this time, the leakage will be harmlessly discharged through the relief valve.

If backpressure on the building side of an RPZ should exceed the supply pressure, as illustrated in Figure 11-21C, both check valves will close. If the second check valve should fail to seal completely under this condition, the leakage will be discharged through the relief valve, as shown in Figure 11-21D.

FIGURE 11-19 Reduced pressure zone backflow preventer
Courtesy of Watts Regulator Co.

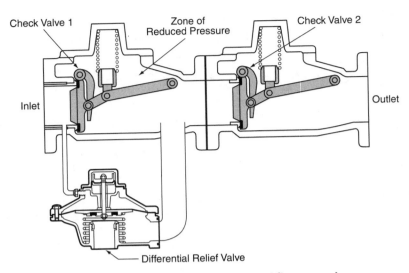

FIGURE 11-20 Cut-away view of a reduced pressure zone backflow preventer
Courtesy of Cla-Val Co., Backflow Preventer Division

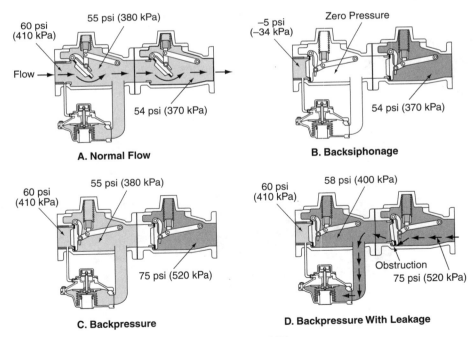

FIGURE 11-21 Valve position and flow direction in an RPZ
Courtesy of Cla-Val Co., Backflow Preventer Division

A definite sign of an RPZ malfunction is continuous drainage from the relief port. Because of the multiple protective systems, the probability of backflow occurring across an RPZ backflow preventer is very small. A typical installation of an RPZ device is illustrated in Figure 11-22.

Even though the RPZ backflow preventer is highly dependable, it is a mechanical device that requires maintenance. Valve faces, springs, and diaphragms deteriorate with age. Also, solids can lodge in or damage the check valves, causing leakage. For these reasons, each installed unit must be periodically inspected, tested, and maintained. RPZ devices should be tested in accordance with the manufacturer's specifications and regulatory requirements.

The backflow preventer's performance can be affected by how and where it is installed. To achieve continued satisfactory performance, these installation procedures should be followed:

- Install the unit where it will be accessible. Accessibility will provide for ease of testing and maintenance. Installation at least 12 in. (300 mm) above the floor is one criterion for accessibility. If the device can be easily inspected, it will increase the probability that any malfunction (leakage) will be noticed and promptly repaired.

- Install the unit so that the relief-valve port cannot be submerged. Submergence of the port creates another cross-connection and may also prevent the unit from operating properly.

- Protect the device from freezing, which will damage the unit.

- Protect the device from vandals. Accessible units—particularly the associated gate valves—are a temptation to vandals.

- If a drain is needed, an air gap must be provided below the relief-valve port. Most manufacturers have designed attachments to fit the unit.

- Install a screen upstream of the preventer to eliminate the possibility of debris becoming lodged in the unit, which could render it inoperative. Screens are available commercially.

- Most models must be installed horizontally. Check with the manufacturer to determine the allowable positions for each model.

Double Check Valve Assemblies

The double check valve backflow preventer is designed basically the same way as the RPZ but without the relief valve (Figure 11-23). The absence of the relief valve significantly reduces the level of protection provided. The unit will not give any indication that it is malfunctioning.

A double check valve assembly is not recommended as protection in situations where a health hazard may result from failure. Before installing such a unit in a potable water line, contact the agency having statutory jurisdiction to determine whether or not its use is permitted.

FIGURE 11-22 Typical installation of reduced pressure zone backflow preventer

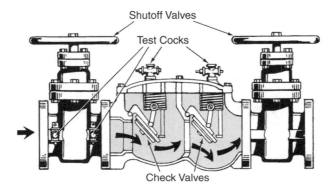

FIGURE 11-23 Double check valve assembly

The installation and maintenance of a double check valve assembly should follow these procedures:

- The unit must be an approved model. Two single check valves in series will not be satisfactory.
- The installation must be protected from freezing and vandals.
- The unit must be periodically tested and maintained. Frequency of testing and maintenance is governed by state and local codes.
- The unit should be installed in a position that will allow testing and maintenance to be performed.
- The unit should be installed in an area that does not flood. The test cocks can form a cross-connection when submerged.

In recent years, many water systems have become interested in installing check valves on customer water services. This is particularly the case where many homes still have old private wells in use. There is always a possibility that a connection within the home plumbing will allow the well water to flow out into the public water system. The principal problem has been finding a location to install one or preferably two check valves for this purpose.

Several manufacturers have recently developed double check valve assemblies for this use. Various configurations are available, each placing the assembly near the meter, as illustrated in Figure 11-24. Locating the check valves on the discharge side of a meter has the advantage of preventing water from draining out of the residence plumbing while a meter is being replaced.

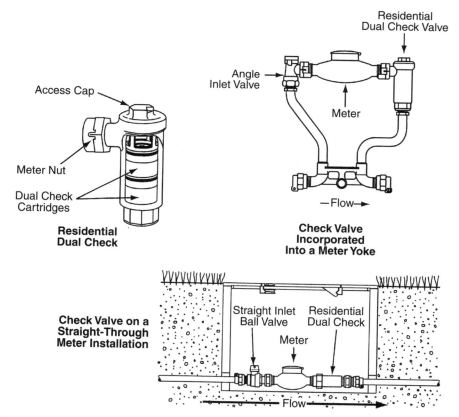

FIGURE 11-24 Examples of residential dual check valves
Courtesy of the Ford Meter Box Company, Inc.

Vacuum Breakers

Backsiphonage occurs when a partial vacuum pulls nonpotable liquids back into the supply lines. If air enters the line between a cross-connection and the source of the vacuum, the vacuum will be broken and backsiphonage will be prevented. This is the principle behind a vacuum breaker.

Figure 11-25 shows an atmospheric vacuum breaker, consisting of a check valve operated by water flow and a vent to the atmosphere. When flow is forward, the valve lifts and shuts off the air vent. When flow stops or reverses, the valve drops to close the water supply entry and open an air vent. A version of the unit designed for a hose bibb is shown in Figure 11-26.

Atmospheric vacuum breakers are not designed to protect against backpressure, nor are they reliable under continuous use or pressure because the gravity-operated valve may stick. Atmospheric vacuum breakers must only be used where there is no possibility of

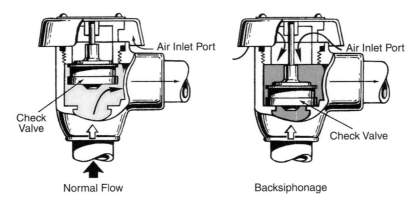

FIGURE 11-25 Atmospheric vacuum breaker

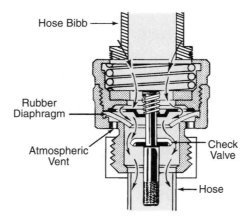

FIGURE 11-26 Hose-bibb type of atmospheric vacuum breaker

backpressure being applied. These breakers must be installed at least 6 in. (150 mm) above the highest point of the downstream outlet.

Figure 11-27 shows a pressure vacuum breaker. The unit illustrated has a spring-loaded check valve that opens during forward flow and is closed by the spring when flow stops. When pressure in the line drops to a low value, a second valve opens and allows air to enter this breaker. With this arrangement, the breaker can remain under supply pressure for long periods without sticking and can be installed upstream of the last shutoff valve. The placement and use of the pressure vacuum breaker is restricted to situations where no backpressure will occur and where it can be installed 12 in. (300 mm) above the highest point of the downstream outlet.

Barometric Loops

The barometric loop is a simple installation that prevents only backsiphonage. To create a barometric loop, an inverted U is inserted into the supply pipe upstream of the cross-connection. Figure 11-12 illustrates how the height of a column of water, open to the atmosphere at the bottom, is limited to about 33.9 ft (10.3 m). If the barometric loop is taller than the 33.9-ft (10.3-m) limitation, siphonage or backsiphonage will not occur.

Although the barometric loop is effective against backsiphonage and requires no maintenance or surveillance (other than to ensure that it remains leak-free), the space requirement is a serious disadvantage. The loop must extend about 35 ft (11 m) above the highest liquid level. In addition, the barometric loop is completely ineffective against backflow due to backpressure. For these reasons, barometric loops are no longer installed.

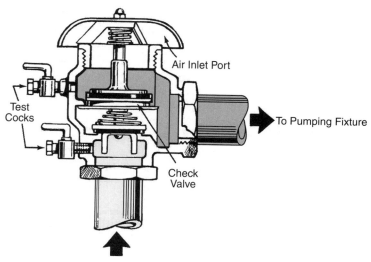

FIGURE 11-27 Pressure vacuum breaker

Other Methods and Devices

One of the first methods for cross-connection control was the complete separation of potable and nonpotable piping systems. When water piping systems from two different sources are located in the same building, they are usually identified by color coding. This is an effective approach when adequate surveillance is provided to prevent connections between systems. In some instances, however, separate systems have been interconnected by such devices as spool pieces, flexible temporary connections, and swing connections. None of these interconnections is recommended for use regardless of the degree of risk involved. Each type should be removed from any system. However, these types of connections (as opposed to cross-connections) may be useful when two *potable* systems need to be connected for emergency or other approved service.

Level of Protection

The type of backflow-prevention device that is to be used for each installation depends on the degree of hazard present and whether or not backflow could result from backpressure or backsiphonage. Table 11-2 provides a listing of the minimum protection suggested for various applications. Local codes or regulations must be followed before a method is selected for protecting an installation. Table 11-3 lists the types of cross-connections discussed in this chapter and suggests the level of protection appropriate for each.

CROSS-CONNECTION CONTROL PROGRAMS

Passage of the Safe Drinking Water Act in 1974 made each water utility responsible for the quality of water at the consumer's tap. Legal proceedings have also established that the utility is responsible for cross-connection control in some jurisdictions. In addition, many states assign the cross-connection control responsibility directly to the water supplier. The size of the utility has no bearing on the degree of risk posed by cross-connections. Therefore, all water systems, large and small, public or private, should maintain an active cross-connection control program.

Developing a Cross-Connection Control Program

A water utility can develop and operate a cross-connection control program or jointly participate in such a program with other municipal agencies. The specific nature of any program depends on state and local laws and regulations, municipal agencies, and the size of the community.

An effective program deals with the two major sources of cross-connection problems: (1) plumbing within the customer's premises and (2) auxiliary water sources. Plumbing within the customer's premises usually falls under the supervision of state and local health departments or local building–engineering departments. Auxiliary water sources can fall under the supervision of a health department or environmental protection agency or its equivalent. When different agencies have an interest, the program must be administered as a cooperative effort.

A utility is more likely to serve as the single program manager in a smaller community. Many small communities have a public water distribution system but still depend on a higher authority, such as a county or district, for health and sanitary support.

A control program will give the water utility (or responsible agency) the legal authority to do the following:

- Take actions to protect the water supply
- Provide a systematic procedure for locating, removing, or protecting all cross-connections in the distribution system
- Establish the procedures for obtaining the cooperation of customers and the public

Anyone attempting to establish such a program should consult the state water program agency or other authorities for legal direction.

The operator's involvement in a cross-connection control program will vary from almost no involvement, to being an inspector, to providing major assistance in establishing and operating the program. Knowledge of program content and procedures are valuable at every level of responsibility.

Program content

An effective cross-connection control program has the following elements:

- An adequate plumbing and cross-connection control ordinance
- An organization or agency with overall responsibility and authority for administering the program, with adequate staff
- Systematic inspection of new and existing installations with formal record keeping
- Follow-up procedures to ensure compliance
- Backflow-prevention device standards, as well as standards for inspection and maintenance
- Cross-connection control training
- A public awareness and information program

Procedures

The procedure for initiating a program begins with planning, which is normally a function of management. Assistance from the health, building, or plumbing inspection departments should be obtained at the beginning. Planning should include identifying a tentative organization, the appropriate authority, and internal procedures for executing the program.

TABLE 11-2 Recommended minimum protection requirements

Type of Hazard on Premises	Minimum Protection at Meter*					Minimum Options to Isolate Area of Plant Affected†					Comments
	AG	RPZ	DCVA	AVB or PVB	None	AG	RPZ	DCVA	AVB or PVB	None	
1. Sewage treatment plant	x	x									RPZ at meter with air gap in plant also
2. Sewage lift pumps	x	x								x	
3. Domestic water booster pumps			x					x			
4. Equipment or containers manufactured for industrial use without proper backflow protection											
A. Dishwashing									x		See comments
B. Clothes washing				x			x		x		Normally machine has built-in AG
C. Food processing		†	†				†				If no health hazard exists, DCVA is acceptable
D. Pressure vessels			x				x				
E. Tank or vat containing a nonpotable or objectionable solution	x	x	x			x	x		x		A vacuum breaker or DCVA may be used if no health hazard exists
F. Sinks with hose threads on inlet									x		
G. Any dispenser connected to a potable water supply									x		
H. Aspirator equipment									x		
I. Portable spray equipment								x			
5. Reservoirs, cooling towers, circulating systems		x				x	x				System where no health hazard or potential for a health hazard exists, a DCVA may be used

TABLE 11-2 Recommended minimum protection requirements (Continued)

Type of Hazard on Premises	Minimum Protection at Meter*					Minimum Options to Isolate Area of Plant Affected†					Comments
	AG	RPZ	DCVA	AVB or PVB	None	AG	RPZ	DCVA	AVB or PVB	None	
6. Commercial laundry	x	x								x	Must be hot-water related
7. Steam-generating facilities and lines			x				x				
8. Equipment under hydraulic test or hydraulically operated equipment	x	x					x	x			
9. Laboratory equipment											
A. Health hazard		x	x							x	
B. Not health hazard				x					x		
10. Plating facilities	x	x				x					
11. Irrigation systems	x	x	x	x		x		x			
12. Fire-fighting systems	x	x	x			x		x			Chemicals are often used in such systems
13. Dockside facilities	x	x				x	x				DCVA may be used on dockside if outlet is protected
14. Tall buildings		x	x								Two devices should be installed in parallel; if a health hazard exists within the building, two RPZs in parallel may be required
15. Unapproved auxiliary supply	x	x				x	x				
16. Premises where inspection is restricted	x	x					x				
17. Hospitals, mortuaries, clinics	x	x				x	x				
18. Laboratories	x	x				x	x				DCVA, if no health hazard exists

Table continued next page

TABLE 11-2 Recommended minimum protection requirements (Continued)

Type of Hazard on Premises	Minimum Protection at Meter*					Minimum Options to Isolate Area of Plant Affected†					Comments
	AG	RPZ	DCVA	AVB or PVB	None	AG	RPZ	DCVA	AVB or PVB	None	
19. Chemical plants using a water process	x	x				x	x				
20. Petroleum processing or storage plants	x	x				x	x				
21. Radioactive material processing plants or nuclear reactors	x	x				x	x				
22. Swimming pools	x	x				x	x				

NOTE: The list is not all-inclusive and does not necessarily conform to state and local codes. These codes must be checked before installation of any protective device.

* The following abbreviations are used:air gap—AG; reduced pressure zone backflow preventer—RPZ; double check valve assembly—DCVA; atmospheric vacuum breaker—AVB; pressure vacuum breaker—PVB.

† In areas where no health hazard or potential for a health hazard exists, a vacuum breaker properly installed in the line to the problem area may be adequate protection (only if there is no backpressure present.)

TABLE 11-3 Suggested backflow protection for situations discussed in this chapter

Application	Reference Figure	Minimum Protection		Notes
		At Meter	At Cross-Connection	
Auxiliary water system	11-1	RPZ or air gap	RPZ	
Sink	11-2	RPZ or air gap[*]	Atmospheric or pressure vacuum breaker	
Water truck	11-3	RPZ or air gap[*]	Air gap, RPZ	See Figure 11-18
				A double check-valve assembly may be acceptable if no health hazard exists
Garden hose	11-4	RPZ or air gap[*]	Atmospheric or pressure vacuum breaker	The hose-bibb type of breaker is used
Ship	11-5	RPZ or air gap	RPZ	
Cooking vessel	11-6	RPZ or air gap[*]	RPZ	If no health hazards exist, double-check valve assembly may be used
Water softener	11-7	RPZ or air gap[*]	RPZ	

[*] Protection at the meter may not be necessary if protection at the cross-connection is ensured. RPZ = reduced pressure zone backflow preventer.

The next step is to inform the municipal government (or other applicable government agency) and the public of the nature of cross-connections and backflow, as well as the steps needed to protect the public. This can be done through newspaper announcements, interviews, and public presentations. The government agency uses this background information to understand and enact the authorizing control ordinance, which is the legal basis of a local program. The public should continue to be informed even after the program has been implemented.

The authorizing control ordinance provides authority for establishing and operating the program. The amount of detail included depends on state laws and codes. Where the state has an extensive and detailed code, the ordinance can be very simple. In other locations, it may be necessary to provide details to describe the program completely. These details should include at least the following:

• Authority for establishment
• Responsibilities and organization for operating the program
• Authority for inspections and surveys
• Prohibitions and protective requirements (if the state code is not specific, this description can become very detailed)
• Penalty provisions for violations

After establishing the program, the water utility or other agency designated by the ordinance begins inspecting customers' premises. Those premises that, by nature of their activity, present the greatest risk to public health should be inspected first. The inspector, who must be trained to identify cross-connections and the actions needed to protect the potable water supply, should be accompanied by the owner or the owner's representative so that there is an understanding of the procedures and results.

The complete plumbing system should be inspected. Particular attention should be given to pipelines leading to process areas, laboratories, liquid storage, and similar facilities. When cross-connections are found during the inspection, they should be pointed out to the owner and their significance discussed. Areas to be covered include: What is the material that could backflow? How hazardous is the cross-connection? Can the cross-connection be eliminated or must it be protected? How can it be protected?

Generally, the water utility is responsible for the water distribution supply to the owner's premises. Therefore, for the purpose of a cross-connection program, the inspector does not need to require that each cross-connection be protected. Instead, protection needs to be provided only at the point of entry to the premises (this tactic is termed *containment*). However, the inspector should recommend that some action be taken on each connection, both for the owner's protection and to fulfill the requirements of the Occupational Safety and Health Administration and local plumbing or building codes.

In addition, local plumbing or health codes may require that each connection be protected. Only when the codes are specific should the inspector specify the item needed for a certain application. Otherwise, a minimum level of protection should be recommended in accordance with the code and Table 11-2. Upon completing the survey, the inspector should prepare a report and notify the owner of the corrective actions needed, the approvals required, and the time limits.

For new installations, jurisdictions normally require building and plumbing plans to be submitted and approved. The plan review stage presents an opportunity to identify potential problem areas. The review should identify cross-connections, with a view toward their elimination or protection. Only after the plans are changed to ensure adequate protection should they be approved. Most jurisdictions inspect new buildings during construction and after completion. A survey for cross-connections should be included in these inspections.

Repeat inspections are needed both to ensure that all corrective actions required from previous inspections have been completed and to look for new cross-connections. These inspections are conducted in the same manner as the initial inspection. Timing of repeat inspections will vary according to their nature and whether compliance checks are routine or special.

Routine inspection intervals will vary according to the degree of hazard involved and human resources available. However, inspections should be conducted as often as feasible. The regulatory code may recommend appropriate inspection intervals, which could be every 3 to 6 months for high-risk installations and annually for others. Special compliance checks should be made immediately after the utility is notified of a plumbing change or a change in activity at an installation (the protection may need to be upgraded if the risk has increased). These checks should also be made when the utility is notified of a violation that could endanger public health.

When a customer is not convinced of the seriousness of the potential hazard or refuses to cooperate, the water purveyor must have the legal authority to shut off the customer's water. Before the water is shut off, the purveyor needs to make a final effort to gain the customer's cooperation and warn of the consequences of noncompliance. Discontinuing water service should be a last-resort measure, and it is recommended that legal counsel be consulted first. All warnings and notifications must be made exactly as specified by ordinance.

One possible corrective action resulting from inspections is the actual elimination of cross-connections. These actions will require only follow-up inspections to ensure that the old connections are not reconnected. A second corrective action is the installation of backflow-prevention devices. If records of testing and maintenance are not maintained by the agency performing the field surveys, repeated inspections should include a check of the owner's records to ensure that the backflow equipment is being tested and maintained.

Backflow-prevention devices

For backflow-prevention devices to operate properly for long periods of time, two primary conditions must be met.

1. Good quality backflow-prevention units must be installed.
2. Inspection and maintenance must be performed periodically.

The first condition is met by setting minimum standards (e.g., ANSI/AWWA C510, C511; ANSI/NSF Standard 61) for design, construction, and performance, followed by tests to evaluate conformance. Testing and approval are done by special laboratories and in some cases by the state or utility. A customer should consult a list of state-approved devices before installing any backflow preventer.

The second condition—periodic inspection and maintenance—is met by systematic inspection and testing of all units that have been installed. Inspection consists of visually checking a unit weekly or biweekly for leaks and external damage. The owner should be required to make the inspections and keep a record of the findings. If an inspection shows defects, qualified repair personnel should be called to perform further examination and repair. Each unit should be tested for correct operation annually by an individual specifically qualified for such testing and repair. Any devices that fail must be repaired immediately, and a report of the test and repair should be made when the job is completed.

Education and training for cross-connection control should be performed at several levels. The first level is a continuous program of public education. This program should describe applicable codes and cross-connections and the hazards they pose. The goal is to obtain the consumer's cooperation and reduce the overall risk. At the next level, water utility personnel, plumbing inspectors, and health personnel need special training to develop and operate the control program. Finally, specialized training in maintenance and testing (certification of testers may be required by regulation) is needed for those individuals assigned to maintain backflow preventers. The names of schools with programs for training in backflow-preventer repair can be obtained from the manufacturers of the devices.

RECORDS AND REPORTS

Records and reports play an essential role in administering a cross-connection control program. Complete records are a utility's first line of defense against potential legal liability in case of public health problems resulting from a cross-connection.

Records should be kept on installing, inspecting, testing, and repairing backflow-prevention devices as a shared responsibility of the water customer, the water utility or agency operating the control program, and personnel performing tests and repairs.

Water Customer Reports

To fulfill their responsibility for maintaining safety within the premises, water customers need to be informed about their building plumbing systems, any hazardous conditions that are creating potential cross-connections, and the condition of any backflow-prevention devices that have been installed. Information about the plumbing system should be available from building plans. Information about cross-connections and their protection comes from field inspection surveys, periodic visual observations, and test reports.

Utility or Agency Operating Records

Effective control dictates a time-based formal record system—that is, records that are organized so that each inspection, survey, test, and corrective action is performed as scheduled. Records should consist of inspection reports, test and repair reports, reports of corrective actions, authorized testing and repair personnel lists, approved backflow-preventer lists, and backflow-preventer installation locations.

Testing and Repair Personnel Reports

To ensure continued satisfactory operation of each backflow preventer, only qualified, authorized (certified testers may be required in many locations) individuals should perform the periodic test or needed maintenance. A report of work performed should be recorded. The distribution of the report by the control agency is specified by local regulations and practices.

Another way to ensure that each preventer is adequately maintained is to assign each unit to a single, qualified individual or firm for testing and repair. The assignment can take the form of a contract between the owner and an independent repair service or it may be a letter of instruction to a qualified individual who is an employee of the owner. A record of the assignment should be kept by the owner and made available to the field inspector.

The agency operating the cross-connection control program should be given the name of the individuals to whom each backflow preventer is assigned. To ensure continued satisfactory performance, repair personnel should have servicing instructions available for all backflow preventers they must maintain. Records of previous servicing and performance tests are also important and may be required by law.

BIBLIOGRAPHY

ANSI/NSF Standard 61: Drinking Water System Components—Health Effects. Ann Arbor, MI: NSF International.

AWWA Standard for Double Check Valve Backflow Prevention Assembly. ANSI/AWWA C510. Denver, Colo.: American Water Works Association (latest edition).

AWWA Standard for Reduced-Pressure Principle Backflow Prevention Assembly. ANSI/AWWA C511. Denver, Colo.: American Water Works Association (latest edition).

Backflow Prevention Assemblies. 1991. Westlake, Ohio: American Society of Sanitary Engineers.

Craun, G.F., and R.L. Calderon. 2001. *Waterborne Disease Outbreaks Caused by Distribution System Deficiencies. Jour. AWWA*, 93(9):64–75.

Cross-Connection Control Manual. 2003. USEPA Publication 816-R-03-002. Washington, D.C.

Cross Connection Control Manual, 6th ed. 1995. Seattle, Wash.: Pacific Northwest Section American Water Works Association.

Manual of Cross-Connection Control, 9th ed. 1993. Los Angeles, Calif.: Foundation for Cross-Connection Control and Hydraulic Research, University of Southern California.

Manual M14, Recommended Practice for Backflow Prevention and Cross-Connection Control. 2004. Denver, Colo.: American Water Works Association.

Potential Contamination Due to Cross-Connections and Backflow and the Associated Health Risk, An Issues Paper, Office of Ground Water and Drinking Water, USEPA, Washington D.C., August 13, 2002.

2000 Survey of State and Public Water System Cross-Connection Control Programs. 2000. Bryan, Texas: American Backflow Prevention Association.

Water-related Disease Outbreaks; Surveillance: Annual Summary. Atlanta, GA: Centers for Disease Control and Prevention. 1982–1999.

CHAPTER 12

Water Main Installation

Water mains must be properly installed to maximize their service life and minimize future maintenance problems. Improper installation of water mains can result in frequent temporary loss of service to customers and of fire protection, as well as unnecessary costs for water system maintenance and repair. Construction operations for installing distribution system mains and accessories can vary depending on the pipe material used (see American Water Works Association [AWWA] installation standards listed at the end of this chapter). The various operations are discussed in general in this chapter without reference to specific materials. Additional details on procedures for specific piping materials are available in the form of published recommendations from the manufacturer and in the references and sources listed in appendixes A and B.

PIPE SHIPMENT

Pipe may be shipped from the factory to the jobsite using various types of transportation. Determining the most economical method depends on several factors, including the distance that the pipe must be transported, the size and weight of the pipe, the amount of pipe to be used on the job, and the availability of various types of transportation. Figures 12-1, 12-2, and 12-3 show pipe being transported by truck, railroad, and barge.

FIGURE 12-1 Pipe being transported by truck
Courtesy of U.S. Pipe and Foundry Company

FIGURE 12-2 Large-diameter pipe on railroad flatcars
Photograph furnished by American Concrete Pressure Pipe Association

FIGURE 12-3 Pipe being delivered by barge
Photograph furnished by American Concrete Pressure Pipe Association

From the user's standpoint, direct truck delivery is the most satisfactory because the pipe can usually be unloaded directly at the jobsite. Many utilities require pipe to be delivered with the ends covered (with plastic wrap or some other material). This procedure keeps the pipe protected from foreign material that could be picked up during transportation or storage. Also this prevents drying and cracking of cement–mortar lining of metallic pipes. When pipe is shipped by rail or barge, it will usually be necessary to stockpile the pipe at an unloading point and then reload it onto trucks to deliver it to the construction site.

PIPE HANDLING

There are four general steps in handling pipe:

1. Inspection
2. Unloading
3. Stacking
4. Stringing

Pipe and Fitting Inspection

Pipe normally receives a final inspection before leaving the factory. It can, however, be damaged in transit, when the carrier still has responsibility for it. Therefore, pipe should be inspected as it is unloaded. All pipe, fittings, gaskets, and accessories should be checked against the shipping list for proper size, class, number, and condition before delivery is accepted. Any missing, damaged, or improper materials should be acknowledged in writing by the driver and reported to the shipping company. Damaged material should be saved, and a claim should be made according to the shipping company's instructions.

Unloading

Pipe handling is extremely important. Any type of pipe can be damaged by rough handling. It is generally best to use appropriate mechanical equipment to unload pipe to protect it from damage. Care should be taken to prevent abuse and damage to the pipe, no matter what method of unloading is used. Pipe should not be dropped or allowed to strike other pipe. Lined and coated pipe must be handled particularly carefully to avoid damaging the lining and coating.

Small-diameter pipe may be unloaded with skids and snubbing ropes, as shown in Figure 12-4. Using a derrick or other power equipment makes unloading both safer and quicker (Figure 12-5). Large-diameter pipe must be unloaded with heavy-duty equipment.

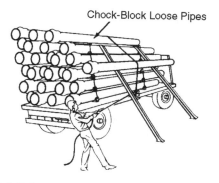
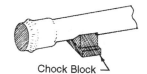

FIGURE 12-4 Unloading PVC pipe with snubbing ropes

FIGURE 12-5 Pipe being unloaded with power equipment
Courtesy of U.S. Pipe and Foundry Company

Use of a forklift for unloading pipe is not generally recommended unless the forks are padded to prevent pipe damage. Rubber-covered hooks can be used for handling pipe, but fabric slings are even better. Plastic pipe is particularly vulnerable to damage from rough handling. Scratches on the exterior can grow into cracks over time. Such scratches can eventually be the cause of pipe failure. Polyvinyl chloride (PVC) pipe is susceptible to impact damage and is more easily damaged in cold weather.

Fittings should be kept as clean as possible while in storage. It is best to store them in a secure location to protect them from vandalism. They may then be brought to the job-site as needed. Gaskets should be protected from dirt, oil, excessive heat, and excessive exposure to sunlight. Gaskets are also subject to damage from ozone, so they should not be stored near electric motors or other operating electrical equipment. Always store gaskets in a clean, secure location until they are needed.

Stacking

If pipe is to be stockpiled, stacking should be done in accordance with the manufacturer's published instructions. Following are some general recommendations:

- Build stockpiles on a flat base and off the ground to minimize contamination (Figure 12-6).
- Support the bottom layer uniformly on timbers so that the bells do not touch the ground. Use secured blocks at each end to prevent rolling.

FIGURE 12-6 Pipe stockpiles kept off ground surface
Courtesy of the Ductile Iron Pipe Research Association

- For ductile-iron and concrete pipe, place boards with blocks at each end between each layer of pipe.
- When bell-end pipe is being stacked, project the bell ends over the end of the barrels in alternate layers.
- Group pipe of the same sizes and classes together.
- Group short lengths, fittings, and adapters separately.
- Leave PVC pipe in shipping units until needed. Do not stack pipe more than three bundles high. If it is loose, do not stack more than 3 ft (0.9 m) high. Protect the pipe from sunlight with a canvas tarpaulin or other opaque material. Do not use clear plastic sheets, and allow air to circulate beneath the cover.

Stringing

When pipes are distributed (strung) along the trench in preparation for installation, observe the following procedures:

- Lay or place pipe as near to the trench as possible to avoid excess handling (Figure 12-7).
- If the trench is already open, string pipe on the side opposite the spoil bank (i.e., the side without the pile of dirt). This way, the pipe can easily be moved to the edge of the trench for lowering into position.
- If the trench is not open, determine which side excavated earth will be thrown to. Then string the pipe on the opposite side, leaving room for the excavating equipment.

FIGURE 12-7 Pipe strung out along a work site
Courtesy of CertainTeed Corporation

- Place pipe where it will be protected from traffic and heavy equipment.
- Place the bells in the direction of installation progress.
- Place and secure each pipe so there is no chance of it rolling into the trench.
- If there is a danger of vandalism or other damage, string out only enough pipe for one day's laying.
- If pipe must be strung in advance in residential areas, there is a particular danger of children playing around the pipe and injuring themselves. The pipe might also roll onto the street. In these cases, take special care to provide secure blocking.
- Place pipe where dirt will not get into the ends. If this is not possible, cover the pipe ends to prevent contamination.

EXCAVATION

This section discusses the many issues that must be taken into consideration before and during the excavation (trench-digging) process.

Preparations for Excavation

As discussed in chapter 1, the project engineer should prepare detailed plans of the project in advance. These plans should include details of the alignment, grade, and depth of mains; the locations of valves, hydrants, and fittings; and details of all known obstructions. Usually, the design engineer submits the plans to the state drinking water control agency for approval before work is started. Additional approval may also be needed from the county, city, state highway department, railroad companies, or other authorities. If it will be necessary to enter onto private property to do the construction work, easements or other property access and usage rights must be obtained from the property owner in advance.

Before any digging is done, all other utilities must be notified to locate and mark their underground pipes and cables. Some utilities may take several days to do this if they are particularly busy. They should be provided with ample notice before construction is to begin. Most states now have an underground excavation alert system in which a single phone call will notify all utilities about the water company's need to know pipe and cable locations.

For the purposes of traffic control and community outreach, it is best to inform the public. It is a good idea to mark the work area in advance with signs to inform the public of what is going to be done. It is also good public relations to write a brief letter to all adjoining property owners to let them know what will be done, why it is being done, when work is expected to begin, and when it will be completed. An apology can also be made for any inconvenience they will suffer as a result of the construction work. The excavation site should be properly marked with barricades, flashing lights, warning or detour signs, and a flagger if necessary to protect the work crew and the public. Work-area protection is discussed in chapter 13. In some locations, erosion and sedimentation control may also be required by regulatory agencies.

The selection and use of appropriate excavation equipment are important considerations. Machines that are too large will probably cause unnecessary damage. Machines that are too small will move the job too slowly, which not only is uneconomical but may also cause problems such as subsequent settlement due to frozen backfill, side-slope instability, and prolonged public inconvenience.

Most trenching for water main installation is done by hydraulic backhoes because they are easy to operate and provide excellent control. The best bucket width for the job should be determined and provided. The width of excavation should be minimized, yet it must be wide enough to maintain side support and allow workers to work properly in the trench. Most buckets can be changed very quickly, and an appropriate size can usually be rented if necessary. Trenching machines may be used for large installations requiring high productivity. Specialized equipment may be required in rocky terrain.

Asphalt roads and driveways should be scored in advance with an air hammer and a flat spade cutter to give a relatively smooth cut. Concrete can also be scored with an air hammer before it is broken out, but this usually leaves a rather jagged edge for patching. A smooth cut can be made with a diamond-edged power saw, which minimizes the amount of pavement replacement. It is usually most economical to employ a firm specializing in concrete cutting to do this work.

Large pieces of concrete and asphalt debris should generally not be used for backfill. They must be hauled away for disposal elsewhere before the excavation starts. The debris from asphalt removed with a rotomill will be small enough that it can be replaced in the upper part of the trench backfill. If traffic cannot be diverted around the area where the hard surface was removed, the hole can be filled with gravel or crushed stone or covered with steel plates to make a temporary driving surface.

Trenching

Several factors control the depth and width of trench excavation. Some of the principal considerations are

- ground frost conditions,
- groundwater conditions,
- traffic load that will be over the pipe,
- soil type,
- size of pipe to be installed,
- economics,
- surface restoration requirements, and
- depth of other utility lines that must be crossed or paralleled by the water main.

The most expensive part of pipe installation is the excavation. The trench should therefore be as shallow and as narrow as safety and soil conditions permit.

Trench depth

In colder climates, the maximum depth of frost penetration that can be expected governs the depth at which mains are buried. During an extremely severe winter, frost may penetrate to twice the average winter frost depth. A section of water main can generally be surrounded by frost for a short period of time without freezing as long as water is moving in the pipe. The greatest danger of installing mains too shallowly is that frost heaving (expansion of the frozen soil) will later increase the frequency of broken water mains. When frost penetrates to the depth of the mains, it is also likely that water services connected to the main will freeze.

Consideration must also be given to the fact that frost usually penetrates much deeper under pavements and driveways because they are kept free of snow. Surfaces that have continual snow cover often have less frost penetration, though this is not always the case. Most areas occasionally have a winter with little snow cover during extremely cold periods. Frost can penetrate quite deeply within a few days.

If adequate depth is not possible, a main may be insulated. Closed-cell Styrofoam insulation board, 2 in. (50 mm) thick and 2 to 4 ft (0.6 to 1.2 m) wide, works well. The insulation should be placed on stable fill, a few inches above the pipe, for as long a distance as necessary. Bead-board-type sheets should not be used. They are less expensive, but they have low insulating value and will absorb water, thereby further reducing their effectiveness.

In warmer climates where there is no frost, pipe must still have a minimum amount of cover to protect it from damage. This cover will also protect the pipe from highway wheel loads and other impact loads at the surface. The minimum required cover is usually 2.5 ft (0.8 m) for mains and 18 in. (0.5 m) for services.

When deciding on trench depth, the project engineer should also consider whether there is any possibility of future changes in ground surface elevation. For instance, if there is any chance of changes in grade due to future road construction or erosion of soil over the installed main, it is far less expensive to bury a main a little deeper initially than to have to re-lay it in the future.

Trench width

The width of trench necessary for pipe installation is generally governed by the following considerations:

- Minimum width to allow for proper joint assembly and compaction of soil around the pipe
- Safety considerations
- Economics
- The need to minimize the external loading on the pipe

Trench width below the top of the pipe should generally be no more than 1 to 2 ft (0.3 to 0.6 m) greater than the outside diameter of the pipe. This width provides workers with enough room to make up the joints and tamp the backfill under and around the pipe.

The pressure from the tire load of vehicles passing over a trench spreads out through the soil. Below a certain depth (a few feet below the surface) the loading is generally not a danger to buried pipe. Maintaining trench width as narrow as possible in paved areas will reduce the possibility of pipe damage due to vehicle loads.

Wide trenches should be avoided for small-diameter pipe if at all possible, particularly in hard clay soils. Tables 12-1 and 12-2 list recommended trench widths for smaller diameter pipe. If a wider trench width is necessary or if exceptionally heavy loads may be imposed over the pipeline, a design engineer or pipe manufacturer should be consulted for recommendations.

When pipe must be laid on a curve, the maximum permissible deflection of the joints limits the degree of curvature. Trench widths for curved pipelines must, of course, be wider than usual.

Trenching operations

As the trenching process proceeds, excavated soil should be piled on one side of the trench, preferably between the trench and the traffic. It must be placed far enough away from the trench so that it will not fall back into the excavation. In addition, it must be placed far enough away that it will not significantly increase the weight on the trench wall. Soil that is too close will increase the likelihood of cave-in. There must also be enough room for workers to walk alongside the trench.

The bottom of the trench must be dug to the specified depth while maintaining the specified grade (the elevation of the bottom of the pipe as specified in the plans). Both depth and grade should be double-checked frequently. Installers usually do this by lining a "story pole"

TABLE 12-1 Trench widths for ductile-iron mains

Nominal Pipe Size		Recommended Trench Widths	
in.	*(mm)*	*in.*	*(m)*
3	(80)	27	(0.69)
4	(100)	28	(0.70)
6	(150)	30	(0.76)
8	(200)	32	(0.81)
10	(250)	34	(0.86)
12	(300)	36	(0.92)
14	(350)	38	(0.96)
16	(400)	40	(1.0)
18	(450)	42	(1.07)
20	(500)	44	(1.12)
24	(600)	48	(1.22)
30	(750)	54	(1.37)
36	(900)	60	(1.52)
42	(1,050)	66	(1.68)
48	(1,200)	72	(1.83)
54	(1,350)	78	(1.98)
60	(1,500)	84	(2.13)

with a string line stretched out along the side of the ditch. This line is set an even number of feet above the grade and provides a reference line for installers. The bottom of the trench should form a continuous, even support for the pipe. A proper trench bottom will usually require some handwork and possibly some special fill material to provide proper bedding.

Excavation of the trench should not extend too far ahead of pipe laying. Keeping the excavation just ahead of the pipe installation will minimize the possibility of cave-ins or trench flooding during rainy weather. An open trench presents a danger not only to the workers but also to traffic, pedestrians, and children. The dangers of an open trench are

TABLE 12-2 Trench widths for PVC pipe

Pipe Diameter		Trench Width			
		Minimum		Maximum	
in.	*(mm)*	*in.*	*(m)*	*in.*	*(m)*
4	(100)	18	(0.46)	29	(0.74)
6	(150)	18	(0.46)	31	(0.79)
8	(150)	21	(0.53)	33	(0.84)
10	(250) and greater	24	(0.31) greater than outside diameter of pipe	36	(0.61) greater than outside diameter of pipe

often greater after working hours than during construction hours. These dangers can be minimized if open sections of trench are kept as short as possible and if written warnings, proper barricades, signals, and flaggers are used.

Local regulations often require that the trench be filled or protected in a specific way overnight. Long sections of open trench may also cause disruptions and inconvenience to local residents, certain municipal services, and emergency vehicles.

Special Excavation Problems

Some possible problems encountered during excavation include rocky ground, poor soil, and groundwater.

Rock excavation

The term *rock* generally applies to solid rock, ledge rock such as hardpan or shale, and loose boulders more than 8 in. (200 mm) in diameter. In any type of rock formation, the rock must be excavated to a level 6 to 9 in. (150 to 230 mm) below the grade line of the pipe bottom. Proper bedding material must then be added for pipe support. Excavated rock should be hauled away and not used for backfill.

Some rock may require blasting. If blasting is necessary, it is best to employ a professional firm that is experienced and specially insured to do this type of work. A detailed record should be kept of dates when blasting is done, the condition of surrounding property prior to blasting, and details of any damages or injuries incurred as a result of the blasting.

Poor soil

Where poor soil conditions exist (e.g., coal mine debris, cinders, sulfide clays, mine tail-ings, factory waste, or garbage), the soil should be excavated well below the grade line, hauled away, and properly disposed of. The excavation should then be filled with more suitable material for a foundation under the pipe bedding.

Groundwater

Groundwater will enter the trench when the trench bottom is below the water table. Pipe should not be laid or joined in groundwater. In many locations, the groundwater level fluc-tuates during the year, so the elevation at which pipe is to be installed may be above the groundwater table only at certain times of the year. If this is the case, the pipe installation should be timed, if possible, for the driest time of the year.

When pipe must be installed below the groundwater table, disposal of the water greatly increases costs. In addition, working in saturated ground can be hazardous because of the danger of trench cave-in. To dewater the ground in advance of excavation, a system of well points can be placed at intervals along the trench, as shown in Figure 12-8. The points have a screen at the bottom and are installed below the trench bottom on one side (and sometimes both sides) of the trench line before excavation. The points are then connected to a pipe manifold and pumping system. It is important to check with local reg-ulatory agencies to ensure proper disposal of the pumped water.

Avoiding Trench-Wall Failure

The following paragraphs discuss the reasons trench walls fail and ways of minimizing the risks.

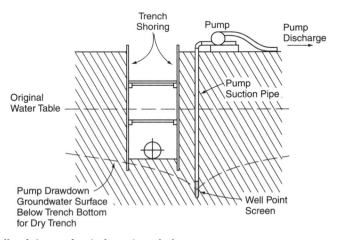

FIGURE 12-8 Well point pumping to keep trench dry

Types of soil

Variations in soil types, water content, and slope stability require modifications in trenching methods. When the digging takes place in firm soil, there is usually a tendency to make trench walls too steep and trenches too narrow in order to minimize the amount of excavation. When the digging occurs in wet silts or free-running soils, workers may dig too deep instead of taking steps to restrain the soil in the excavation.

Soils are generally characterized as clays, tills, sands, and silts. Firm clay and tills with low moisture content can usually be excavated easily and safely, but they require careful control during backfill. Dry silt, which is fairly uncommon, behaves in the same manner. Operating in dry sand requires special care during excavation because sand can slip or run easily. Wet silts, which are common, often require special treatment because of their unpredictability and their potential for caving in.

Causes of trench failure

The possibility of trench-wall failure and cave-in is undoubtedly the greatest danger for workers on a main-installation job. The primary causes of failure include

- the hydrostatic pore pressure,
- external loads caused by construction equipment operating near the edge of the trench,
- the weight (load) of excavated soil that is piled too close to the trench,
- trench walls that are too steep for the type of soil being excavated, and
- cleavage planes (fissures or cuts) in the soil caused by previous excavations.

Failures occur more often in winter and early spring when ground moisture is higher. These times should be considered particularly dangerous for excavation. Failures usually give little warning and occur almost instantaneously.

Danger signs

Some of the danger signs workers should look for are

- tension cracks in the ground surface parallel to the trench, often found a distance of one half to three quarters of the trench depth away from the edge;
- material crumbling off the walls;
- settling or slumping of the ground surrounding the trench; and
- sudden changes in soil color, indicating that a previous excavation was made in the area.

Trench-wall failures can be prevented if problem areas and conditions are recognized before they occur.

Methods of preventing cave-in

Four basic means of cave-in prevention and protection are sloping, shielding, shoring, and sheeting.

Sloping involves excavating the walls of the trench at an angle. This approach means that the downward forces on the soil are never allowed to exceed the soil's cohesive strength. For any section of an excavation, there will be a certain angle, called the angle of repose, for which the surrounding earth will not slide or cave back into the trench. The angle varies with the type of soil, the amount of moisture the soil contains, and the surrounding conditions, especially vibration from machinery. Figure 12-9 shows the approximate angles of repose for different common soil conditions. The angle ratios shown in the figure indicate the horizontal distance compared with the vertical. For instance, an excavation at a 2:1 ratio indicates that the cut is 2 ft (0.6 m) out from the edge of the ditch for each foot of excavation depth. Figure 12-10 shows an excavation with sloped trench walls.

Shielding involves the use of a steel box—open at the top, bottom, and ends—that is placed into the ditch so workers can work inside it, as illustrated in Figure 12-11. As the work progresses, the protective box is pushed or towed along the trench to provide a constant shield against any caving of the trench walls. This box is also called a trench shield, portable trench box, sand box, or drag shield. The shield is constructed of steel plates and bracing, welded or bolted together. It is important that the shield extend above ground level or that the trench walls above the top of the shield be properly sloped.

Shielding does not prevent a cave-in. The shield cannot fit tightly enough in the trench to hold up the trench walls—otherwise, it would be impossible to move. However, if a cave-in does occur, workers within the shield are protected.

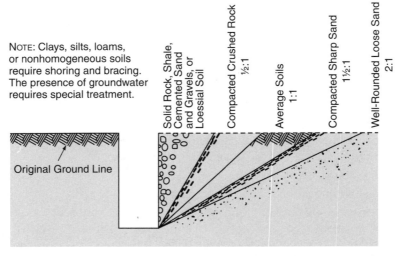

FIGURE 12-9 Approximate angles of repose for different soil types being excavated

A major disadvantage of a shield is that workers have a tendency to leave its protection in order to check completed work, to help adjust pipe placement, or just to get out of the way of the job in progress. The shield offers protection in only a limited area. As soon as work in an area is completed, the trench should be progressively excavated, the shield moved forward, and the trench backfilled.

Shoring, if properly installed, will actually prevent the caving in of trench walls. It is basically a framework support system of wood, metal, or both that maintains pressure against both trench walls. It is important that the type and condition of the soil be checked before excavation begins. This allows the correct shoring to be selected. The type and condition of the soil also need to be checked for changes as construction progresses. Soil conditions can vary greatly within a few feet, and conditions may also be changed by weather or vibration.

FIGURE 12-10 Pipe excavation with sloped trench walls
Photograph furnished by American Concrete Pressure Pipe Association

FIGURE 12-11 Pipe being installed using a shield
Photograph furnished by American Concrete Pressure Pipe Association

A shoring assembly has three main parts, as illustrated in Figure 12-12.

1. Uprights are the vertically placed boards that are in direct contact with the faces of the trench. The spacing required between them will vary depending on soil stability.
2. Stringers are the horizontal members of the shoring system to which the braces are attached. The stringers are also known as whalers, because they are similar to the reinforcements in the hulls of wooden ships.
3. Trench braces are the horizontal members of the system that run across the excavation to keep the uprights separated. They are usually either timbers or adjustable trench jacks.

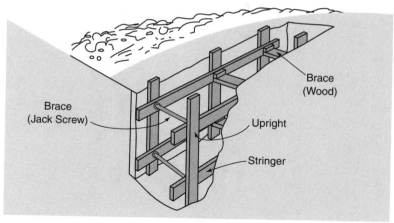

FIGURE 12-12 Principal parts of a shoring assembly

To prevent movement and failure of the shoring, there should not be any space between the shoring and the sides of the excavation. After the shoring has been installed, all open spaces should be filled in and the material compacted. A basic principle of shoring operations is that shoring and bracing should always be installed from the top down, then removed from the bottom up after the operation has been completed. This will give maximum protection to workers constructing or dismantling the shoring. A shoring installation is shown in Figure 12-13.

When a pipe installation within a shored excavation is finished, it may be necessary to leave the shoring in place while the trench is backfilled and compacted. The recommended procedure for removing shoring is to raise the shoring a few feet, then place and compact a layer of backfill, then raise the shoring a few feet more and place more backfill. In some installations, uprights that are above the top of the pipe can be left in place until all or most of the backfill has been placed. Then they can be removed with power equipment.

When no compacting of backfill is needed and the shoring is to be removed, ropes should be used to pull jacks, braces, and other shoring parts out of the trench. This way, workers do not have to enter the trench after the protection has been removed.

Sheeting is the process of installing tightly spaced upright planks against each other to form a solid barrier against the faces of the excavation, as illustrated in Figure 12-14. Under normal soil conditions, the sheeting can be installed as the excavation progresses. When soil conditions are very poor or for very deep excavations, steel sheet piling may have to be driven into the ground before excavation begins. Sheeting is generally removed as backfilling progresses, much the same as with shoring.

Local regulations and the Occupational Safety and Health Administration regulations require shoring or shielding in certain excavation situations. Regulatory requirements must be satisfied for all pipeline installation procedures.

FIGURE 12-13 Typical shoring installation

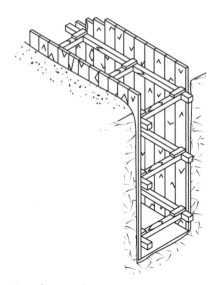

FIGURE 12-14 Tight sheeting in soft ground

Avoiding Other Utilities

To avoid the pipes and cables of other utilities, a water utility must either adjust the line and grade of a new water main or arrange for the other utility to relocate its lines. The grade of a sewer line is usually fixed, so the water main grade must be adjusted. If gas, electric, telephone, or existing water services are in a location where a new water main unavoidably must go, those services can usually be relocated. However, there will be delays in the main installation unless the conflicts can be anticipated and the adjustment made in advance. The other utility may charge a fee for making the changes. A conflict in grade with a pipe that cannot be relocated can also be avoided if several bends are installed in the water main (Figure 12-15).

Working around other utilities requires careful work with machines and by hand. The pipe must not touch or rest on other pipes or be used to support another structure. Accidentally cutting another utility line can be costly and dangerous. If gas from a punctured gas main is ignited, a serious fire can result. A buried electric cable often looks much like a tree root, but a worker in a damp ditch can be electrocuted by breaking the insulation with a shovel. Cutting a fiber-optic communication cable can require repairs costing thousands of dollars. In addition, each time another utility is damaged, it is often necessary to stop the water main construction job until the repair is completed.

Potentially serious problems can result when sanitary sewers and water mains are buried close together. Sewer lines may leak, and a water main could be surrounded by sewage-contaminated soil. If there is also a leak in the water main at the same location, it is possible for sewage to be drawn into the main if a main break or fire creates a vacuum in the main.

In general, potable water pipes and sanitary sewers should not be laid in the same trench. Wherever a water main crosses a sewer line, the water main should be at least 18 in. (0.45 m) above or below the sewer line. For 10 ft (3 m) on either side of the crossing, the sewer line should be made of either (1) ductile iron with mechanical or push-on joints or (2) PVC with mechanical-joint couplings. If water and sewer lines are parallel to each other, the distance between the two should be at least 10 ft (3 m). Refer to local or regulatory agency rules for minimum separation distances and specific requirements such as pipe materials, joints, and connections.

Bedding

The bottom of the trench excavated for water main installation must be properly leveled and compacted so that the barrel of the pipe will have continuous, firm support along its full length. A leveling board should be used to ensure that there are no voids or high spots

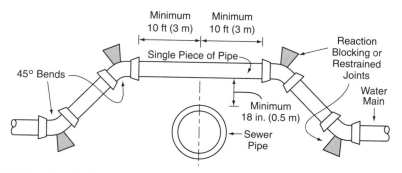

FIGURE 12-15 Bends installed in a water main to avoid a sewer pipe

and that the grade is correct. Any high spots should be shaved off, and voids should be filled with well-tamped soil. The practice of laying pipe on blocks or earth pads to allow room and position for joining pipe is not recommended.

There are some situations in which special pipe bedding may be specified by the design engineer. This usually occurs where the local soil contains many large rocks or is soft or unstable. If special bedding material is required, it should be a clean, well-graded, granular material up to 1 in. (25 mm) in size. It should contain no lumps or frozen ground and have no more than 12 percent clay or silt that can be sensitive to water. The bedding material should be spread over the trench bottom to the full width of the trench.

When natural trench bedding is unable to structurally support the pipe, such as in very soft soils, it may be necessary to dig an extra 12 to 24 in. (300 to 600 mm) deeper. Then coarse granular material in well-mixed sizes up to 3 in. (75 mm) can be used as backfill on top of geotextile fabric. This material should be compacted, and more added if necessary, until the trench bottom has been brought up to the proper grade. This approach is especially important in areas where there is an upward flow of groundwater or in muddy material. The trench bottom must be stabilized before the pipe can be laid. In extreme cases, pilings or timber foundations may be required.

"Haunching" (compacting backfill beneath the pipe curvature) is necessary to increase the load-bearing capacity of the pipe (Figure 12-16). If the pipe is supported only over a

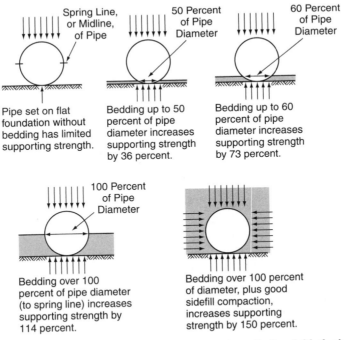

FIGURE 12-16 Effect of bedding on a pipe's load-bearing strength (for rigid pipe)

narrow width, as with a round pipe on a flat-bottom trench, there could be a very intense load at the bottom of the pipe. Failure is then possible. Distributing the load over a wider area will reduce the load intensity beneath the pipe. Proper haunching tamped up to the pipe spring line (middle of the pipe) greatly increases the load-bearing strength of the pipe.

As illustrated in Figure 12-17, it is also important that there is not undue weight placed on the pipe bell. Normal practice is to excavate a "bell hole" at the proper place so that the bell is completely free for assembly of the joint.

LAYING PIPE

This section gives guidelines for laying pipe once the length of trench is ready. It discusses inspection and placement, jointing, connecting to existing mains, tunneling, thrust restraints, and air vents.

Inspection and Placement

After the trench bottom has been prepared, the pipe may be set in place. The proper procedure varies somewhat with the type of pipe, but the following general directions for laying pipe apply to all types.

Inspection

Before the pipe is lowered into the trench, it should be inspected for damage. Any unsatisfactory sections should be rejected.

Cleaning

The inside of each pipe length should also be inspected for dirt, oil, grease, animals, and other foreign matter that must be removed before installation. If the pipe has been tapped before being placed in the trench, the holes or corporation stops in the pipe should be covered to keep out dirt during pipe placement.

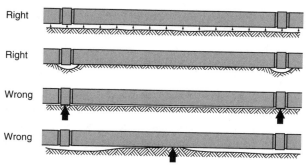

FIGURE 12-17 Good and bad pipe bedding
Courtesy of J-M Manufacturing Co., Inc.

If mud and surface water have been permitted to stand or flow through strung-out pipe, the inside should be swabbed with a strong hypochlorite solution. This will save time and expenses later when the pipe is disinfected. All gaskets should be kept clean and dry until they are ready for use.

Placement

The pipe should be lowered into the trench by hand or mechanical equipment. It should never be rolled into the trench from the top. Smaller diameter pipe may be lowered into the trench by two people using ropes, similar to the method illustrated in Figure 12-4 for unloading pipe from a truck. Larger pipe sizes must be handled with power equipment.

When pipe is lowered by machinery, it is usually supported by a sling in the middle of the pipe. The sling must be removed once the pipe is down, which usually requires some hand excavation to free the sling. The space created under the pipe should be backfilled and tamped. The use of a lifting clamp greatly facilitates handling pipe (Figure 12-18).

Pipe that is joined by couplings may be laid in either direction. Bell-end pipe is normally laid with the bells facing the direction in which the work progresses; however, this approach is not mandatory. When a main is being laid on a slope steeper than 6 percent, contractors often find it is easier to lay the pipe with the bells facing uphill.

Installers should make a conscious effort to keep the inside of the pipe clean during laying. When laying is not in progress, the open ends of installed pipe should be plugged to prevent animals, dirt, and trench water from entering. A piece of plywood placed over the end of the pipe at the end of the day will not always be enough to keep out animals, dirty water, or children who wish to throw rocks into the pipe. The most satisfactory method of plugging is with a standard pipe plug made for the type of pipe joint being used.

Jointing

Jointing pipe is an important, sensitive part of pipe installation. It must be done correctly to minimize leakage and ensure long and satisfactory service. The method for making up a joint depends on the type of pipe material and the type of joint. In all cases, sand, gravel, dust, tar, and other foreign material should be carefully wiped from the gasket recesses in the bells. Otherwise the joint may leak. The spigots must be smooth and free of rough edges.

Bell holes must be dug by hand where pipes will be joined. This will allow room for the joint to be installed while the remainder of the pipe rests on the bed. Bells or couplings should not be allowed to support the pipe. The size of the bell hole will vary with the type of pipe and joint used. Bell holes that are too large result in undue stress on the pipe.

Published instructions from the pipe manufacturer for making up joints should be followed. General directions for the more common joint types are given in this section.

FIGURE 12-18 Pipe being lifted with a pipe clamp
Courtesy of the Ductile Iron Pipe Research Association

Push-on joints

The four general steps in making up ductile-iron or PVC pipe push-on joints are illustrated in Figure 12-19. The most common cause of joint failure is not having the joint completely clean. For instance, a small leaf caught between the spigot end and gasket will usually leak enough that the installation will not pass the pressure test. But at the same time, the leak can be so small that it is difficult to locate. Finding a small leak and correcting it can be quite time-consuming. It is well worth the time to make up joints carefully.

Another potential problem is not pushing the spigot end the full distance into the bell. The spigot ends of all pipe sections are provided with a painted line to indicate when the pipe is fully "home." If a short piece of pipe is cut on the job, a similar mark should be made on the new spigot end with a piece of chalk or felt marker to ensure that the joint is properly made up.

The spigots of all push-on joint pipe are provided with a beveled end to allow for easy slipping into the joint gasket. If special pipe lengths are cut from full lengths in the field, the new spigots must be provided with a similar bevel. Although the bevel can be made with a rasp file or hand grinder, a special tool for beveling will save a lot of time if there are many cut lengths used on a job.

If a push-on joint is to be deflected, it is best to first push it home in line with the previous pipe before deflecting it. If installers try to push the pipe home on an angle, the pipe may resist going in or may not make a good joint. Table 12-3 lists the maximum deflections that can be obtained for full lengths of standard ductile-iron push-joint pipe.

1. Thoroughly clean the groove and the bell socket of the pipe or fitting; also clean the plain end of the mating pipe. Using a gasket of the proper design for the joint to be assembled, make a small loop in the gasket and insert it in the socket, making sure the gasket faces the correct direction and that it is properly seated. NOTE: In cold weather, it is necessary to warm the gasket to facilitate insertion.

2. Apply lubricant to the gasket and plain end of the pipe in accordance with the pipe manufacturer's recommendations. Lubricant is furnished in sterile containers, and every effort should be made to protect against contamination of the container's contents.

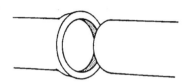

3. Be sure that the plain end is beveled; square or sharp edges may damage or dislodge the gasket and cause a leak. When pipe is cut in the field, bevel the plain end with a heavy file or grinder to remove all sharp edges. Push the plain end into the bell of the pipe. Keep the joint straight while pushing. Make deflection after the joint is assembled.

4. Small pipe can be pushed into the bell socket with a long bar. Large pipe requires additional power, such as a jack, level puller, or backhoe. The supplier may provide a jack or level puller on a rental basis. A timber header should be used between the pipe and jack or backhoe bucket to avoid damage to the pipe.

FIGURE 12-19 Push-on joint assembly

Small-diameter pipe may be pushed home by hand or using a pry bar, as illustrated in Figure 12-20. A block of wood should always be used between the bar and the pipe bell to avoid damaging the bell. Larger joints should be pulled together using a come-along or chain hoists, as illustrated in Figure 12-21.

Mechanical joints

The assembly of a mechanical joint (MJ) is illustrated in Figure 12-22. Although these joints are more expensive and take longer to assemble, they make a very positive seal and allow some deflection. Table 12-4 indicates the proper torque that should be used in tightening MJ bolts of various sizes.

TABLE 12-3 Maximum joint deflection,* full-length pipe—push-on type joint pipe

Nominal Pipe Size		Deflection Angle θ	Maximum Offset—S† in. (m)				Approx. Radius of Curve— R† Produced by Succession of Joints ft (m)			
in.	(mm)	deg.	$L^\dagger = 18$ ft (5.5 m)		$L^\dagger = 20$ ft (6 m)		$L^\dagger = 18$ ft (5.5 m)		$L^\dagger = 20$ ft (5 m)	
3	(76)	5*	19	(0.48)	21	(0.53)	205	(62)	230	(70)
4	(102)	5*	19	(0.48)	21	(0.53)	205	(62)	230	(70)
6	(152)	5*	19	(0.48)	21	(0.53)	205	(62)	230	(70)
8	(203)	5*	19	(0.48)	21	(0.53)	205	(62)	230	(70)
10	(254)	5*	19	(0.48)	21	(0.53)	205	(62)	230	(70)
12	(305)	5*	19	(0.48)	21	(0.53)	205	(62)	230	(70)
14	(356)	3*	11	(0.28)	12	(0.30)	340	(104)	380	(116)
16	(406)	3*	11	(0.28)	12	(0.30)	340	(104)	380	(116)
18	(457)	3*	11	(0.28)	12	(0.30)	340	(104)	380	(116)
20	(508)	3*	11	(0.28)	12	(0.30)	340	(104)	380	(116)
24	(610)	3*	11	(0.28)	12	(0.30)	340	(104)	380	(116)
30	(762)	3*	11	(0.28)	12	(0.30)	340	(104)	380	(116)
36	(914)	3*	11	(0.28)	12	(0.30)	340	(104)	380	(116)
42	(1,067)	3*	11	(0.28)	12	(0.30)	340	(104)	380	(116)
48	(1,219)	3*			12	(0.30)			380	(116)
54	(1,400)	3*			12	(0.30)			380	(116)
60	(1,500)	3*			12	(0.30)			380	(116)
64	(1,600)	3*			12	(0.30)			380	(116)

* For 14-in. (356-mm) and larger push-on joints, maximum deflection angle may be larger than shown above. Consult the manufacturer.
† See figure below.

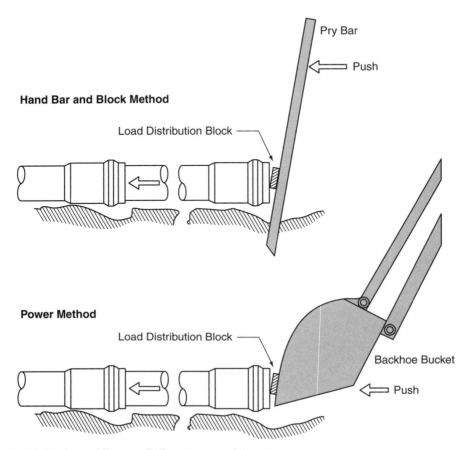

FIGURE 12-20 Assembling small-diameter push-joint pipe

Connecting to Existing Mains

Installers can connect new mains to existing mains either by inserting a new tee or by making a pressure tap.

Tee connections

To insert a tee fitting in an existing main, the main must be out of service for several hours. The valves on either side of the location should be tested in advance to be sure they will work, and customers with water services connected within the section must be notified. The type of pipe and outside diameter must also be determined in advance; the correct fittings and gasket sizes are sure to be available once the main is shut down. Two methods are commonly used for inserting a tee.

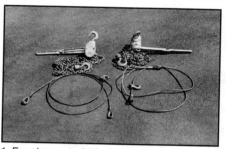

1. For pipe up to 24 in. (610 mm) in diameter, use two 1-ton (900-kg) chain hoists with 25 ft (7.6 m) of chain.

   For larger size pipe, use 2.5-ton (2,300-kg) hoists.

2. Wrap two chocker slings around the pipe just behind the bell, with one on each horizontal center line.

3. Double-wrap the chains of the two chain hoists around the pipe barrel about 6 ft (2 m) from the spigot end.

4. Attach the hook of each chain hoist into the eye of each bell choker sling.

5. Operate the chain hoists evenly to pull the spigot into the bell.

FIGURE 12-21 Using chain hoists to assemble push-on joints
Courtesy of U.S. Pipe and Foundry Company

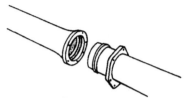

1. Clean the socket and the plain end. Lubrication and additional cleaning should be provided by brushing both the gasket and the plain end with soapy water or an approved pipe lubricant meeting the requirements of ANSI/AWWA C111/A21.11 just prior to slipping the gasket onto the plain end for joint assembly. Place the gland on the plain end with the lip extension toward the plain end, followed by the gasket with the narrow edge of the gasket toward the plain end.

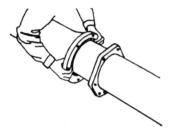

2. Insert the pipe into the socket and press the gasket firmly and evenly into the gasket recess. Keep the joint straight during assembly.

3. Push the gland toward the socket and center it around the pipe with the gland lip against the gasket. Insert bolts and hand-tighten nuts. Make deflection after joint assembly but before tightening bolts.

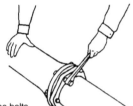

4. Tighten the bolts to the normal range of bolt torque while at all times maintaining approximately the same distance between the gland and the face of the flange at all points around the socket. This can be accomplished by partially tightening the bottom bolt first, then the top bolt, next the bolts at either side, finally the remaining bolts. Repeat the process until all bolts are within the approximate range of torque. In large sizes (30–48 in. [762–1,219 mm]), five or more repetitions may be required. The use of a torque-indicating wrench will facilitate this procedure.

FIGURE 12-22 Mechanical-joint assembly

TABLE 12-4 Mechanical-joint bolt torque

Joint Size		Bolt Size		Range of Torque	
in.	*(mm)*	*in.*	*(mm)*	*ft•lb*	*(N•m)*
3	(76)	⅝	(16)	45–60	(61–81)
4–24	(102–610)	¾	(19)	75–90	(102–122)
30–36	(762–914)	1	(25)	100–120	(136–163)
42–48	(1,067–1,219)	1 ¼	(32)	120–150	(163–203)

The first method is illustrated in Figure 12-23. After a section of main is cut out, a cutting-in sleeve is slipped all the way to one side over the existing pipe. The tee is then installed on the other side of the cut, and the sleeve is slid back so that the spigot end engages the joint of the tee. The length of existing main that must be cut out is quite specific. Instructions on how to compute this length are included with the inserting sleeve. The normal procedure for cutting out a piece of main is to make a cut or snap at the outside edges of the section to be removed. Then a third cut is made midway between them. When the pipe is hit with a sledge in the center, it will easily break out.

The second method, illustrated in Figure 12-24, uses two short pieces of new pipe (one on either side of the tee) and two rubber-joint sleeves. The tee and new pieces of pipe can be made up in advance. The section of old main is then cut out to match the same laying length. It is essential to verify that proper-size sleeve gaskets are available before the existing main is broken out. For example, if the old main is sand-cast pipe and the short pieces on either side of the tee are ductile-iron pipe, each sleeve will need two different gasket sizes.

Pressure taps

Today most connections of new mains to existing mains are made with a pressure tap, which offers the following advantages:

- Customers are not inconvenienced by having their water turned off.
- There is a much lower probability of contaminating the water system, particularly if working conditions in the excavation are poor.
- Fire protection in the area remains in service.

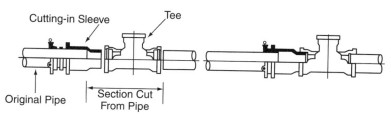

1. Cut a section from the existing pipe equal to the length of the tee plus the length specified by the sleeve manufacturer.

   Place the sleeve gland and gasket on the pipe. Slide the sleeve as far as it will go on the main.

   Install the tee with one bell mounted on the opposition side of the cut-out section.

2. Pull the sleeve spigot end toward the valve, and place it all the way into the bell of the tee.

   Assemble the gaskets and glands, and tighten all bolts.

FIGURE 12-23 Steps in using a cutting-in sleeve to install a tee

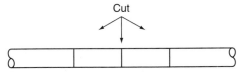

1. Cut pipe in three places and break pieces out.

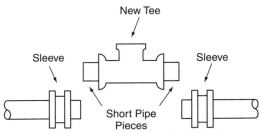

2. Assemble tee with two short pieces of pipe.
   Slide sleeves onto existing pipe.

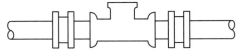

3. Place tee in line and slide sleeves back to cover joints.

FIGURE 12-24 Steps in using short pieces of new pipe and rubber-joint sleeves to install a tee

- There are likely to be fewer complaints of discolored water.
- There is no water loss.
- Based on the time required to close valves and notify customers before inserting a tee is taken into account, making a pressure tap is much faster.

Once the appropriate excavation has been prepared, a pressure tap is completed in the following steps:

1. The surface of the existing main is cleaned. The tapping sleeve is installed on the main, with the face of the flange vertical.
2. The special tapping valve is attached to the sleeve's projecting flange.
3. The tapping machine is installed on the valve, with the shell cutter attached. The machine is supported on temporary blocks (Figure 12-25).
4. With the tapping valve open, the cutter is advanced to cut a hole in the main.
5. The cutter is retracted to beyond the valve, and the valve is closed.
6. The tapping machine is removed, and the new main is connected (Figure 12-26).

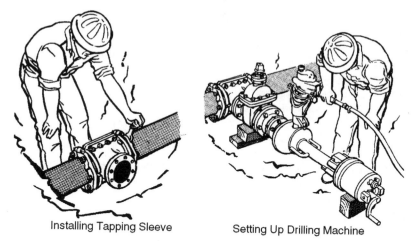

Installing Tapping Sleeve   Setting Up Drilling Machine

FIGURE 12-25 Preparing to tap a large connection
Courtesy of Mueller Company, Decatur, Ill.

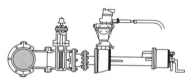

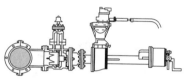

1. The tapping sleeve and valve are first attached to the main. Then the drilling machine, with a shell cutter fastened to its boring bar, is attached to the tapping sleeve and valve by an adapter. The assembly should be pressure-tested before the cut is made.

2. With the tapping valve open, the shell cutter and boring bar advance to cut the main.

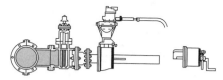

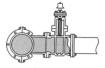

3. The boring bar is retracted and the tapping valve closed to control the water pressure.

4. With the machine removed, the lateral is connected and the tapping valve opened to pressurize the lateral and place it in service.

FIGURE 12-26 Using a tapping sleeve to install a large lateral main
Courtesy of Mueller Company, Decatur, Ill.

As shown in Figure 12-27, the shell cutter cuts a round "slug" or "coupon" from the existing main and retains it on the cutter. These coupons should be tagged with the date and location of the tap because they provide a good record of the condition of the interior of water mains in the system. Figure 12-28 shows a pressure tap being made on large-diameter concrete pressure pipe.

When workers are making direct pressure taps on PVC mains, they should strictly follow the pipe manufacturer's published recommendations.

Tunneling

When pipelines are installed under railroad tracks, major highways, and other obstructions, the following requirements are usually imposed on the installation:

- If the pipe should break or leak, the water must be carried to one side of the roadway. If water from a broken main were to come up in the middle of a railroad track or major highway, it could cause a serious accident.
- If the water main should require repair or replacement, such action must be possible without the need for excavating in the roadway.
- The pipe must be protected from excessive dead and impact loads, which might otherwise cause the pipe to fail.

FIGURE 12-27 Coupon cut from main
Courtesy of T.D. Williamson, Inc.

The most common method of installing a water main under these conditions is to install a casing pipe and then place the water main inside it. When a casing pipe is used for highway or railroad crossings, the project must meet applicable federal, state, and local regulations, as well as any requirements of the highway department or railroad company.

The casing pipe is usually made of steel or concrete pipe. For casing diameters up to 36 in. (920 mm), boring is done to create the casing hole, with a maximum length of about 175 ft (55 m). Jacking is used for diameters of 30 to 60 in. (760 to 1,500 mm) and for lengths of about 200 ft (60 m). Tunneling is done for pipes 48 in. (1,200 mm) and larger for longer lengths. The casing pipe diameter should be 2–8 in. (50–200 mm) larger than the outside diameter of the water main bells.

After the casing is in place, runners or skids are attached at each pipe joint with steel bands. The chocks must be of sufficient size that the bells will clear the casing. The water main is then pushed or pulled through the casing.

FIGURE 12-28 Tap being made on a large-diameter main
Photograph furnished by American Concrete Pressure Pipe Association

Thrust Restraints

Water under pressure and water in motion can exert tremendous forces inside a pipeline. The resulting thrust pushes against fittings, valves, and hydrants and can cause joints to leak or pull apart. All tees, bends, reducers, caps, plugs, valves, hydrants, and other fittings that stop flow or change the direction of flow should be restrained or blocked.

Thrust locations

Thrust almost always acts perpendicular to the inside surface it pushes against. As illustrated in Figure 12-29, the thrust acts horizontally outward, tending to push the fitting away from the pipeline. If uncontrolled, this thrust can cause movement in the fitting or pipeline that will result in leakage or complete separation of a joint. Although upward or downward bends are more unusual than horizontal bends in water pipelines, the same type of thrust should be anticipated and controlled to prevent the pipe joints from separating under a surge situation.

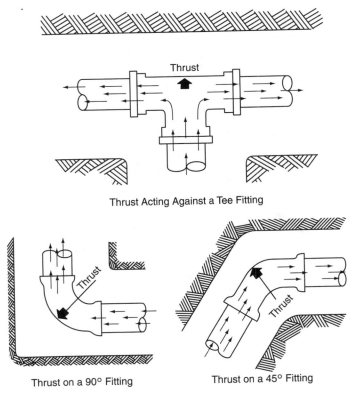

Thrust Acting Against a Tee Fitting

Thrust on a 90° Fitting

Thrust on a 45° Fitting

FIGURE 12-29 Thrust actions

Plastic pipe has a particularly smooth surface. It is especially prone to sliding out of a push-on joint if not firmly restrained. In addition, when polyethylene bags are placed around ductile-iron pipe for corrosion protection, the surrounding soil provides much less restraint for the pipe. In other words, if there is water hammer in the line, the pipe may slide within the plastic if it is not well restrained. Installers must be particularly careful to provide good blocking of joints in plastic pipelines or plastic-encased pipelines.

Thrust control

Thrust can be controlled by restraining the outward movement of a pipeline by four general methods: installation of thrust blocks, installation of thrust anchors, use of restraining fittings or joints, or use of batter piles. Thrust control is covered in additional detail in the hydraulics section of *Basic Science Concepts and Applications*, another book in this series.

Thrust blocks are masses of concrete that are cast in place between a fitting being restrained and the undisturbed soil at the side or bottom of the pipe trench. Figure 12-30 illustrates thrust blocks installed for various pipeline bend situations. It is important that the block be centered on the thrust force. The block should also partially cradle the fitting to distribute the force, but it should not cover the joint fittings.

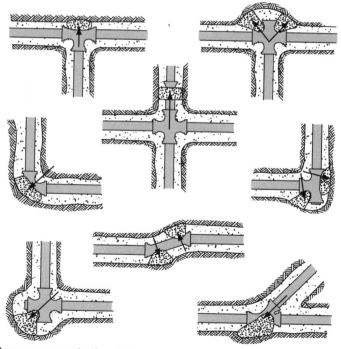

FIGURE 12-30 Common concrete thrust blocks
Courtesy of J-M Manufacturing Co., Inc.

It is also extremely important to cast the block against undisturbed soil. Disturbed soils cannot generally be expected to provide adequate support. For this reason, it is important not to over excavate a trench at bends in the pipeline.

Thrust anchors may be used when a thrust block cannot be used because there is an obstruction or there is no undisturbed soil to block against. They are also used on vertical bends. As illustrated in the typical designs illustrated in Figure 12-31, steel rods connected to relatively massive blocks of concrete restrain the piping bends.

Tie rods are frequently used to restrain mechanical-joint fittings that are located close together. As shown in Figure 12-32, threaded rods are usually used. Nuts on either side of each joint connection take the place of the MJ bolt that they replace. If nonalloyed carbon steel is used for tie rods, the rods should be properly coated or covered for protection against corrosion.*Restraining fittings*, which make use of clamps and anchor screws, are coming into more general use for restraining joints because they are so easy to use. They are also particularly useful in locations where other existing utilities or structures are so numerous that thrust blocks are precluded.

Air Vents

Whenever a water main is laid on uneven ground, there will be air trapped at each high point as the main is filled. If the "bubbles" of air are allowed to remain at these locations,

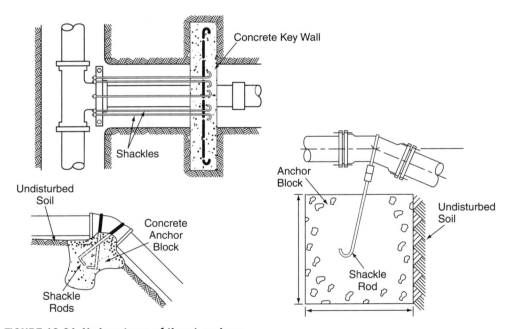

FIGURE 12-31 Various types of thrust anchors
Courtesy of J-M Manufacturing Co., Inc.

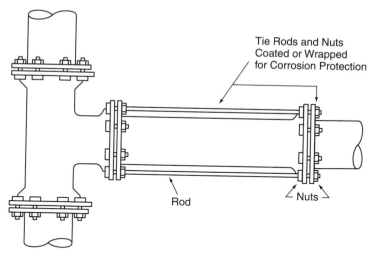

FIGURE 12-32 Tie rods used to secure piping where blocking cannot be used

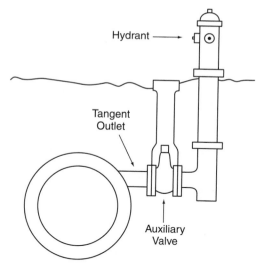

FIGURE 12-33 Fire hydrant installed on a tangent outlet located on a high point of a large-diameter water main

they will constrict the flow much the same as if there were a throttled valve in the line. For small-diameter mains, opening one or more fire hydrants will ordinarily produce enough velocity in the water to blow out the air, but with large mains, this is not possible. One method of providing for a release of air is to install fire hydrants at the high points. As illustrated in Figure 12-33, a special tangent tee or connection should be used so that all

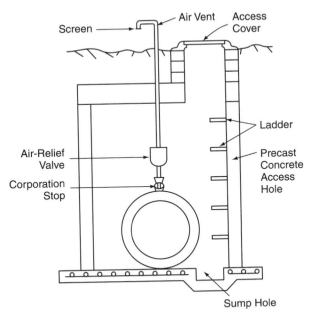

FIGURE 12-34 Installation of an air-relief valve at a high point on a large-diameter water main

air will be released from the top of the pipe. The only problem with this method is that there is some entrained air in most water that will gradually refill the air pockets. A regular schedule should be set up for manually operating the hydrants to release accumulated air. Another method of releasing air is to provide automatic air-relief valves at each high point. Figure 12-34 illustrates one design. Although the valves operate automatically, they must periodically be inspected to make sure they are operating properly.

SELECTED SUPPLEMENTARY READINGS

AWWA Standard for Disinfecting Water Mains. ANSI/AWWA C651. Denver, Colo.: American Water Works Association (latest edition).

AWWA Standard for Installation of Ductile-Iron Water Mains and Their Appurtenances. ANSI/AWWA C600. Denver, Colo.: American Water Works Association (latest edition).

AWWA Standard for Underground Installation of Polyvinyl Chloride (PVC) Pressure Pipe and Fittings for Water. ANSI/AWWA C605. Denver, Colo.: American Water Works Association (latest edition).

Manual M9, Concrete Pressure Pipe. Denver, Colo.: American Water Works Association (latest edition).

Manual M11, Steel Pipe—A Guide for Design and Installation. 2004. Denver, Colo.: American Water Works Association.

Manual M23, PVC Pipe—Design and Installation. 2002. Denver, Colo.: American Water Works Association.

Water Distribution Operator Training Handbook. 2005. Denver, Colo.: American Water Works Association.

CHAPTER 13

Backfilling, Main Testing, and Installation Safety

After a section of new water main has been installed, it must be carefully backfilled to protect and support the pipe and fittings. Finally, the main must be tested for leaks, flushed, disinfected, and tested for bacteriological quality before the water can be used by the public. Safety for the workers as well as the public is also a major concern for a construction project; it must be kept in mind from the time the project is planned until the final work is completed.

BACKFILLING

Backfill material is placed directly around and over the pipe to

- provide support for the pipe,
- provide lateral stability between the pipe and the trench walls, and
- carry and transfer surface loads to the side walls.

Placing Backfill

Backfill is usually placed using mechanical equipment to meet the requirements of the appropriate authority. Granular material or selected soil should be used for the first layer of backfill. If soil is used, it must be either carefully selected excavated soil or imported material. Native soil can be used if it does not contain large rocks, roots, organic material. Backfill material should contain enough moisture to permit thorough compaction. The backfill should have no large rocks, roots, construction debris, or frozen material in it.

Compacting

The pipe embedment and backfill placed in a trench are generally compacted by a process of tamping, vibrating, or saturating the soil with water, depending on the type of soil or material used.

Tamping

The first layer of backfill should be placed equally on both sides of the pipe, joints, valves, and fittings up to the center line of the pipe. Then it should be compacted. This process is sometimes referred to as haunching. A hand tamper, which is often used for this purpose, is illustrated in Figure 13-1. Pneumatic tampers are also used.

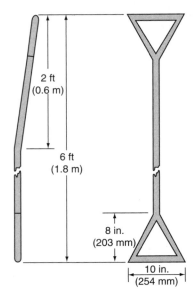

FIGURE 13-1 Hand tamper

Above the spring line (the pipe midline), backfill practices vary considerably depending on local conditions and regulatory requirements. In general, an initial backfill should be placed around the upper half of the pipe. This will prevent damage to the pipe and prevent the pipe from moving or becoming buoyant until the remaining backfill is placed. The compacted covering layer should be 6 to 12 in. (150 to 300 mm) for pipe smaller than 8 in. (200 mm) and 12 to 24 in. (300 to 610 mm) for larger size pipe.

If the final layer of backfill is to be mechanically compacted, workers should backfill the remainder of the trench by placing the material in layers and compacting it thoroughly. This backfill does not need to be selected, placed, or compacted quite as carefully as the material placed adjacent to the pipe—unless the specifications or regulatory agency say otherwise. However, the fill should be uniformly dense, and unfilled spaces should be avoided.

Saturating with water

Where water is available at a reasonable cost and the soil drains relatively freely, water settling can be used to compact the backfill. However, simply flooding the backfilled or partially backfilled trench will result in good compaction only when granular material with few fines is being used. If the native soil is relatively dense, only the upper backfill may be compacted by the initial saturation. The backfill may then continue to settle over a period of years.

A process known as jetting involves repeatedly pushing a pressurized water pipe vertically to near the bottom of the loose fill at intervals along the excavation, as shown in Figure 13-2. This approach provides good compaction if done thoroughly. The use of water will generally compact the backfill to within 5 percent of the maximum density.

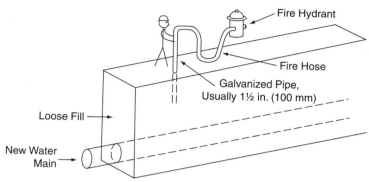

FIGURE 13-2 Settling a ditch with water: A worker plunges the pipe into the loose fill repeatedly to completely saturate the new fill

In areas where the surface doesn't need to be restored immediately to its original condition, it may be adequate to simply mound excess fill over the ditch and let the fill be settled by rain and gravity. It will then be necessary for someone to return periodically to provide additional fill material to compensate for the material that has settled.

Granular backfill

When trenches are located in areas that must be repaved—such as near roads, sidewalks, or driveways—excavated soil must be very well compacted to minimize future settlement. A common practice in many areas is to backfill these sites with imported clean granular material or processed material, instead of native soil. If cost is the primary consideration, the expense of purchasing the new backfill and disposing of the excess excavated material must be considered. Some engineers or local authorities specify that granular or processed material must be used under surfaces that are to be repaved. This allows the pavement to be restored to use more quickly. It also avoids any future pavement settlement resulting from inadequately compacted soil.

Shoring removal

When soil must be compacted in shored trenches, the shoring timbers and braces should be withdrawn in stages to match the layers of earth being placed. When shoring extends below the spring line of the pipe, workers may not be able to withdraw it without disturbing the pipe bedding. In these cases, the only option is to cut the shoring off near the top of the pipe.

Where poor ground conditions exist from the bottom of the trench up to the surface, it may be necessary to leave shoring in place for the entire depth to maintain the installation's stability during backfilling. In this case, the top of the shoring is usually cut off about 2 ft (0.6 mm) below the surface. Regardless of which compaction method is used, all voids caused by removing the shoring must be filled. In some locations, any shoring that will be left in place must be made of treated lumber.

Backfilling and Tamping Equipment

The goal in backfilling and tamping is to have a good balance in the sizes and types of equipment on the job. A small bulldozer or a loader works well in combination with a self-propelled vibratory roller for clay-like or silty soils. Vibratory compactors work better for granular materials. Trying to use the same machine both for backfilling and as a crane for lowering pipe is not usually a good idea if the job needs to move at a good pace.

The following are some types of mechanical compacting equipment and their appropriate applications:

- Irregular drum tampers are used for clays, tills, and silts.
- Hand-controlled plate tampers (Figure 13-3) are used for sand in shallow lifts.
- Boom-mounted plate tampers are used for clays or sands in deep, narrow trenches that prevent workers from going into the trench.

PRESSURE AND LEAK TESTING

After the trench has been partially backfilled, water mains should be tested for their ability to hold pressure and to determine if there is any leakage. Testing can be done between valves after individual pipe sections are completed or it can be done after the entire pipeline has been laid. In either case, these tests should be done before the trench is completely closed so that any leakage can be observed and easily repaired.

FIGURE 13-3 Mechanical plate tamper
Courtesy of The WACKER Group

Years ago, when only lead joints were used, joints were rarely perfect. It was assumed that there would be a little leakage from many of them. Accordingly, testing standards were written to specify just how much leakage should be acceptable. However, experience has shown that if a pipeline has all mechanical, push-on, or other rubber-gasket joints that have been properly made up, there should be very little joint leakage. If there is some leakage, it is most likely from another source.

Testing Procedure

Leakage is defined as the volume of water that must be added to the full pipeline to maintain a specified test pressure within a 5-psi (34-kPa) range. The water added to the pipeline will be equal to the amount of water that leaks out. This test process assumes, of course, that all air has been expelled from the line. The specified test pressure and the allowable leakage for a given length and type of pipe and joint are given in American Water Works Association (AWWA) standards and manuals.

Although pressure and leakage tests can be performed separately, the usual practice is to run one test combining both. The following procedure is used:

1. Allow at least 5 days for the concrete used for thrust blocks to cure, unless high-early-strength cement was used.
2. Install a pressure pump equipped with a makeup reservoir (a container of additional water to be pumped into the pipeline during the test), a pressure gauge, and a method for measuring the amount of water pumped.
3. Close all appropriate valves.
4. Slowly fill the test section with water while expelling air through valves, hydrants, and taps at all high points. If chlorine tablets are to be used for disinfection, take particular care to fill the pipe slowly or the tablets will be dislodged.
5. Start applying partial pressure with the positive-displacement pump. Before bringing the pressure to the full test value, bleed all air out of the mains by venting it through service connections and air-relief valves. Corporation stops may have to be installed at high points in the pipe to release all air properly.
6. Once the lines are full and all air has been bled, leave partial pressure on and allow the pipe to stand for at least 24 hours to stabilize.
7. For pressure testing, subject the test line to the hydrostatic pressure specified in the applicable AWWA standard. A pressure of either 1.5 times the operating pressure or 150 psi (1,030 kPa) for a period of 30 minutes is usually the minimum. Sometimes a pressure chart recorder is connected to the main to keep a record of pressure changes during the test period.
8. Examine the installed pipe and fittings for visible leaks or pipe movement. Any joints, valves, or fittings that show leakage should be checked, adjusted, or repaired as needed. Repeat the test after any adjustments or repairs.

9. After the test pressure has been maintained for at least 2 hours, conduct a leakage test by using the makeup reservoir and measuring the amount of water that has to be pumped into the line in order to maintain the specified test pressure.

10. Compare the amount of leakage to the suggested maximum allowable leakage given in the appropriate AWWA standards and manuals. A swift loss of pressure is likely due to a break in the line or a major valve being open. A slow loss of pressure may be due to a leaking valve or leaking joint.

Measuring Makeup Water

The method used to measure makeup water volume usually involves one of the following:

- A calibrated makeup reservoir (the preferable method)
- A calibrated positive-displacement pump
- A very accurate water meter (not normally recommended)

The test setup can be installed at the end of the pipe, at a service connection, or at a hydrant, as shown in Figure 13-4. Alternatively, a hose with a pressure gauge, pump, and sensitive water meter could be connected from an existing operable fire hydrant on the distribution system to the new pipeline being tested. In this case, a double-check valve should also be placed in the connection to prevent any backflow into the distribution system.

Failed Pressure Tests

If a newly installed water main fails the leakage test, it is necessary to find where the excessive leakage is taking place. Steps can be taken to determine if the leak is in a pipe or at a fitting. First, leave the line under normal pressure overnight and repeat the test the next day. If the leakage measured the next day is greater than before, the leak probably is in a pipe joint or a damaged pipe. If the leakage is the same, it is probably in a valve or service connection.

The following is a useful checklist of causes for a failed leak test:

- A hydrant valve may be held open by a piece of rag, wood, or some other foreign object.
- There may be some dirt or foreign material under a coupling or a gasket or a worker may have forgotten to tighten the bolts on a mechanical joint.
- Fittings, valves, and hydrants in the test section may not have been sufficiently restrained and have moved.
- Corporation stops may not have been tightly closed.
- There may be leakage through a valve at either end of the test section. This is particularly likely if there is an old valve at the connection of the new pipeline with the existing water system.

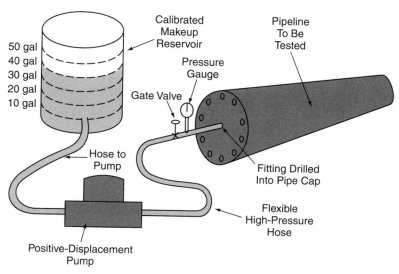

50 gal
40 gal
30 gal
20 gal
10 gal

Calibrated
Makeup
Reservoir

Pipeline
To Be
Tested

Pressure
Gauge

Gate Valve

Hose to
Pump

Fitting Drilled
Into Pipe Cap

Flexible
High-Pressure
Hose

Positive-Displacement
Pump

FIGURE 13-4 Equipment for leakage and pressure testing of a pipeline

- The packing on a valve may be leaking.
- The test pump may be leaking. This could be the check valve or the gate valve.
- The test may include too long a section of pipe.
- The saturation time may have been too short—it should be 24 hours.
- There may be a break in the pipe—either a crack or a blowout. A lateral crack may, on occasion, not leak at line pressure but will open up and leak at higher pressure.
- There may be some faulty accessory equipment—possibly a valve, fitting, hydrant, saddle, corporation stop, or relief valve.
- The test gauge may be faulty.
- The test pump suction line may be drawing air.
- Air may be entrained in the test section.

A leak that is large enough to cause a failed test, yet small enough that water does not immediately come to the surface, can be particularly frustrating. In this event, the only solution is first to isolate the section with the leak and make sure the cause is not a leaking valve. Then arrange to continuously subject the line to the highest pressure that can be safely applied, and wait for water to come to the surface. Leak detection equipment can also be used to locate a leak.

Any leaks in the line should be repaired. The line should be retested until the measured leakage is less than or equal to the allowable amount.

FLUSHING AND DISINFECTION

Any new or repaired water main must be thoroughly flushed, disinfected, and tested for bacteriological quality before it can be put into use for customers. Flushing is primarily necessary to remove any mud and debris that were left in the pipe from the installation.

Flushing

One or more fire hydrants should be used to perform the flushing. A blowoff connection, if one has been installed, can also be used. A velocity of at least 2.5 ft/sec (0.8 m/sec), and preferably 3.5 ft/sec (1.1 m/sec), should be obtained in the pipe. This velocity should be maintained long enough to allow two or three complete changes of water for proper flushing action. Table 13-1 provides information on the rate of water flow necessary to flush various pipe sizes adequately. If the pipeline is large or if there is insufficient water for flushing a new main, the pipe can be cleaned with polypigs.

Disinfection

Chlorine compounds are the most common chemicals used to disinfect large pipelines. Calcium hypochlorite and sodium hypochlorite solutions are generally used for smaller pipelines.

Application point

The chlorine solution is usually injected through a corporation stop at the point where the new main connects to the existing system. Water utility personnel must ensure that chlorine-dosed water does not flow back into the potable water supply. All high points on the main should be vented to make sure there are no air pockets that would prevent contact between the chlorinated water and portions of the pipe walls.

Chlorine dosage

The chlorination requirement should normally be in conformance with AWWA Standard C651, *Disinfecting Water Mains* (most recent edition), unless there are other local and state requirements. In general, the rate of application should result in a uniform free chlorine concentration of at least 25 mg/L at the end of the section being treated. Under certain conditions, higher chlorine dosages may be required. It should be kept in mind that too much chlorine for an excessive time can damage mortar linings, brass, and other fittings.

Calculating the amount of chlorine and water needed for proper disinfection involves determining the

* capacity of the pipeline,
* desired chlorine dosage,
* concentration of the chlorine solution,
* pumping rate of the chlorine-solution pump, and
* rate at which water is admitted to the pipeline.

TABLE 13-1 Flow rate and number of hydrant outlets required to flush pipelines (40-psi [280-kPa] residual pressure in water main)*

Pipe Diameter		Flow Required to Produce Velocity of Approx. 2.5 ft/sec (0.76 m/sec) in Main		Number of 2 ½-in. (65-mm) Hydrant Outlets*	Size of Tap, *in. (mm)*		
					1 (25)	1½ (40)	2 (50)
in.	*(mm)*	*gpm*	*(L/sec)*		Number of Taps on Pipe†		
4	(100)	100	(6)	1	1	—	—
6	(150)	200	(13)	1	—	1	—
8	(200)	400	(25)	1	—	2	1
10	(250)	600	(38)	1	—	3	2
12	(300)	900	(57)	2	—	—	2
16	(400)	1,600	(100)	2	—	—	4

* With a 40-psi (280-kPa) pressure in the main and the hydrant flowing to atmosphere, a 2½-in. (65-mm) hydrant outlet will discharge approximately 1,000 gpm (60 L/sec) and a 4½-in. (115-mm) hydrant nozzle will discharge approximately 2,500 gpm (160 L/sec).

† Number of taps on pipe based on no significant length of discharge piping. A 10-ft (3-m) length of galvanized iron (GI) piping will reduce flow by approximately one-third.

Tables 13-2 and 13-3 provide basic guides to the amount of hypochlorite required for proper disinfection. Additional information on dosage computations is included in *Basic Science Concepts and Applications*, another book in this series, and AWWA Standard C651, *Disinfecting Water Mains* (most recent edition).

Procedures

Three of the most commonly used methods of applying disinfectant are the continuous feed method, the slug method, and the tablet method.

In the continuous feed method, water from the distribution system is slowly admitted to the new pipe section while a concentrated chlorine solution is simultaneously forced into the main. Chlorine solution may be injected into the main with a solution-feed chlorinator or by a booster pump.

The concentration used is usually at least 50 mg/L available chlorine. The residual should be checked at regular intervals to ensure that the proper level is maintained. Chlorine application should continue until the entire main is filled and a chlorine residual of at least 25 mg/L can be measured in water being bled from the end of the line. The water should then remain in the pipe for a minimum of 24 hours. During this time, all valves and hydrants along the main must be operated to ensure that they are also properly disinfected.

TABLE 13-2 Ounces of calcium hypochlorite granules to be placed at beginning of main and at each 500-ft interval

Pipe Diameter[*]		Calcium Hypochlorite Granules	
in.	*(mm)*	*oz*	*(g)*
4	100	1.7	48
6	150	3.8	113
8	200	6.7	200
10	250	10.5	300
12	300	15.1	430
14 and larger	(350 and larger)		

* Where D is the inside pipe diameter in feet D = $d/12$.

TABLE 13-3 Number of 5-g calcium hypochlorite tablets required for chlorine dose of 25 mg/L[*]

Pipe Diameter		Number of 5-g Calcium Hypochlorite Tablets for Length of Pipe Section *ft (m)*				
in.	*(mm)*	13 (4) or less	18 (5.5)	20 (6)	30 (9)	40 (12)
4	(100)	1	1	1	1	1
6	(150)	1	1	1	2	2
8	(200)	1	2	2	3	4
10	(250)	2	3	3	4	5
12	(300)	3	4	4	6	7
16	(400)	4	6	7	10	13

* Based on 3.25 g available chlorine per tablet; any portion of tablet rounded to next higher number.

In the slug method, a long slug of water is fed into the main with a constant dose of chlorine to give it a chlorine concentration of at least 300 mg/L (make sure that the pH is within acceptable range). Water is then slowly bled from the end of the line at a rate that will cause the slug to remain in contact with each point on the pipe for at least 3 hours as it passes through the main. As the slug passes tees, crosses, and hydrants, the adjacent valves must be operated to ensure that any dead-end sections of pipe are disinfected. This method is used primarily for large- diameter mains, where continuous feed is impractical.

With the tablet method, calcium hypochlorite tablets are placed in each section of pipe, in hydrants, and in other appurtenances. These tablets are usually glued to the top of the pipe. The main is then slowly filled with water at a velocity of less than 1 ft/sec (0.3 m/sec) to prevent dislodging the tablets and washing them to the end of the main. The final solution should have a residual of at least 25 mg/L. It should remain in contact with the pipe for a minimum of 24 hours. It is also advisable to occasionally bleed a little water from the system to ensure that chlorinated water is distributed within the pipe.

The tablet method cannot be used if the main needs to be flushed before it will be disinfected. If the tablet method is used, workers must take extra care to keep pipe free of dirt, debris, and animals during installation.

Large-diameter pipes or very long pipelines are usually best disinfected with chlorine gas, but special equipment, procedures, and safety precautions are required. Water utility personnel should obtain advice from the manufacturer of the chlorination equipment to be used for the application.

Contact period

Chlorinated water should normally be left in a pipeline for 24 hours. If unfavorable or unsanitary conditions existed during pipe installation, the period may have to be extended to 48 or 72 hours. If shorter retention periods must be used, the chlorine concentration should be increased, but pH must be monitored. High pH reduces the disinfection effectiveness.

At the end of the contact period, chlorinated water should be flushed to an acceptable location (storm sewer, storage pond, or flood control channel, as allowed by local regulation) until the chlorine residual in water leaving the pipe approaches normal. Highly chlorinated water may kill grass, so it may be necessary to use a fire hose to duct water from a hydrant to the discharge point. Local and state regulatory agencies should be consulted on how to dispose of the highly chlorinated water. The discharge may have to be dechlorinated before being released to the environment.

Bacteriological Testing

After a new pipe has been flushed of chlorinated water and refilled with water from the system, bacteriological tests must be performed. These tests must meet the requirements of the applicable regulatory agencies. Samples must be analyzed by a certified laboratory, and results of the analysis must show that the samples tested negative (no coliform present) before customers are allowed to use the water.

If the results fail to meet minimum standards, the water should be tested again. If the water fails the second test, the entire disinfection procedure must be repeated, including sampling and bacterial testing.

FINAL INSPECTION

Before a new water main is put into service, as-built plans should be completed and used as the basis for the final inspection. The plans should then be recorded and filed for future use. Detailed records should be completed identifying the types and locations of valve boxes on the new line, as well as the locations of hydrants and other appurtenances.

All valves on the line should be operated and left in the fully open position. The number of turns needed to close and open each valve should be counted and recorded, along with the direction of opening. Each fire hydrant should also be flow-tested to determine its flow capacity and to verify that it is in good operating condition.

SITE RESTORATION

Good restoration requires common sense. It should take whatever form local conditions, owners, and regulatory agencies require. Since restoration is performed at the end of the project, there is often a tendency to give it a low priority. In some cases, this means that some restoration may never get done unless adjacent property owners make strenuous demands. However, operators must remember that the community judges the water utility in great part by what takes place after the work is completed.

Before-and-after photographs or videotapes of the jobsite are useful in evaluating the quality of restoration. Photographs taken before the job begins show the original condition of trees, shrubs, sidewalks, and fences. They can be particularly useful if someone alleges that the work caused inordinate damage.

The following observations on restoring jobsites to their original condition are generally applicable.

Backfilling Trenches

Properly compacted or settled backfill in trench cuts will reduce settling. If the trench is not well compacted, settlement results, and water utility or street maintenance departments often have to add fill continually to compensate for the settlement, which may extend over several years.

Pavement Repair

Many municipalities require utilities to make temporary repairs of paving cuts by using cold patch. These municipalities assume that additional settlement will occur. They often specify a delay of at least 6 months before a permanent repair is made.

Grass Replacement

A good grade of sod or quick-growing grass placed over a layer of topsoil should be used to replace grass that was disturbed. Sod is usually used to restore lawns and boulevards in urban areas. Depending on the time of year, sod may have to be watered periodically until the roots are established, which may take a month or so. In addition to creating a good image for the water utility, replacing grass is also often necessary to control erosion of the exposed earth.

For larger jobs, grass seed may be sprayed on with a mulch that will help prevent erosion and retain moisture until the grass becomes established. Sprayed seed does not generally require routine watering, but it is slower to grow than sod. Areas that do not grow may require respraying at a later date. Grass matting material is also available to prevent erosion and promote seed growth without much need for maintenance.

Ditches and Culverts

Ditches and culverts in the project area should be checked for proper drainage. Any excessive silt or debris should be removed. A plugged ditch in a heavy storm can flood the surrounding area. Culverts should be checked for any damage caused by the construction, and arrangements should be made for repairs if necessary.

Trees and Shrubs

Damaged roots can cause some species of trees to die. In some cases, the trees might not die until 5 years after the project is completed. A qualified expert should be consulted on the repairs or feeding needed to help save any trees that were damaged. Some locations require that utilities avoid disturbing valuable or historic trees by jacking or boring pipelines rather than performing open-cut trenching adjacent to trees.

Utilities

Any underground pipes or cables uncovered during construction should preferably be properly supported and protected during the work phase. In some cases, it is impossible to restore these pipes and cables while main installation is in progress. It may be necessary to return later and complete proper restoration.

Curbs, Gutters, and Sidewalks

Concrete and asphalt surfaces removed during excavation should be replaced to match the old surfaces as nearly as possible, including the texture of the finish. In some cases, facilities or private property that are not directly in the line of the work may be damaged during construction. In some cases, this damage may be unavoidable but it may also be due to the carelessness of workers and machine operators. One common example is for sidewalks not directly in the line of work to be damaged by the movement of heavy equipment. If sufficiently damaged, some of these facilities may have to be replaced—even though these replacements weren't originally part of the job plans.

Machinery and Construction Sheds

All construction machinery and structures should promptly be removed from the site when the job is completed. The ground where they were located must then be restored to its original grade and condition.

Watercourses and Slopes

If a stream or wetland must be disturbed during construction, it is important to check with local, state, and federal regulatory officials regarding the need for a permit. Areas that a few years ago may have been considered an unimportant "swamp" may now be considered a "wetland" that must be protected under new environmental protection laws. Some localities now require erosion and sedimentation control during construction.

Slopes should be structurally restored and sodded, seeded, and/or riprapped as necessary. Stream beds should be inspected for excessive mud and debris for a short distance downstream from the actual construction site. Cleanup operations should be conducted promptly to limit the amount of silt and debris sent down the stream.

Roadway Cleanup

Almost any type of construction will track some dirt and dust over the roads being used by construction equipment. Many local governments now have strict laws requiring that roadways be maintained clean and free of dust. If the ground is muddy, arrangements may have to be made to wash or scrape mud from the wheels of trucks before it is tracked onto roadways. During dry weather, dust forms on the roads. It may be necessary to use a power sweeper to clean the roads periodically. On gravel roads, applying calcium will usually help reduce the amount of dust.

Traffic Restoration

When construction is completed, restoration may involve removing signs and filling holes, removing any temporary roads, and putting the area back into its original condition by a process of grading and sodding.

Restoration of Private Property

All construction debris must be removed from the site. Private driveways, walkways, fences, lawns, and private appurtenances must be returned to their original condition.

WATER MAIN INSTALLATION SAFETY

During water main installation, the following types of activities require special precautions:

- Material handling
- Working near trenches

- Traffic control
- Work requiring personal protection equipment
- Chemical handling
- Use of portable power tools
- Vehicle operation

Material-Handling Safety

Back injuries are a common and debilitating type of injury related to material handling. Lifting heavy objects by hand can be done safely and easily if common sense is used and a few basic guidelines are followed. Refer to *Safety Practices for Water Utilities* (AWWA Manual M3) for detailed descriptions of labor-related safety procedures.

Trench Safety

Trenches can be made safe if proper excavation and shoring rules are followed and if proper equipment is used. Proper trench shoring cannot be reduced to a standard formula. Each job presents unique problems and must be considered under its own conditions. Under any soil conditions, cave-in protection is required for trenches or excavations 5 ft (1.5 m) deep or more. Where soil is unstable, protection may be necessary in much shallower trenches.

Traffic Control Safety

Barricades, traffic cones, warning signs, and flashing lights are used to inform workers and the public of when and where work is going on. These devices should be placed far enough ahead of the work so the public has ample opportunity to determine what must be done to avoid obstructions. If necessary, a flagger should be used to slow traffic or direct it. Everyone involved in the work should wear a bright reflective vest.

Approved traffic safety control devices that should be used for obstructing roadways are described in detail in *Manual on Uniform Traffic Control Devices for Street and Highways,* prepared by the US Department of Transportation, Federal Highway Administration.

Most states have also prepared simplified booklets describing work area protection that should be used during street and utility repairs. The state highway department or department of transportation should be contacted for available information. Water utility operators should be aware that they could be held liable for damages if an accident occurs as a result of utility operations that were not guarded in conformance with state or federal procedures. Figure 13-5 illustrates the directed guarding procedure for utility work in one lane of a low-traffic volume roadway.

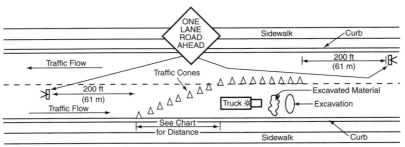

(If traffic is heavy or construction work causes interference in the open lane, one or more flaggers should be used.)

Speed Limit mph (kph)	Lane Width						Number of Cones Required
	10 ft (3 m)		11 ft (3.5 m)		12 ft (3.7 m)		
	Taper Length,		Taper Length,		Taper Length,		
	ft	(m)	ft	(m)	ft	(m)	
20 (32)	70	(21)	75	(23)	80	(24)	5
25 (40)	105	(32)	115	(35)	125	(38)	6
30 (48)	150	(46)	165	(50)	180	(55)	7
35 (56)	205	(62)	225	(69)	245	(75)	8
40 (64)	270	(82)	295	(90)	320	(98)	9
45 (72)	450	(137)	495	(151)	540	(165)	13
50 (81)	500	(152)	550	(168)	600	(183)	13
55 (89)	550	(168)	605	(184)	660	(201)	13

FIGURE 13-5 Recommended barricade placement for working in a roadway

Personal Protection Equipment

Most utilities must provide a broad range of personal protection equipment for workers. Hard hats, safety goggles, and steel-toed shoes are probably the most widely used safety equipment. The equipment is usually issued to all workers so that they have no excuse for not using proper protective gear. Each person must be responsible for maintaining his or her equipment in good condition and having it available when it is needed.

Gloves are also necessary for protection from rough, sharp, or hot materials. Special long-length gloves provide wrist and forearm protection. Workers should wear rubber gloves when handling oils, solvents, and other chemicals. Gloves should not, however, be used around revolving machinery. The machine might catch onto a glove and injure the worker.

Hard hats, which are necessary whenever an operator is working in a trench or near electrical equipment, have been very successful in reducing serious injuries or deaths due to head injuries. However, metal hard hats should never be used where there is an electrical hazard.

Chemical Safety

Chlorine, in the form of chlorine compounds, is the principal chemical used in distribution system operations. Chlorine must be treated, stored, and used carefully to prevent accidents. Even with proper use, accidents can still happen, so the operator should know how to react to them. Depending on the concentration and the length of exposure, chlorine can cause lung irritation, skin irritation, burns, and a burning feeling in the eyes and nose.

Chlorine is most often used in the distribution system in the form of either calcium hypochlorite (HTH) or sodium hypochlorite. HTH is available as a dry powder, crystals, or tablets. It is corrosive in small amounts of water and can support combustion. Sodium hypochlorite (bleach) is a strong acid that can cause similar problems. Workers should wear proper protective equipment when handling hypochlorite, including eye goggles, gloves, and a coat.

If chlorine gas is used, supplied-air masks (air packs) should be available, and operators should be trained in their use. Any individuals caught around a chlorine gas leak without a mask should leave the area immediately. They should keep their heads high and mouths closed and avoid coughing and deep breathing. Detailed safety precautions for chlorine gas are covered in *Water Treatment*, another book in this series.

Portable Power Tool Safety

Electric power tools should always be grounded. A ground-fault interrupter circuit should be used whenever the tool is used outside or near water. The cord should be in excellent condition and should never be used to lift the tool or as a line. The cord or tool should never be left where it could trip someone.

Air tools can be dangerous if the hoses and connections are not correctly maintained. Workers should not point air tools at anyone. They should not clean off any part of their bodies or clothing with compressed air.

Vehicle Safety

Records indicate that the greatest number of accidents in the water utility industry involve vehicles. Workers should be made particularly aware of the potential of accidents while they are operating large trucks and handling heavy construction equipment.

BIBLIOGRAPHY

AWWA Standard for Disinfecting Water Mains. ANSI/AWWA C651. Denver, Colo.: American Water Works Association (latest edition).

AWWA Standard for Installation of Ductile-Iron Water Mains and Their Appurtenances. ANSI/AWWA C600. Denver, Colo.: American Water Works Association (latest edition).

Chlorine Basics, 7th ed. 2008. Washington, D.C.: The Chlorine Institute.

Connell, G.F. 1996. *The Chlorination/Chloramination Handbook*. Denver, Colo.: American Water Works Association.

Manual M3, Safety Practices for Water Utilities. 2002. Denver, Colo.: American Water Works Association.

US Department of Transportation, Federal Highway Administration. 2009. *Manual on Uniform Traffic Control Devices for Streets and Highways.* Washington, D.C.: US Government Printing Office.

CHAPTER 14

System Operations

Operating and maintaining a water distribution system is an involved process. There are two major objectives for drinking water distribution system operational policies: (1) maintain water quality from the point of entry into the distribution system to the point of use and (2) maintain adequate pressure and deliver adequate flow to satisfy customer demands and protect from fire losses. This chapter discusses operational practices and procedures that are designed to address these two objectives.

MAINTAINING WATER QUALITY

System operators have traditionally focused their attention on achieving adequate pressure and flow. Customers (and regulators) are now demanding an increasing emphasis on the water quality.

The distribution and storage system is one component of a multiple-barrier approach to preventing contamination (Table 14-1). Each component must be optimized to provide the maximum level of protection from contamination as water is delivered to consumers. Several groups (departments) within a utility have key roles in maintaining and operating distribution systems. To make sure that all groups can provide their input, standard operating procedures (SOPs) should be jointly developed.

Five steps are suggested to optimize water quality in the distribution system:

1. Gather information to understand your system and define the cause of any problems.
2. Set water quality goals that go beyond regulatory requirements and establish preliminary performance objectives.
3. Evaluate alternative solutions to address problems and satisfy the preliminary performance objectives.
4. Implement good management practices and monitor effectiveness of these practices in maintaining water quality.
5. Prepare final performance objectives and SOPs with input from all affected operating groups (or departments). These steps should result in a distribution system that is optimized to provide the highest quality water possible.

Water Quality Monitoring

A comprehensive, well-designed water quality monitoring program is necessary to optimize distribution system operation to deliver high-quality water to consumers. The goal of a well-managed distribution system should be to provide water to customers' taps that has not changed from the point of entry into the system. Attaining this goal will result in increased customer satisfaction and compliance with all regulatory requirements.

TABLE 14-1 Multiple protection barriers

1. Source protection and management

2. Treatment (may include a number of internal barriers)

3. Disinfection

4. Distribution system operation and maintenance

5. Monitoring and response

Routine monitoring

The analysis of water samples collected from the distribution system provides evidence that the water being delivered to customers is safe to drink and desirable to use. The tests ensure that water treatment processes are functioning properly and that the water has not objectionably degraded while in the distribution system. Testing will also give an indication of whether there is corrosion or scale accumulation in the distribution piping.

Development of a sampling plan

The purpose of a routine sampling plan is to examine characteristics of water quality in the distribution system as the water moves from source to tap. The plan must include all samples to meet regulatory requirements. Of particular interest to distribution system operators is the US Environmental Protection Agency (USEPA) Lead and Copper Rule. This regulation requires special testing of samples collected from residences and in the distribution system. The sampling plan should also include sites, parameters, and frequencies necessary to fully describe the distribution system. Key components of a water quality sampling plan are listed below.

Sampling plan components.

Site Selection Criteria
- Age of the water
- Locations where multiple sources mix
- Storage facilities
- A selection of main materials and conditions
- Locations of booster disinfection stations
- Critical facilities (hospitals, high-usage customers)

Test Parameter Selection
- All regulated parameters
- Disinfectant residual
- Color, turbidity, pH

- Heterotrophic plate count bacteria
- Total coliform bacteria
- Other targeted parameters (corrosion inhibitors, coagulant residuals, nitrite, ammonia, etc.)

Sampling Frequency Criteria

- Satisfy regulatory requirements
- Continuous (may be useful at some strategic locations, e.g., storage facilities)
- Daily (disinfectant residuals, bacteria, pH, temperature, at some locations)
- Weekly (storage facility grab samples, dead ends, targeted parameters)
- Monthly (long-term trends or special surveillance samples)

Nonroutine monitoring

Special sampling studies are needed at times to help manage the system, investigate water quality problems, respond to emergencies, and evaluate alternatives for system improvements. Operators frequently establish monitoring programs that go beyond regulatory requirements. Special sampling is often necessary to identify areas where water quality is deteriorating. Some examples of nonroutine monitoring studies for specific purposes are given below.

Finished water storage facilities. Many distribution system water quality problems can be traced to finished water storage facilities. Water quality monitoring at these sites can be used as a quality control check. In addition to more routine parameters, additional parameters such as *Cryptosporidium*, nitrite, ammonia, odor, total trihalomethanes, and iron may be examined.

Customer complaint investigations. Water quality testing is an important part of an onsite investigation. Coliform bacteria, temperature, pH, and disinfectant residual should normally be tested. In addition, other tests may be necessary depending on the complaint (odor, color, iron, particle identification). The results of water quality tests should be communicated to the customer as the inquiry is resolved.

Construction activities. Main rehabilitation and replacement activities usually require disinfection. The mains must be tested and found acceptable prior to a return to service. Test parameters may include coliform bacteria, pH, turbidity, and disinfectant residual. Other possible tests are odor, color, and volatile organic compounds.

Emergency monitoring. Main breaks, treatment upsets, and backflow events are only a few possible emergency situations that may occur in distribution systems. Water quality monitoring is important to help determine the extent of the problem and verify when corrective measures have been successful. Generally, quick field tests are needed to provide information in a timely manner. It may be useful to monitor the following parameters in an emergency: bacteria, disinfectant residual, turbidity, color, pH, conductivity, alkalinity, and fluoride.

System Design for Water Quality Enhancement

Drinking water distribution systems have, historically, been designed to satisfy potable and fire flow demands. Little consideration was given to the impact of distribution system design on water quality. Many potential water quality problems can be reduced or eliminated by including design features that are specific to water quality.

Planning considerations

As utilities consider distribution system improvements, it is critical that they include factors such as water quality regulations, customer expectations, customer growth, budget limitations, and financial impacts. Long-range master plans that include a capital improvement program must incorporate water quality changes as well as growth demands.

Hydraulic and water quality modeling

Water distribution system models simulate the behavior of physical facilities and water use patterns within the system. These computer models (Figure 14-1) are essential tools for designing distribution system improvements or extensions. The model can simulate actual or proposed operating practices such as operation of pumps, opening or throttling valves, fire flow events, and major main breaks.

Water quality models are extensions of hydraulic models and can estimate water quality changes in a distribution system. An accurate, calibrated, hydraulic model is a necessary building block for a water quality model since the flows, pressures, and volumes in storage are major contributing factors that influence water quality. These computer simulations of a distribution system can be used to illustrate the effect of blending between sources for parameters such as hardness, total dissolved solids, or conductivity. Water age, pH, chlorine residual, trihalomethane formation, or the spread of a waterborne contaminant can also be estimated using sophisticated calculations to account for changing characteristics of the parameter. Information from water quality models can provide valuable information that can be used to establish water quality monitoring sites based upon system characteristics, thus providing an opportunity to better diagnose the cause of changes in the system.

Pipeline design

Water quality considerations need to be included in the design of pipeline networks. By integrating these concepts with volume and pressure requirements, many future water quality problems can be avoided. Special attention should be given to issues affecting pipeline water main sizing. Fire flow requirements generally determine the need for larger mains. This requirement may lead to stagnation, bacterial regrowth, and depletion of disinfectant residual. When designing a pipeline network, a balance between flow requirements and water quality consequences must be considered. The pipeline network design should strive to eliminate (or at least reduce) the number of dead ends,

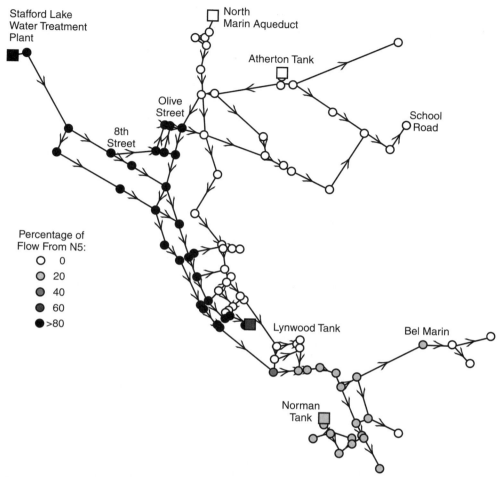

FIGURE 14-1 North Marin hydraulic calibrations for 6:00 a.m. showing direction of flow and percentage of flow from N5 (Stafford Lake Water Treatment Plant)

which are notorious for producing water quality problems. The network design should avoid low or negative pressure areas, prevent uncontrolled pipe scouring, and also reduce the collection of suspended sediments during routine operations.

By selecting the best pipeline materials of construction for the given water quality, future problems can be avoided. Generally, lined ductile iron, concrete pressure, and polyvinyl chloride pipe are ranked by utilities as the most favorable materials for maintaining water quality. Selection of the most suitable material is dependent on many factors including water quality, pressure, size of main, structural loads, soil conditions, and cost.

Pressure zone adjustments

In some systems it is necessary to divide the service area into defined pressure service zones. These zones may be connected in series (water must flow through one zone to reach the next). This practice may increase the age of the water and thus incur the water quality problems associated with this issue. Pressure zone boundaries may also form barriers, in effect establishing artificial dead ends in the system. The dead ends are often the cause of multiple water quality problems: tastes and odors, decay of disinfectant residual, bacterial regrowth, increased corrosion, changes in pH, and collection of sediment.

Finished water storage facilities

Monitoring, inspection, maintenance, and mixing to enhance water quality (Table 14-2) must be considered when designing storage facilities. Many water quality problems are directly associated with water retention in storage facilities. The size and type of facility is normally heavily influenced by hydraulic considerations such as maximum hour demand, fire flow and emergency supply considerations, and future growth demands. Major factors in storage facility design that can mitigate water quality problems are selecting and maintaining the correct materials of construction, providing adequate and appropriate cathodic protection to prevent or control corrosion, and preventing dead zones (Figure 14-2), thus reducing any problems that are associated with stagnant water.

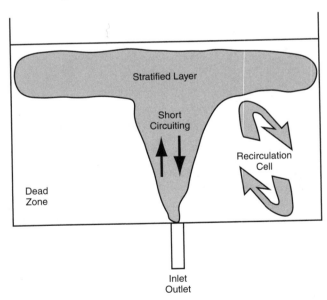

FIGURE 14-2 Nonideal flow patterns in a storage tank

TABLE 14-2 Examples of water quality design considerations for storage facilities

1. Conform to latest American Water Works Association (AWWA)/National Science Foundation standards.

2. Provide for isolation.

3. Provide for bypass.

4. Minimize dead storage.

5. Protect drain from cross-connections.

6. Provide redundant units for maintenance.

7. Isolate altitude valve.

8. Ensure roof is watertight.

9. Ensure vents and overflow are tamper resistant, lockable, and screened.

10. Be sure that overflow pipe is sized equal to or greater than maximum inflow rate.

11. Be sure that inlet and outlet pipe are sized and positioned to promote mixing.

12. Ensure access hatches are lockable, hinged, and constructed to avoid stormwater inflow.

13. Check that safety ladders conform with regulations.

14. Install lockable security fence if not already in place.

15. Provide cathodic protection if needed.

16. Provide monitoring instruments for level, flow, and water quality parameters.

17. Identify sampling locations for inflow and outflow.

18. Provide security alarms.

Operations and Maintenance Practices to Maintain Water Quality

Operators can influence water quality through their system operating procedures. These practices need to focus on three issues:

- Minimize bulk water detention time (control water age).
- Maintain positive pressure.
- Control the direction and velocity of the bulk water.

Most water quality problems in the distribution system can be controlled by effectively addressing these concerns.

Hydraulic detention time

Water quality deterioration is often proportional to the time the water is resident in the distribution system. The longer the water is in contact with pipe walls and is held in storage facilities, the greater the opportunity for water quality changes. SOPs should be developed to minimize the detention time in the system. Hydraulic models can be used to help define water age throughout the system and to help evaluate ways to reduce this value. Monitoring for disinfection by-products (DBPs) and disinfectant residual can also provide information leading to the identification of areas where detention time may be excessive.

Storage facility operation

Finished water storage facilities have been sized and operated to provide reserves for emergency service and fire-fighting needs and to satisfy peak demands. These requirements often lead to oversizing the facility for optimum water quality. SOPs should include practices that will promote mixing and reduce water age.

Water quality in storage facilities should be carefully monitored. Routine sampling of the facility outlet is a minimum monitoring step. Operators may also need to periodically examine the water quality at other locations inside the facility, including taking samples at various depths. It is possible for the water to become stagnant or stratified within the facility. Disinfectant residual can be a good parameter to use to evaluate water quality changes in storage facilities.

SOPs should establish a minimum turnover goal based on water quality at the facility. Generally, one complete turnover of the contents of a finished water storage facility is recommended (seasonal adjustment may be necessary) about every 5 days. A fluctuating water level promotes mixing. However, the process used to fluctuate water level may not always result in the removal of stagnant water. An effective method is to lower the water level in a continuous operation rather than in small increments throughout the day.

Some relatively rare storage facilities need special operational considerations. Facilities with floating covers must take precautions to avoid damaging the fabric. Ice formation is a particular concern. Mechanical recirculation systems are provided to mix water in some facilities. These installations need to be monitored to verify their effectiveness and the mechanical equipment must be constructed to prevent possible water contamination. Storage facilities are a convenient place to redisinfect the water. This may be done by adding the disinfectant to the inlet, outlet, or inside the facility (batch disinfection). Chlorine is most commonly used for this purpose. If the water is chloraminated, special care must be taken so that the desired residual (free chlorine or combined chlorine) can be attained.

Flushing programs

The velocity of flow in most mains is normally very low (hydraulic detention time can be long) because mains are designed to handle fire flow, which may be several times larger than domestic flow. As a result, corrosion products and other solids tend to settle on the pipe bottom. The problem is especially bad in dead-end mains or in areas of low water consumption.

These deposits can reduce the carrying capacity of the pipe. They can also be a source of color, odor, and taste in the water when the deposits are stirred up by an increase in flow velocity or a reversal of flow in the distribution system. These sediments provide an environment for future biofilm growth and can result in a high disinfectant demand. In some cases, the disinfectant residual can be completely consumed. In areas where other operations cannot improve the hydraulic detention time, flushing can restore the water quality and help avoid the need for reactive maintenance procedures.

Flushing programs can be reactive (emergency), routine, or systemwide. Reactive flushing is often a response to a customer complaint (Figure 14-3). An example of routine flushing would be regular maintenance of known dead-end mains or other trouble spots. A scheduled, systematic, systemwide flushing program can result in long-term water quality improvements.

Extensive flushing programs can become expensive and time consuming. However, the cost of emergency repairs or reactive flushing over a large area can also become costly. Each system must evaluate the scope and cost of its flushing programs. Some costs can be saved by coordinating the scheduled flushing with other maintenance activities (like fire hydrant maintenance or valve testing). Most systems will incorporate all three types of flushing programs in their overall system maintenance plan.

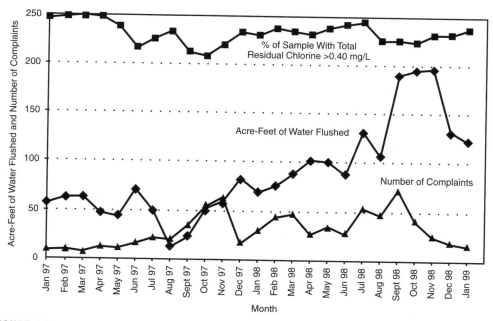

FIGURE 14-3 Southern California Water Company flushing performance measurement data analysis

Courtesy of Southern California Water Company

Flushing procedures. Flushing involves opening a hydrant located near the problem area. The hydrant should be kept open as long as needed to flush out the sediment; this may require up to three water changes. The amount of water needed can be estimated by calculating the volume in the pipeline and the flow from the hydrant. Only through experience will an operator be able to tell how often or how long certain areas should be flushed. Some systems find that dead-end mains must be flushed as often as weekly to avoid customer complaints of rusty water. It is not normal practice for larger utilities to flush the entire system, but problem areas are often flushed in response to customer complaints. Large areas of the system may receive systematic flushing on a rotational basis, thus resulting in an entire system flushing over a number of years.

The following points should be considered when developing a flushing program:

- A map of the system and past experience should be used to plan (Figure 14-4) the flushing schedule. Emphasis should be placed on areas where there has been a high incidence of customer complaints.

- If the complete system is to be flushed, the process should begin in the area of the well or treatment plant and then progress outward. If only one area is to be flushed, the process should be started at the point closest to the source.

- Flushing the system late at night will achieve greater flows through the line and cause fewer customer complaints. In addition, fewer customers will see how bad the water sometimes looks coming out of the system.

- If possible, media announcements should be made in advance to explain the flushing schedule. This will alert customers that there may be a temporary condition of discolored water. Posting signs in the area to be flushed may also assist in customer relations.

- After flushing is completed, customers should be advised in advance to run enough water (cold water from the bathtub is best) to flush their service lines before using the water. They should also be told that the water is safe to drink but warned not to wash clothes while the water is turbid because it may discolor white clothing.

- The work of flushing crews should be coordinated to avoid flushing too many hydrants at once. If too many hydrants are opened at the same time, a reduced pressure situation could be created, which could increase the chances of backflow through any existing cross-connections.

- Before flushing begins, the area in which the flushed water will drain should be inspected to ensure that the water will not flow into basements, excavations, or buildings. In addition, storm drains should be checked to make sure they are open. Diffusers (e.g., water retention dams) may be needed to reduce erosion.

- Local regulations may require dechlorination of flushed water. Devices that are used to control erosion may also be useful to ensure adequate dechlorination. Dechlorination procedures and chemicals are described by Kirmeyer et al. (2000b).

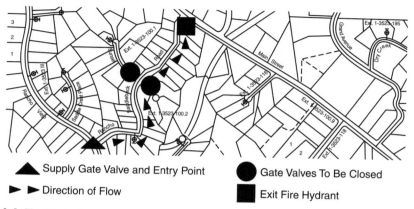

Supply Gate Valve and Entry Point

► ► Direction of Flow

Gate Valves To Be Closed

Exit Fire Hydrant

FIGURE 14-4 Flushing plan map

- The flow required for effective flushing usually is at least 2.5 ft/sec or 220 gpm in a 6-in. main (0.76 m/sec or 830 L/min in a 150-mm main). A flow of 3.5 ft/sec or 310 gpm (1.07 m/sec or 1,173 L/min) is considered better. Care should be taken to avoid too high a velocity that could cause scouring of pipelines (flow should generally be limited to no more that 10–12 ft/sec [3.1–3.7 m/sec]).

- It may be necessary to turn on additional wells or booster station pumps during flushing to ensure that adequate water quantity and pressure are available.

- Hydrants used for flushing must be opened fully—the hydrant valve is not intended for throttling flow. (Hydrants may be fitted with a gate valve on one of the connections if a low flow is desired.)

- Hydrant valves must be opened and closed slowly to prevent water hammer.

- A nonrigid diffuser, screen, length of fire hose, or other means of reducing the force of the water stream is recommended, especially in unpaved areas (Figure 14-5). Flushing should be stopped if the water is damaging a road or parkway. If damage has already been done, it should be marked with a lighted barricade. The location should be recorded so that the damage can be repaired as soon as possible.

- Flowing hydrants should not be left unattended.

- Flushing should be continued no longer than necessary to remove sediment and ensure that the hydrant is operating properly. Monitoring water quality parameters (e.g., chlorine residual, turbidity, pH) may be useful in determining when flushing has achieved the desired results and can be terminated. In some cases, flushing at a low velocity (using the proper throttling valve) may be adequate for the intended purpose.

- When hydrant use is completed, the hydrant should be checked to ensure that the barrel drains properly. An operator can usually determine this either by placing a hand over one nozzle to feel if a slight vacuum is formed or by listening for air being drawn in when an outlet cap is screwed on loosely.

FIGURE 14-5 Diffuser used to reduce force of water from hydrant during flushing
Courtesy of Pollardwater.com

- Hydrants that have had their drains plugged must be pumped out if there is a possibility of freezing.
- Nozzle caps should be tightened to the point that unauthorized persons cannot remove them by hand.
- The code number of each hydrant flushed, the length of time flushed, the water condition at the start and end of flushing, and other special concerns should be recorded.
- Water quality tests should be conducted to document the results of the flushing. Total/free chlorine residual, temperature, bacteria, iron, color, and odor are examples.
- Hydrants found to be defective should be noted, flagged if inoperative (notify fire department if out of service), and reported for immediate repair.

Directional flushing. This procedure involves a systematic approach to direct the flow, at the desired velocity, from a clean source to the area to be flushed. It is necessary to map the locations of valves that will need to be operated to direct the flow before beginning this procedure. A hydraulic model of the system can be very helpful in planning directional flushing. Proper valve operation is key to this approach; therefore, it is common to conduct valve inspections and maintenance when performing this procedure. Although time is needed to plan the flushing strategy for directional flushing, some utilities have found that this process is actually less expensive than the traditional (open hydrants without operating valves) procedure since water quality is maintained and there are fewer instances of needing to return to repeat the procedure.

Water main cleaning

Pipelines that exhibit flow restrictions and/or water quality problems should be cleaned. Utilities should first try to clean the mains by flushing. If flushing proves inadequate, air purging or cleaning devices such as swabs or pigs may need to be used. In addition to removing objectionable material from a main, the cleaning operation can increase the flow rate through the pipe.

Mechanical cleaning may be necessary in areas where excessive tuberculation and deposits are found in older cast-iron pipes or where iron bacteria and slime growth are a severe problem. In some cases, removing encrustation or tuberculation may cause leaks that must be repaired. The cleaning process is not always a permanent solution to dirty water problems. Unless the cleaned pipe is lined or the corrosiveness of the water reduced, the condition will probably come back. However, experience has shown that, in many cases, leaving just a thin coating of iron oxide on the smooth interior of the pipe wall delays the occurrence of red water and the regrowth of encrustation.

Main-cleaning preparations. Thorough planning must precede a pipe-cleaning operation. The section of main or system to be cleaned should first be mapped. The order of work, the water source, the entry and exit points, and the disposal method for the flushed water should all be determined. The vehicles to be used, the size of the crew, and the necessary equipment and materials can then be listed and arrangements made for them to be available.

Before cleaning begins, valves and hydrants should be checked to ensure they are operable. Customers should be notified concerning the date and time the system will be out of service. Temporary water service should be arranged for any customers who must have water for medical reasons.

Other utilities and agencies that will be affected by the planned operation must also be notified, including police and fire departments. Regulatory agencies should be consulted concerning any special requirements, and any necessary safety procedures should be planned.

Before the utility work crew flushes or cleans any main, a way to control pressure surges should be provided. The sudden stopping of flow if a line valve is operated too rapidly or if a pig suddenly stops moving can cause a large pressure surge. Such surges can raise system pressure 20 to 60 psi (450 to 1,350 kPa) for each foot-per-second (meter-per-second) of velocity change, which can destroy water mains and appurtenances.

Air purging. In the air-purging process, air mixed with water is used to clean mains. Generally, small mains, up to 4 in. (100 mm) in diameter, are cleaned using this procedure; however, some utilities have reported using this process for larger mains. Before the procedure is performed, all services must be shut off. Air from a compressor is then forced into the upstream end of a main after the blowoff valve is opened at the downstream end. Spurts of the air–water mixture remove all but the toughest scale.

Swabbing. The swabs used in pipe-cleaning operations are polyurethane foam plugs. These plugs are somewhat larger than the inside diameter of the pipe to be cleaned, and they are forced through the pipe by water pressure. Swabs can remove slime, soft scale, and loose sediment without the need for high-velocity water flow. However, they will not significantly remove hardened tuberculation. Swabs wear out quickly in heavily encrusted mains and must be frequently replaced.

Swabs can be purchased commercially either in specific sizes or in bulk (bulk swabs can then be cut to size). The swab material is available in soft and hard grades. Soft swabs are typically used for mains that have an undetermined cross section or condition. They are also used in mains where the reduction in pipe diameter is expected to be 50 percent or more. They may also be used in mains where there is severe encrustation but where a pig cannot be used because of bends or partial obstructions in the pipe. Hard swabs are commonly used in newer mains, in mains that have minor reductions in diameter, and in mains where deposits need continuous hard-swab pressure.

An experienced crew can swab up to several thousand feet of main per day if the operation is planned properly. Swabbing procedures vary with each job, but a typical procedure is as follows:

1. Notify the public and shut off the system.
2. Install the necessary equipment at the entry and exit points for launching and retrieving swabs.
3. Isolate the water main or portion of the system to be cleaned. Be sure all valves on the main to be cleaned are open.
4. Open the valve on the upstream water supply to launch the swab and to control speed.
5. Operate swabs at the speed recommended by the manufacturer. If they travel too fast, they remove less material and wear out more rapidly.
6. Estimate the flow rate by using a Pitot gauge at the exit or a meter on the inlet supply.
7. Note the entry time and estimate the time of exit. If the travel time is too long, the swab may have become stuck. Reverse the flow and time the return to determine the location of blockage.
8. Perform enough swab runs that flushing water becomes clear within 1 minute following the run.
9. Account for all the swabs to make sure none were left in a main. A typical cleaning operation may take from 10 to 20 swabs.
10. Do a final flush until the water is clear of swab particles.

Pigging. Pipe-cleaning pigs are stiff, bullet-shaped foam plugs that are forced through a main by water pressure. They are similar to swabs but are harder, less flexible, and more durable. These differences allow the pigs to remove harder encrustations. However, pigs have more limited flexibility, which somewhat reduces their ability to change direction at fittings and at points where there are significant changes in pipe cross section.

Pigs are purchased commercially in various sizes, densities, grades of flexibility, and external roughness. A number of different types of pigs are available for use in system cleaning. Special pigs can be made for most situations. The ones most commonly used are classified as bare pigs, cleaning pigs, and scraping pigs.

Bare pigs, made of high-density foam, are usually the first pigs sent through a tuberculated main to determine the inside diameter of the obstruction. *Cleaning pigs* have a tough coat of polyurethane synthetic rubber applied in a crisscross pattern. When sent through a main, a cleaning pig removes most types of encrustation and growths. A bare pig made of low-density foam or a swab is sometimes sent behind an undersized cleaning pig to maintain the seal. *Scraping pigs* have spirals of silicon carbide or flame-hardened steel-wire brushes. These pigs are used to remove harder encrustations and tuberculation. Cleaning or scraping pigs of increasing size may be sent through a main to remove layers of encrustation gradually.

Figure 14-6 shows three methods of launching pigs with permanent or portable launchers. During a launch from a fire hydrant base, an external source of water is required to force the swab or pig down the hydrant and into the main. This water is supplied either by a fire hose connected to a hydrant not in the isolated section or by a small, high-pressure pump with an independent water supply. After a pig has been forced into the main, the hydrant branch can be isolated. The pig will then be pushed along the main by water from the distribution system. Operators may be able to launch a pig into larger mains by removing the gates from a gate valve and inserting the pig into the main through the body of the valve.

Pigging procedures vary with the anticipated condition in the pipeline, the location of the section that needs cleaning, and the type of pig to be used. The utility should determine the procedure to be used at each location independently, especially for larger pipe sizes. The first time pigs are used, the utility's crew should work with someone experienced in using the equipment and performing the cleaning operation. Assistance is available from pipeline-cleaning firms and manufacturers of cleaning devices. A general procedure for using pigs to clean a line is as follows:

1. Notify the public and shut off the system.
2. Install the necessary equipment at the entry and exit points for launching and retrieving pigs. In many cases, it will be necessary to cut into the main to install the equipment.
3. Isolate the water main to be cleaned. Be sure all gate valves on the main to be cleaned are open.
4. Make provisions to control surges.
5. Open the upstream water supply to launch the pig.
6. Time the passage of the pig in order to gauge the valve setting required to achieve the desired speed.
7. Control the speed of the pig with the downstream hydrant or blowoff valve. Typical speeds are 1 to 5 ft/sec (0.3 to 1.5 m/sec). If pigs travel too fast, they remove less material and wear out more rapidly.
8. Avoid sudden changes in speed or stops in pig movement, which will cause destructive surges in water pressure.
9. Run the final flush until the water turns clear.

Metal scrapers. Cleaning units that are forced through a main by water pressure are also available in designs that have metal scrapers. They consist of a series of body sections on which high-carbon spring-steel blades of various shapes are mounted to provide scraping and polishing. The sections are free to rotate and are pushed through the main by water pressure acting against pusher cones.

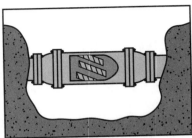

Oversize Spool Inserted in Line

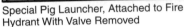

Special Pig Launcher, Attached to Fire Hydrant With Valve Removed

Y-Section Inserted in Line

FIGURE 14-6 Launching methods for pipe-cleaning pigs
Courtesy of Girard Industries

Power-driven cleaning. When deposits in a water main are particularly thick or dense, mechanical cleaning with a power drive unit is needed. In this process, a rod similar to a sewer rod is used to pull a cutter through the main to dislodge the built-up material. This type of work must generally be performed by specialized contracting companies.

Final cleaning procedures. After a water main has been cleaned, it should be flushed until the water runs clear. The main should also be chlorinated before it is returned to service. All valves should be checked to make sure that they have been left in the open position and that all customer water services have been reactivated.

Before and after the cleaning is performed, a flow test should be conducted on mains that are to be cleaned. The results of the test after cleaning should indicate if the procedure was successful or if further cleaning is necessary.

Chlorine treatment. When the carrying capacity is reduced by slime growths in pipelines, chlorine may be used to solve the problem. The section of the system affected is usually given a "slug" dose of chlorine to kill the bacteria causing the problem (usually 25 mg/L for 24 hours; specific procedures are detailed in AWWA Standard C651, most recent edition). This slug should be followed by thorough flushing to get the slime and chlorine out of the system. The precautions discussed previously concerning flushing are even more important in this situation, *especially* those that relate to informing customers to flush their service lines thoroughly before using the water.

A second dose of chlorine may be needed to complete the job. A disinfectant residual should be maintained throughout the system after the treatment is done to prevent the bacterial slime from regrowing.

Lining water mains

Cleaning can usually restore interior pipe surfaces to a condition close to that of a newly laid main. However, experience has shown that cleaning iron pipe without lining it is only a temporary solution. If the water quality remains unchanged, tuberculation will occur again at an even faster rate after cleaning. The flow coefficient will decline back to its previous level. For this reason, cleaning alone does not accomplish much.

Cement–mortar lining. After cleaning is finished, a thin layer of cement mortar can be applied to the pipe walls to line them in place. This not only prevents interior surface deterioration from recurring but also results in improved water quality, volume, and pressure to customers.

The cost of mortar lining in place depends on the following:

*   Pipe diameter and length
*   Condition of the pipe
*   Layout and profile of the line
*   Number of bends
*   Locations and types of valves
*   The need to provide temporary bypass lines
*   Type and depth of soil cover
*   Accessibility
*   Traffic conditions

The longer the length of pipe that can be lined in one operation, the greater the production rate and the lower the cost per unit length. As illustrated in Figure 14-8, small-diameter pipe is lined by remote-controlled equipment that is pulled through the main. Large-diameter pipe can be lined using equipment that allows a worker to enter the pipe to control the operation.

When distribution mains are lined, temporary water service must be provided to all customers until the operation is completed. Water utilities usually provide this service by running a 2-in. (50-mm) pipe on the surface over the main. All services must be connected to this pipe through meter pits or to building sill cocks. After the lining has been placed, each service must be dug up and the mortar removed from corporation stops before service can be restored to customers.

FIGURE 14-7 A cement-mortar lining machine for use in small-diameter pipe

Sliplining. Another method of lining cleaned water mains in place involves slip lining with a high-density polyethylene pipe. The plastic pipe is lightweight and flexible and can be either pulled or pushed into the existing main from access points cut into the system. Valves, tees, and services must be recut, the same as with cement–mortar lining.

Booster disinfection

Most systems perform disinfection only at the water treatment plant or wellhead. The dosage applied is adequate to provide a measurable residual throughout the distribution system. In some cases, this is not possible without adding a prohibitive amount at the source. Booster disinfection (redisinfection) is needed in this instance. Some utilities have employed booster disinfection as a strategy to maintain a more uniform disinfection residual at a lower level, reduce DBPs, and lower total disinfection cost (in some cases).

Where chloramines are used to provide the disinfectant residual, booster disinfection requires special consideration. If free chlorine is applied for booster disinfection, it is possible for free chlorine from this process to blend with chloramine residual in the system in an uncontrolled manner. The potential problems of blending are related to the breakpoint curve (Figure 14-9) for free and combined chlorine. Various resultant concentrations of mono-, di-, and free chlorine are possible. This may lead to undesirable tastes and odors or possible reduction of chlorine residual in some areas. Some utilities have dealt with this situation by measuring the ammonia residual remaining in the water at the point of free chlorine addition. They have then added the correct ratio of free chlorine to reform predominately monochloramine. In this case, blending with residual chloramine is not an issue. Other utilities have found that free chlorine booster disinfection in a chloraminated supply does not cause problems. The best strategy for each utility is dependent on many factors, some of which are unique to the site. Therefore, each situation should be fully evaluated before implementing any disinfection strategy.

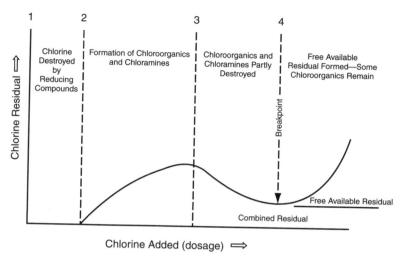

FIGURE 14-8 Breakpoint chlorination curve

Customer complaints

Many of the operational practices and system design features that are described in this chapter and throughout the book are directed toward promoting customer satisfaction. A valid customer complaint is evidence that 100 percent satisfaction has not been achieved. In this way, customer complaints indicate a failure by the water supplier to *prevent* complaints.

No distribution system is completely free of situations that occasionally create water quality problems and thus generate customer complaints. Water quality complaints should be tracked separately from other customer issues. The information from a complaint investigation can be useful in identifying the source of a deteriorating water quality condition. Persistent complaints in one area may lead to a change in operational practices or system alterations to permanently address the problem. The utility operator should establish a goal to reduce customer complaints to improve customer satisfaction.

Source water blending

When a utility has access to more than one source of water, a detailed blending analysis should be conducted. Blending can have beneficial or detrimental consequences on distribution system water quality and pipe materials. Conservative parameters such as iron, phosphate, fluoride, and manganese may be estimated using blending models that are based on concentration and flow. Parameters that decay or grow under certain circumstances may need more complex models to predict system concentrations.

Blending may result in conditions that affect the stability of pipeline materials and the films that adhere to internal surfaces. For example, changes in water chemistry can reduce or increase the rate of corrosion. Alternatively, microbiological counts have increased in pipelines subject to blended supplies. Many of these situations can be predicted and adverse consequences can be avoided by planning and modeling blending practices. It may be necessary to avoid blending by isolating areas of the distribution system with valves or by using pressure zones (this practice may create dead-end areas that could develop water quality problems). Care should be taken to notify customers if changes in water quality are anticipated.

Source water treatment

The quality of water entering the distribution system can significantly impact the receiving system and customers' satisfaction with their drinking water. The system operator's goal is to maintain water quality at the point of entry to the customers' service (the extent of the system operator's responsibility may be defined by local regulations). This goal can only be achieved if the source water treatment is optimized to match the system conditions to prevent water quality changes from occurring.

Several major treatment practices (discussed in more detail in the *Water Treatment* book in this series) can greatly aid system operators in their quest to maintain water quality, including

- stabilize pH,
- control corrosion,
- optimize primary disinfection,
- optimize turbidity removal,
- reduce organic compounds, and
- minimize iron and manganese concentrations.

Most water quality changes that can be attributed to source water treatment can be addressed with one or more of these practices.

The pH of the water affects many common reactions. Therefore, pH stability is critical to maintaining water quality throughout the distribution system. Corrosion of system components, the rate of DBP formation, disinfectant decay, and tastes and odors can all be influenced by pH. The buffer intensity is a measure of the resistance to a change in pH. This measure is better than alkalinity for determining the pH stability of a given water. The effect of pH (and alkalinity) on buffering intensity is shown in Figure 14-9. The minimum buffering intensity occurs at pH 8.3, which is why it is very difficult to maintain a stable pH at this value. The buffering intensity can be increased by adding carbonate and raising the dissolved inorganic carbon (DIC) level.

Many natural and treated waters are corrosive and will dissolve some pipe and plumbing materials. Corrosive waters can deteriorate both the distribution system piping and domestic plumbing systems. This deterioration (or corrosion) can, in turn, cause the quality of the water delivered to the customer to be significantly degraded. Corrosion can also reduce flow capacity and shorten the life-span of the distribution system.

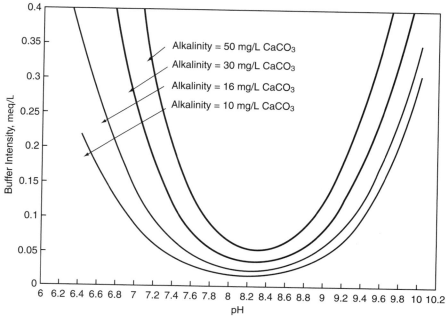

FIGURE 14-9 Theoretical buffer intensity versus pH

Figure 14-10 illustrates a corrosion cell and shows how corrosion can cause pitting and the formation of rough tubercules. Even slight corrosion and tuberculation can increase the roughness of a pipe's surface, thus significantly reducing the pipe's carrying capacity. Major corrosion can also weaken pipes. Corrosion products may break off of mains, causing clogged services and customer complaints of rusty water.

The principal concern with internal corrosion in the past was its effect on unlined cast-iron or steel mains, steel tanks, and other metal surfaces in the distribution system. As a result, unlined cast-iron and steel pipe are no longer installed, but there are still hundreds of miles of old unlined pipe in use in older systems. Corrosion of the interior of steel tanks is generally controlled by a protective coating or by cathodic protection.

Internal pipeline corrosion control is usually implemented to reduce lead, copper, and iron leaching into treated water. Strategies used for this purpose involve pH and alkalinity adjustment, the use of corrosion inhibitors, and calcium adjustment. Lead control in the pH range of 6–9 generally requires the DIC to be greater than 2 mg C/L (2 mg carbon/liter). Carbon dioxide, soda ash, and sodium bicarbonate may be used to adjust the DIC. Orthophosphate may be used as a corrosion inhibitor in the pH range of 7.4–7.8 at a typical residual of 1–5 mg PO_4/L. Copper and iron corrosion are reduced by using orthophosphate or by adjusting pH and DIC. The chemistry of corrosion and ways to reduce internal corrosion are detailed in *Water Treatment*, another book in this series.

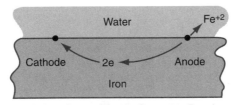

A. Minor Variations Cause Electric Current to Develop

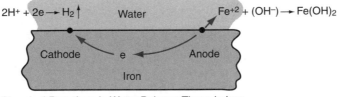

B. Chemical Reactions in Water Balance Those in Iron

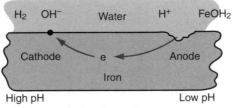

C. Rate of Corrosion is Accelerated

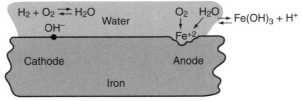

D. Rust Forms

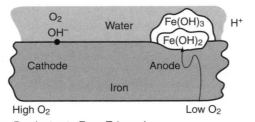

E. Rust Precipates to Form Tubercules

FIGURE 14-10 Chemical and electrical reactions in corrosion cell

Effective primary disinfection protects the distribution system from pathogens that may be present in the source water. The USEPA specifies the disinfection requirements for drinking water systems in the Microbial and Disinfection Byproduct (MDBP) Rules (pubweb.epa.gov/ogwdw/mdbp.html).

New Rules
- Ground Water Rule
- Long Term 2 Enhanced Surface Water Treatment Rule
- Stage 2 Disinfectants & Disinfection By-products Rule

Older MDBP Rules
- Surface Water Treatment Rule
- Total Coliform Rule
- Interim Trihalomethanes Rule
- Information Collection Rule
- Interim Enhanced Surface Water Treatment Rule
- Stage 1 Disinfectants & Disinfection By-products Rule
- Filter Backwash Recycling Rule
- Long Term 1 Enhanced Surface Water Treatment Rule.

Most systems provide primary disinfection with free chlorine or ozone. However, monochloramine, chlorine dioxide, and ultraviolet light may also be used under specific conditions.

Water treatment plants are designed to remove particles. Turbidity is a measure of particles that can scatter light. Lower-turbidity water can be disinfected more efficiently and disinfectant demand in the distribution system may be decreased. Water treatment optimization programs such as the Partnership for Safe Water recommend the lowest turbidity possible. This procedure may lower the risk of possible exposure to pathogens such as *Cryptosporidium*.

The removal of organic compounds by activated carbon filtration, ozone biofiltration, membrane filtration, or enhanced coagulation may reduce the occurrence of bacteria in the distribution system. Assimilable organic carbon and natural organic matter (NOM) have been shown to support regrowth of bacteria. Chlorination of NOM has also led to increases in DBPs. Both the regrowth of bacteria and the development of DBPs are problems that may be difficult to overcome in the distribution system. Source treatment is therefore recommended if these conditions are encountered.

High levels of iron and/or manganese can cause discoloration of water, fixtures, and laundry and adverse taste. Removal of these elements from water may lead to improved customer satisfaction. Lime or caustic soda softening or oxidation/filtration are used for removal. Many plants that previously used free chlorine for oxidation have changed to potassium permanganate due to concerns over the production of chlorinated DBPs. A common treatment process is oxidation with potassium permanganate and greensand filtration (a specialized filtration media suited for manganese removal). A discussion of iron and manganese removal treatment is included in the *Water Treatment* book in this series.

Seasonal considerations

Numerous water quality conditions are affected by seasonal changes either in water use or availability. Source water quality often changes seasonally. As water use decreases, hydraulic detention time may increase, unless the system operator takes action to prevent this consequence (e.g., removing excess storage from service in the winter). In many cases, this situation occurs in the winter when the water temperature is lower. This relationship may be beneficial since many adverse water quality conditions are aggravated by increased temperature.

Some utilities may augment water supplies during high-demand periods. Uncontrolled blending of waters of different water qualities can occur in the distribution system. The operator should consider the consequences of this practice and take steps to maintain uniform water quality at least in confined areas of the system (e.g., by creating temporary pressure zones). Flow reversal and pressure surges should also be avoided when introducing supplemental water sources.

Nitrification is the process by which ammonia-oxidizing bacteria convert ammonia to nitrite and nitrate. Most nitrification episodes are associated with warmer water conditions and occur in locations of low water turnover. Overdosing ammonia in chloraminated systems is a common cause of nitrification. Chloraminated systems should therefore monitor free ammonia as a control parameter. Systems may employ seasonal operations such as practicing periodic free chlorination, breakpoint chlorination, and increasing the turnover rate in storage facilities to prevent nitrification. Symptoms of nitrification are loss of chlorine residual and increases in nitrite and/or nitrate concentration. The loss of chlorine residual may also lead to the occurrence of coliform bacteria in the system, thus leading to the possibility of regulatory noncompliance. Minimizing detention time and maintaining a chlorine/chloramine residual are key steps in preventing nitrification. The system operator should assess storage facilities in particular for the possibility of nitrification in isolated sections of the facility (the entire water volume may not be affected). Spot treatment or cleaning may be required to eliminate this condition.

Pressure requirements

An important operational requirement for any distribution systems is to maintain a continuous positive pressure at all locations. Backflow or backsiphonage from cross-connections may occur when there is negative or zero pressure in pipelines. Contaminants may be drawn into the system through leaking pipes, submerged air-and-vacuum relief valves, blow off valves, or faulty check valves. Standards for the minimum pressure in the distribution system vary. However, 20 psi (138 kPa) is the lowest minimum pressure listed in many existing US state standards and water system guidelines (e.g., "Ten States Standards" [Great Lakes 1997] and National Fire Protection Association).

Excessively high system pressures can also cause water quality problems. Higher incidence of main breaks and service repairs can increase the water quality disturbances caused by these procedures. The effects of water hammer are increased by higher pressure. The resultant hydraulic surges can cause material on pipe walls to become dislodged, thus degrading water quality.

Cross-connection control

Backflow from nonpotable water sources that are cross connected to the potable system can cause serious water contamination incidents. Both backsiphonage and backpressure can be the mechanism for the backflow. Where cross-connections exist, backflow may be prevented by the use of approved control measures including

- air gap,
- reduced pressure zone backflow-prevention device,
- double check valves,
- pressure vacuum breaker, and
- atmospheric vacuum breaker.

The application and acceptability of each measure depends on the degree of hazard and the conditions of the potential backflow situation. In many areas, local regulations define the appropriate measure.

All water systems should implement cross-connection programs. The program should

- define authority and responsibility,
- require systemwide inspection and testing of all prevention devices,
- define device maintenance and record requirements, and
- educate all parties.

A more complete description of cross-connection control and backflow prevention can be found in chapter 11 of this book.

Emergency operations

Water quality problems can cause an emergency situation, and emergencies can cause water quality problems. Natural disasters, main breaks, and power outages are examples of emergency situations that can degrade water quality. Likewise, earthquakes, floods, and violent storms can lead to contaminated water incidents.

Dealing with these situations at the time they occur without sufficient planning can cause inefficient and sometimes incorrect responses. Utilities should develop an emergency contingency plan that includes

- an emergency response strategy,
- interconnections with other utility supplies,
- sufficient storage for emergency operations,
- a plan for natural disasters, and
- a conservation plan to help with insufficient supplies.

The plan should address the recovery period that leads the utility back to normal operations. Operators should be familiar with the plan and should participate in periodic simulation drills to test and refine the plan.

Energy management

The focus of most energy management strategies is to operate the system to minimize power expense. The effect on water quality should also be included when considering options regarding energy usage. Strategies to reduce costs of energy use include treatment plant scheduling, pump scheduling, water demand forecasting, and maintenance scheduling. Water quality can be affected when employing these strategies since hydraulic detention time can be changed and water velocity or direction can be influenced.

Redundant power supply. Improving power supply reliability (redundant power supply lines, portable generators, or permanent emergency generators) can reduce the frequency of system shutdowns. Pipeline disturbance due to on–off operation is therefore lessened.

Pumping strategies. Pumping operations should consider velocity in pipelines and the effect of on–off operations and include sequences that effectively increase the turnover rate in storage facilities.

Pipeline friction loss. Friction losses are the primary factor affecting power requirements for pumps. Regular flushing, corrosion control, and pipeline cleaning can reduce power requirements and improve water quality at the same time.

Direction and Velocity Control

Water quality can be affected by rapid changes in velocity or the direction of flow in pipelines. Operators should employ procedures that will avoid or reduce the impact of these changes.

Fire flow testing

Fire flow requirements are the determining factor in many systems when sizing mains, storage facilities, and pumps. Fire flow tests are necessary to ensure that the system is adequate to provide the necessary flow for fire fighting. The flow requirements are specific to each community and are dependent on a number of factors.

One or more fire hydrants are used for these tests. The hydrants are operated from zero to full flow while measuring the pressure at a nearby hydrant. The rapid change in velocity in the main feeding the hydrants can cause water quality problems. Water hammer created from opening and closing hydrants can add to these problems. Customer complaints are often a direct result of fire flow testing. It is best to alert customers that this procedure will be conducted in their area and advise them that they may notice a temporary change in their water quality. The notice should suggest how they can help clear the water following the procedure.

System operators can also take advantage of these tests and coordinate other procedures at the same time. For example, if the tests are performed in a manner consistent with the needs of the routine flushing program (discussed earlier in this chapter), both procedures can be accomplished at the same time. Hydrant valve inspection and maintenance can be conducted. It may also be a good time to inspect the function of air-and-vacuum relief valves and check backflow-prevention devices.

Pump startup, shutdown, and valve operation

Resuspension of sediment and scouring of scale and other attached material on pipe walls may occur when there is a variation in flow velocity or direction. Water quality can be negatively affected by these operations, resulting in customer complaints. Therefore, operations that cause flow variations should be minimized.

Valves and hydrants should be opened slowly. Pumps should be started with the discharge valve closed. As the motor reaches full speed, open the valve slowly. The reverse procedure will help avoid problems upon pump shutdown. Automatic pump control valves can perform this sequence without operator attention. Variable frequency drives can be used on pumps to slowly change the pump speed upon startup and shutdown. If this is not possible, relief or surge chambers may be installed to absorb the pressure shock from shutdown.

Pump operation may also cause a change in the direction of flow. This procedure can result in many of the same water quality problems as changes in flow velocity. System operators should evaluate alternatives to minimize this condition.

MAINTAINING FLOW AND PRESSURE

The second objective of distribution system operators is to ensure system reliability. This is achieved by following practices and procedures that result in positive pressure throughout the system and the delivery of adequate flow for all intended purposes. This section discusses some of the many concerns in keeping a system in working order.

Distribution System Inspection

The performance of distribution system piping depends on the pipe's ability to resist unfavorable conditions and to operate at or near the capacity and efficiency that existed when the pipe was laid. This performance can be checked using flow measurement, fire flow tests, loss-of-head tests, pressure tests, simultaneous flow and pressure tests, and tests for leakage. These tests are an important part of system maintenance. They should be scheduled as part of the regular operation of the system.

Pressure and flow tests

Some state regulatory agencies and local codes require public water utilities to maintain normal pressure in the distribution system between 35 psi (240 kPa) and 100 psi (590 kPa), with a minimum of 20 psi (140 kPa) under fire flow conditions (National Fire Protection Association 2001).

Many changes can occur in a distribution system to reduce (or occasionally increase) pressure and flow to unacceptable levels. These changes include

- system expansion that does not provide additional feeder mains;
- new water lines installed at elevations higher or lower than the original system;
- additional customer services added to existing mains;
- unintentionally closed or partially closed valves;
- undetected leaks in mains or services;
- changes to water storage tanks; and
- reductions in pipe capacity due to corrosion, pitting, tuberculation, sediment deposits, or slime growth.

Minimum pressures must be maintained to ensure adequate customer service during peak flow periods or while water is being used to fight a fire. A minimum positive pressure must be maintained in mains to protect against backflow or backsiphonage from cross-connections.

Excessive pressure is objectionable for use by customers and decreases the life of water heaters and other plumbing fixtures. It can also increase the chance of distribution system damage in the event of water hammer. Some plumbing codes (e.g., Uniform Plumbing Code) and state or local regulations require the installation of pressure-reducing valves on high-pressure services.

Checking pressure. Operators can test water distribution system pressure using a pressure gauge connected to either a fire hydrant or a building faucet. Hydrant pressure gauges (Figure 14-11) can be purchased or made from a standard hydrant cap that has been tapped. After the gauge is installed on a hydrant nozzle, the hydrant is opened far enough to ensure that the drain valve has closed. Air is then bled off through the petcock before the reading is taken. The pressure is read in pounds per square inch (psi) or kilopascals (kPa).

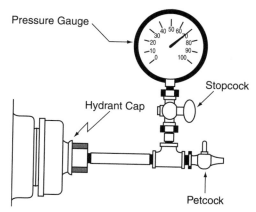

FIGURE 14-11 Fire-hydrant pressure gauge

If a hydrant is not available, operators may need to check pressure using a sill cock on the exterior of a residence or public building. When this is necessary, the operator should first obtain the permission of the resident or property owner. Water cannot be used in the building while the reading is taken. Pressure gauges made for use on a hose thread faucet are readily available from plumbing supply houses.

Checking loss of head. Operators can determine loss of head by isolating a section of the distribution system from all branch lines and services. All water entering the test section of pipe must exit the downstream end. As shown in Figure 14-12, pressure is then measured at points along the main or on customer services to determine how much pressure drop is occurring.

Checking flow. Flow readings are usually taken at hydrants. When locations on the distribution system are found to have reduced flow capacity, the cause may be an increase in pipe interior roughness. This problem is often found in old, unlined cast-iron pipe.

Flow tests conducted on a main over a period of years may gradually indicate decreasing flow rates. This problem is probably due to the added friction or reduced open flow area caused by tuberculation of the pipe interior walls. The calculated flow coefficient (or *C* value) should be compared with the value for the pipe when it was new. If the flow in a main is reduced to the point where adequate fire flow is no longer available, the utility must either clean the main to restore its capacity or install new feeder mains to reinforce the system.

Routine inspection

Fire hydrants and distribution system valves should be regularly inspected and operated to determine whether repair is required. Inspection and operation also keep the hydrants and valves "loose" so they will operate freely when needed.

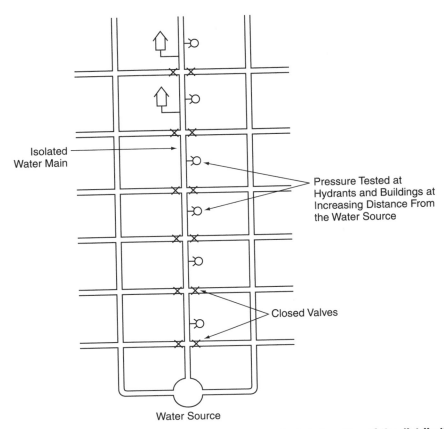

FIGURE 14-12 Locations where head loss is checked in an isolated section of the distribution system

Water utility employees should be specially trained to watch for potential water system problems and notify management of anything suspicious. Meter readers are in a particularly good position to watch for suspected leaks, vandalism, damaged equipment, and unauthorized use. Police, street crews, firefighters, and other municipal employees should also be encouraged to report anything about the water system that appears unusual. The utility may already know about the suspected problem or the situation may not actually be a problem, but the person making the report should nonetheless be thanked for his or her interest. Occasionally, these reports can be very important.

Leak Detection and Repair

Distribution system leaks and the resulting water loss fall into two general categories: emergency leaks and nonemergency leaks.

Emergency leaks require immediate attention. Nonemergency leaks include known leaks that may be repaired when time permits, as well as nonvisible leaks that are located as part of a leak detection survey. Although the repair procedures may be the same for both leak conditions, the method of detection and the sense of urgency involved are often quite different.

Locating leaks

If all water services are metered and all meters are accurate, the difference between the amount of water pumped to the system and the total water metered to customers is known as unaccounted-for water (nonrevenue water). A large portion of unaccounted-for water is due to the various hydrant uses: fire fighting, main flushing, and water leaks. If hydrant use is closely controlled and estimates are made of authorized hydrant use, this amount can be added to the total water metered to customers to create a total of known use. The difference between this figure and the amount of water metered to the system provides an estimate of the amount of leakage in a system.

Large leaks. Unless the ground is very porous, large leaks are usually easy to find. The water either comes to the surface or finds its way into a crack in a sewer pipe or access hole. The starting point for reducing leakage in a system is to urge all police and public works employees to promptly report any unusual puddles or running water. Sewer crews should be asked to report any sewers that seem to have unusually heavy flow.

Small leaks. Unfortunately, small leaks do not always come to the surface. Either they are absorbed into the soil or they flow into sewers, where their flow is too small to be noticed. It can be seen from Table 14-3 that a relatively small leak can pass a surprisingly large amount of water over a period of time. And many small leaks can add up to millions of gallons (or liters) of wasted water.

All water systems generally repair leaks as soon as possible after the leaks are noticed. However, most systems spend little effort looking for hidden leaks. How much effort a utility should make is principally a function of the availability and cost of water. Systems located in water-short areas should have very active leak detection programs to conserve water. Systems that must pay a high cost to obtain or treat water should also be concerned about leaks because the cost of lost water can be substantial.

Leak detection methods

A water utility can control underground leakage only by conducting a painstaking survey of the entire system. The two basic methods are listening surveys and a combination of listening surveys and flow rate measurements, sometimes referred to as water audits.

Listening surveys. An acoustic listening device can be systematically used to locate leaks. It can detect the sound waves created by escaping water. Sound waves are picked up by sensitive instruments and amplified so that a technician can hear them. In the

TABLE 14-3 Water loss versus pipe leak size

| Pipe Leak Size | Water Loss* | | | |
| | Per Day | | Per Month | |
	gal	(L)	gal	(L)
●	360	(1,360)	11,160	(42,241)
●	3,096	(11,720)	95,976	(363,270)
●	8,424	(31,890)	261,144	(988,430)
●	14,952	(56,593)	463,512	(1,755,392)

* Based on approximately 60-psi (410-kPa) pressure.

hands of an experienced operator, the instruments can help locate a leak with remarkable accuracy.

The equipment used to detect the sound may be either mechanical or electronic. Two mechanical devices that have been available for many years and are still used today are the aquaphone and the geophone. The aquaphone resembles an old-fashioned telephone receiver with a metal spike protruding where the telephone wire would go. The spike can be placed against a water pipe, against a fire hydrant, or on a valve key that is placed on a valve. It amplifies the sound surprisingly well. The listening end of the geophone looks like a medical stethoscope, and the listening tubes are connected to two diaphragms. When the diaphragms are placed on the ground over a leak, the sound is amplified and the operator gets a stereo effect that aids in determining the direction of the leak.

Powerful and versatile electronic amplifiers are now available (Figure 14-13). The kits generally include equipment for listening for sounds at the ground surface, as well as probes and direct-contact sound-amplification devices. Normal sounds such as wind and traffic will be amplified along with the leak sounds. For this reason, adjustable filters are provided in the amplifier to reduce the band of unwanted frequencies and enhance the frequency band of escaping water.

Dead ends, crosses, tees, and partially closed valves may make locating and verifying a leak more difficult. Another problem with the listening method is that not all leaks can be detected by listening on surface appurtenances. In general, the smallest leaks are the loudest. On the other hand, many very large leaks make no detectable noise at all until a metal rod is driven through the ground to the pipe wall to provide a direct connection for the listening device.

FIGURE 14-13 Portable laptop leak correlator

Correlator method. Firms specializing in contractual leak detection use much more sophisticated equipment. A leak technician surveys a segment of pipe using transducers, and the data are fed into a computer (called a correlator) for analysis. The computer takes into account the pipe size, type of pipe material, and other factors that affect the speed at which sound travels. It is generally able to accurately pinpoint a leak location. Figure 14-14 illustrates how transducers connected to two valves on a water main provide data to identify a leak location. The procedure for a listening survey can vary from just listening on selected hydrants (a process known as skimming) to completely covering a system by listening on all hydrants, valve stems, and services.

Statistical noise analyzer. This system is useful in areas where there may be excessive interference. A hydrophone sensor is attached to a hydrant or tap. The system "listens" for a continuous period (e.g., 2 hours), usually at night. The units are usually installed in the system at intervals up to about 1,500 ft (460 m). A noise signature is produced at each point, and the statistical variance indicates a leak and its approximate location. The system has advantages over human listening systems, but the cost of the equipment may be a barrier for some utilities.

Acoustic leak detection procedures. Steps in conducting a listening survey usually progress in the following sequence:

1. The system survey is best performed at night, when there is relatively little surface noise and low water flow in mains. Operators use a system map to locate all hydrants

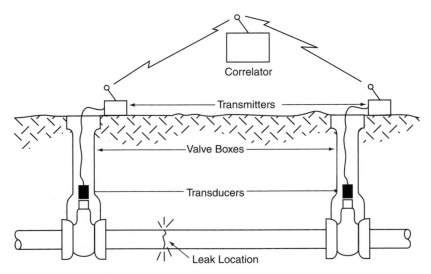

FIGURE 14-14 Pinpointing a leak from data provided by two transducers

in the system. The hydrants are then sounded (checked with a listening device), and those that are noisy are marked on the map.

2. Later, personnel from the transmission and distribution department are consulted for any information that might account for the noisy hydrants.

3. Files are reviewed for any information on abandoned services, previous repairs, and construction that might be the source of a leak.

4. Sewers are checked for possible flow from the potable system. If excessive flow is found in a storm sewer, it may be possible to test a sample for the presence of chemicals that have been added to the treated water, such as fluoride. Finding these chemicals helps confirm the existence of a leak somewhere in the area.

5. Before searching further, the utility should prove that the noise noticed on a hydrant is not from the hydrant valve. Workers can prove this by exercising the valve several times and seeing if the noise stops or the pitch changes.

6. The hydrant auxiliary valve should also be operated to see if doing so makes any difference in the noise.

7. Intersection valves on the mains, the surrounding curb stops, and the customers' outside sill cocks or faucets are then checked for noise level. One of these locations is likely to have significantly louder leakage noise, indicating that the leak is nearby.

8. If the noise seems to be coming from a customer's service line, the service is shut off at the curb stop and the service pressure is relieved by opening an operable faucet. If the noise disappears, the leak is on the customer's side of the service.

9. If the noise is still present after the service is valved off, the leak is on the utility side of the service—most likely on the main near the corporation tap. In this case, the road

surface over the line is sounded with a ground microphone. A mark is then placed on the road where the loudest sound is located to indicate the possible leak. The outside limits where the noise is observed to change or disappear are also indicated on the road surface. In this way, a working area can be identified.

10. To further pinpoint the leak with as little damage to the road surface as possible, the following procedure is used. First, a 3-in. (75-mm) hole is drilled through the pavement over the main. Then, a compressed air supply is used to blow a hole down through the subsoil until the main is reached. If the leaking area has been located, water will usually come to the surface through the probe hole. In areas where probe holes cannot be made, metal rods are driven down to contact the main, and listening devices are applied to the rods.

The major advantages of the listening procedure for locating leaks are that no excavations are needed, distribution crews are not required to operate valves and hydrants, and the work can be performed easily by utility personnel.

Factors affecting leak detection. The following factors affect how well listening devices perform:

- Mains with rubber gasket joints often do not transmit sound much beyond the pipe section that has the leak.
- Copper transmits sound best, followed in order by steel, cast and ductile-iron, plastic, asbestos cement, and concrete.
- Smaller pipes transmit sound well, but as the diameter increases, the quality of the sound diminishes.
- Tees, elbows, and other fittings often amplify sounds, making it difficult to locate a leak.
- Dry, sandy soils produce the best noise transmittance, whereas sounds are dissipated somewhat in loamy soils. Sounds are reduced even further in clay soils.
- Gravel roads and lawn areas are poor transmitters of sound compared with paved surfaces.
- The noise contributed by other buried utilities can also make it difficult to pinpoint a leak.

Water audits

The purpose of a water audit is to select and implement a program to reduce distribution system water losses. The audit identifies the quantity of water lost and the cost. An added benefit is that records and control equipment (such as meters) are checked for accuracy. Before starting the audit, a worksheet to gather the information should be developed and the study period for the audit should be defined. Most utilities select a 12-month calendar year for this purpose. A final step before starting is to select the appropriate unit of measure. Many US utilities use gallons as the preferred measure.

A distribution system water audit consists of eleven main tasks (discussed in detail in AWWA Manual M36, *Water Audits and Loss Control Programs*):

1. Collect distribution system description information
2. Measure water supplied to the distribution system
   - Compile the volume of water from own sources.
   - Adjust figures for total supply including: (a) verify metered accuracy; (b) adjust supply totals; (c) adjust reservoir and tank storage; (d) other adjustments; (e) total all adjustments; and (f) determine the adjusted volume of water from own sources.
   - Compile the volume of water imported from outside sources or purchased from other water utilities.
   - Calculate system input volume.
   - Compile the volume of water exported to outside water utilities or jurisdictions.
   - Calculate the volume of water supplied into the distribution system.
3. Quantify billed authorized consumption
   - Compile the volume of billed authorized consumption (metered water) including: (a) maintain customer accounts data; (b) maintain customer meter and AMR data; (c) compile metered consumption volumes for the water audit period; and (d) adjust for lag time in meter readings.
   - Compile the volume of billed authorized consumption (unmetered water).
4. Calculate nonrevenue water
5. Quantify unbilled authorized consumption
   - Compile the volume of unbilled authorized consumption (metered water).
   - Compile the volume of unbilled authorized consumption (unmetered water) which includes water for (a) fire fighting and training; (b) flushing water mains, storm inlets, culverts, and sewers; (c) street cleaning; (d) landscaping irrigation in public areas; (e) decorative water facilities; (f) construction sites; (h) water quality and other testing; (i) water consumption at public buildings not included in the customer billing system; (j) other unmetered but verifiable uses; (k) sum of all components of unbilled authorized consumption that is unmetered.
6. Quantify water losses
7. Quantify apparent losses
   - Estimate customer meter inaccuracy. This includes checking for proper installation, testing residential meters, and calculating total customer consumption meter error.
   - Estimate systematic data handling error which includes systematic data transfer errors from customer meter reading, systematic data analysis errors, and policy and procedure shortcomings.
   - Estimate unauthorized consumption.
   - Calculate total apparent losses.

8. Quantify real losses
9. Assign costs of apparent and real losses
   - Determine cost impact of apparent loss components.
   - Determine cost impact of real loss components.
10. Calculate performance indicators
    - Calculate the financial performance indicators.
    - Calculate the operational performance indicators which includes apparent losses normalized, real losses normalized, infrastructure leakage index (ILI).
11. Compile the water balance

After the audit, utilities analyze the value of the losses and corrective measures, evaluate potential corrective measures, update the audit, and update their master plan. An example summary from a water audit is shown in Table 14-4.

TABLE 14-4 Typical water audit results

	Water Volume		
	_ML/d_	_mgd_	_%_
Water sold through domestic meters	9.4	2.5	50
Water sold through industrial meters	4.7	1.25	25
Underground leakage located	1.8	0.5	10
Underregistration of industrial meters	0.38	0.1	2
Unauthorized consumption	0.37	0.1	2
Total accounted-for water	16.65	4.45	89
Unmetered use (sewer flushing, fire fighting, street washing)	0.18	0.05	1
Loss through unavoidable leakage at 200 gpd per mile (470 L/d per kilometer) of main	0.07	0.2	4
Unmetered use and underregistration of domestic meters	1.1	0.3	6
Total unaccounted-for water	1.35	0.55	11
Total water produced	18	5	100

Emergency leak repairs

An emergency leak is usually a broken main or a severe service leak. An action plan to deal with main breaks should be established and coordinated with police, fire, and street department personnel. Trained personnel, records, maps, and repair parts should be available at all times in preparation for a major leak.

Preliminary steps. When a major leak is reported, it should be investigated at once to determine the severity of the problem. Utility personnel must determine if immediate protection of private and public property is needed, as well as whether there is any potential danger to pedestrians and traffic. Next, it should be determined, if possible, whether the repair is the responsibility of the utility or a property owner. If the leak is serious or in an area where it could quickly do extensive property damage, it must be repaired immediately.

The first consideration in dealing with the leak should be to get the water loss under control by complete or partial shutdown of the line. Some repairs can be accomplished while the pipeline is still under pressure. Maintaining a positive pressure in the line will ensure that backflow from cross-connections will not occur. Where damage to property is occurring or is likely to occur, water must usually be completely shut off.

If there is no immediate danger to life or property, all valves on that line except one should be closed to reduce the flow. Customers who will be without water should then be notified of approximately when the water will be shut off. They should also be advised to store some drinking water if they will need it during the outage. Special attention should be given to buildings with sprinkler systems, stores with water-cooled refrigeration units, industrial water users, and large users such as hospitals. These customers will have to take special steps to prepare for the water being off. If buildings are unoccupied, the owner, manager, or agent should be notified.

If the leak is at the bottom of a hill, customers at higher elevations should be notified to shut off the inlet valve at their meter to prevent siphoning out of hot-water tanks or softeners. If the occupants are not at home, the curb stops should be shut off.

In some instances, the utility may find it difficult to locate valves and mains because of incomplete records or snow cover. Having electronic locators on hand to locate metallic boxes and pipe can greatly facilitate a water main repair job (Figures 14-15 and 14-16).

The availability of information from a regular valve-locating and valve-exercising program will be helpful in an emergency situation. The valves that are operated and their operating conditions should be recorded. The section of the system that is out of service should be noted, and the locations of all hydrants affected should be reported to the fire department. After customers have been notified, the final valve can be shut down. Record all valve position changes so that they can be returned to service following the repair. An up-to-date system map showing accurate valve locations and main sizes is valuable at this time.

Locating the leak. When water is flowing out of the ground, it may seem obvious where to look for a leak, but time and money can be wasted digging in the wrong spot. The leak is sometimes not directly below where the water comes to the surface. Occasionally, the water

FIGURE 14-15 Electronic pipe detector
Courtesy of Fisher Research Laboratory,
Los Banos, Calif.

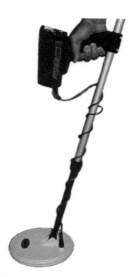

FIGURE 14-16 Valve and box locator
Courtesy of Fisher Research Laboratory,
Los Banos, Calif.

may show up a long way from the leak. A utility employee should pinpoint the leak using a listening device and probing the ground before starting to dig. Some water pressure will have to be left on or temporarily restored to the affected area during the leak-locating phase.

A large leak can undermine roads, walks, railways, or other utilities. In addition, basements may be flooded and slippery driving conditions may result from flooding or freezing water. It may be necessary to arrange for police assistance in controlling traffic and for street department assistance in salting or sanding roadways.

Before a leaking line is excavated for repair, other utilities must be contacted to determine where their underground lines are located. In some areas, all utilities cooperate in a single-call system that immediately alerts all utilities of the need for their assistance.

Excavation. In addition to excavating equipment, the following items should be on hand in preparation for making the leak repair:

* A pump for dewatering the excavation
* A sturdy ladder for entering and leaving the excavation
* Shovels, wrenches, and other hand tools
* Planks and timbers for shoring trench walls
* Traffic control equipment, including barricades, flashers, and cones
* Pipe cutters or saws, as well as proper-size repair clamps, couplings, pipe, and sleeves

- An air compressor and hammer if pavement has to be cut or a clay spade for use in hand excavation
- A generator, lights, and flashlights if the work will be done at night
- Safety ropes or ladders if personnel must work where sudden flooding or a cave-in is possible
- A 5-gal (19-L) bucket with rope, to lower tools into the excavation

The excavation for repairing a leak should normally be parallel to the main or service. It should be located so that the pipe is to one side of the ditch. This will allow a worker to stand next to the pipe while making the repair. A sump hole should also be dug in a far corner of the excavation. The pump suction is placed in this hole to keep the bottom of the excavation as free of standing water as possible. The excavation should not be made too small in an attempt to save time. The extra time taken to make a neat hole that is large enough to work in will usually be more than made up for by the time saved in making the pipe repair quickly.

Safety procedures should not be ignored in the haste to handle the emergency. Accidental injuries are especially likely in unplanned situations. If soil conditions are poor or if the soil is unstable from being saturated by the leak, it may be necessary to install sheeting or shoring to protect workers from a trench cave-in.

Leak repair. If a break is severe, the damaged section may have to be cut away with a saw or pipe cutter and a new section installed. When a new section or appurtenance is inserted into a dewatered pipeline, the materials are disinfected with chlorine according to procedures described in AWWA Standard C651 (most recent edition). Most main breaks are straight across and can be repaired with a flexible clamp or coupling. Two types of repair devices are shown in Figure 14-17. The manufacturer's recommendations should be carefully followed when installing the repair device.

FIGURE 14-17 Pipeline repair clamps

Several preliminary procedures should be observed to help ensure a quick and efficient repair. The diameter of the pipe should be checked to make certain the correct-size clamp is being installed. After the pipe is uncovered, the area where the clamp will be installed should be scraped and washed to remove as much dirt and corrosion as possible.

When applying the clamp, the workers should make sure that no foreign material sticks to the gasket as the bottom half is brought around the pipe. No material should become lodged between the gasket and pipe as the bolts are tightened. A properly sized ratchet wrench should be selected in advance so that the sleeve bolts can be quickly tightened once the clamp is installed.

After the clamp is installed, some utilities apply a tar coating for rust prevention. The utility should then restore line pressure by partially opening one valve to test the repair for leaks before the excavation is backfilled. When the repair has been completed and tested, the main should be flushed to remove any air or dirt that may have entered while it was under repair. The entire line should be chlorinated to reduce the danger of contamination from soil, dirty water, and backsiphonage. The line is then flushed to remove high-chlorine water and bacteria testing is conducted (follow requirements of AWWA Standard C651, most recent edition).

Restoration. All unsuitable excavated material should be hauled away and replaced with suitable material for backfilling, with proper compaction. It is generally best to use sand, road base, crushed stone, or processed material for backfilling under pavement so that permanent street repairs can be made at once. Coordinate with the street department to facilitate repairs. The area should be checked to make sure no equipment or traffic hazards, such as ice or erosion, remain. All valves should be returned to operating position. Customers, the police, and the fire department should be notified that the repair has been completed. Customers may need to clear their service line by running cold water from an outside faucet.

Record keeping. A record of every water main break or leak repaired should be kept for future reference. The cause should be identified, such as shear break, pipe split, blowout, or joint leak. If any old pipe is removed during the repair, a piece should be kept and tagged with the location and date. This piece can be used for future reference on the condition of the pipe interior.

External pipeline corrosion

The soil surrounding buried pipe can cause external corrosion of some pipe materials if the pipe is not properly protected. Soil characteristics vary greatly from one area of the country to another or even from one part of a community to another. In general, asbestos–cement and concrete pipe will suffer harmful external corrosion only under aggressive soil conditions. Cast-iron and ductile-iron pipe do not need any additional protection in most soil conditions, but polyethylene wrap or other technologies may be required in aggressive soil. Steel pipe should be well coated and provided with cathodic protection under essentially all soil conditions.

The electrochemical corrosion process that takes place on the outside of metal pipes is very similar to internal corrosion. It involves a chemical reaction and a simultaneous flow of electrical current. Differences in the soil and the pipe surface create differences of electrical potential on the buried metal to form a corrosion cell. A direct electrical current flows from the anodic area to the cathodic area through the soil and returns to the anodic area through the metal.

Corrosion cells are generally created in metallic pipe by the surface impurities, nicks, and grains that are present. Oxygen differentials in the soil can also create corrosion cells. Severe corrosion of buried metal pipe can be serious enough to cause leakage, breaks, and a shortened service life for the pipe.

Corrosion frequently takes the form of pits in an otherwise relatively undisturbed pipe surface. These pits occur at the anodic areas and may eventually penetrate the pipe wall. Cathodic areas are generally protected in this process. This type of corrosion will occur on steel, cast-iron, and ductile-iron pipe.

Factors affecting external corrosion. The degree of corrosion and the size of the area affected by corrosion depend on the quantity, size, and intensity of the corrosion cells. This intensity, in turn, depends on the corrosivity of the soil. Some soil characteristics and conditions that are likely to increase the rate and amount of corrosion include

- high moisture content;
- poor aeration;
- fine texture (e.g., clay or silty materials);
- low electrical resistivity;
- high organic material content (such as in swamps);
- high chloride and/or sulfate content, which increases electrical conductivity;
- high acidity or high alkalinity, depending on the metal or alloy;
- presence of sulfide; and
- presence of anaerobic bacteria.

Most of these characteristics affect the electrical resistivity of a soil, which is a fair measure of a soil's corrosivity. The amount of metal removed from the pipe wall is proportional to the magnitude of the current flowing in the corrosion cell. The amount of current is, in turn, inversely proportional to the electrical resistivity of the soil or directly proportional to the soil conductivity. Therefore, a soil of low resistivity or high conductivity fosters corrosion currents and increases the amount of metal oxidized.

It has also been found that certain microorganisms in the soil can accelerate corrosion. Under anaerobic conditions (i.e., no oxygen present), iron bacteria or sulfate-reducing bacteria can accelerate corrosion.

Methods of preventing external corrosion. Several solutions to the problem of corrosion have been tried. The solution that will be most effective in any situation will vary depending on the pipe material and the chemical and electrical conditions of the surrounding soil. Methods that have been found effective under certain conditions include

- specifying extra thickness for pipe walls,
- applying a protective coating,
- wrapping pipe in polyethylene plastic sleeves, and
- installing cathodic protection on the pipe.

Adding extra thickness to a pipe as a corrosion allowance is not an effective long-term remedy. Pipe corrosion is not uniform. Specifying a corrosion allowance for buried pipe only postpones the problem. Corrosion-prevention methods are much more cost effective.

Applying a hot bitumastic or coal-tar coating to the pipe exterior may not totally reduce soil corrosion. If there are any pinholes or breaks in the coating, bare metal at these locations will be anodic to the surrounding coated pipe. Corrosion will be concentrated in these areas. Coated pipe must be inspected and handled carefully to ensure that the coating is not damaged.

One way to control corrosion of cast-iron or ductile-iron pipe in corrosive soils is to encase the pipe in a polyethylene sleeve, as shown in Figure 14-18. When a tap is installed on pipe with a polyethylene sleeve, care must be taken to not expose the bare pipe around the completed tap. The method frequently used to easily prevent corrosion around the tap is to apply several wraps of heavy tape around the pipe in the area to be tapped before the tap is made. There is controversy on the effectiveness of poly-wrapping, as some corrosion engineers believe it makes things worse when water is present.

Cathodic protection stops corrosion by canceling out, or reversing, the corrosion currents that flow from the anodic area of a corrosion cell. There are two principal means of applying cathodic protection: sacrificial anodes and impressed-current systems.

Sacrificial anodes, also called galvanic anodes, consist of magnesium-alloy or zinc castings that are connected to the pipe through insulated lead wires, as shown in Figure 14-19. Magnesium and zinc are anodic to steel and iron, so a galvanic cell is formed that works opposite to the corrosion being prevented. As a result, the magnesium or zinc castings corrode (sacrifice their metal), and the pipe, acting as a cathode, is protected.

Impressed-current cathodic protection systems use an external source of direct current (DC) power that makes the structure to be protected cathodic with respect to some other metal in the ground. In most cases, rectifiers are used to convert alternating current (AC) power to DC. A bank of graphite or specially formulated cast-iron rods is connected into the circuit as the anode (corroding) member of the cell. The distribution system component is the protected cathode (Figure 14-20).

Cathodic protection is a very complicated and technical subject. A competent consultant should be involved early in any problem analysis or system revision.

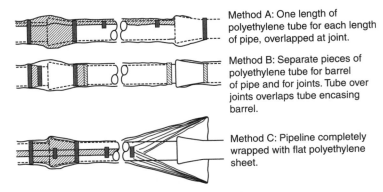

Method A: One length of polyethylene tube for each length of pipe, overlapped at joint.

Method B: Separate pieces of polyethylene tube for barrel of pipe and for joints. Tube over joints overlaps tube encasing barrel.

Method C: Pipeline completely wrapped with flat polyethylene sheet.

FIGURE 14-18 Polyethylene encasement of ductile-iron pipe
Courtesy of Farwest Corrosion Control Company

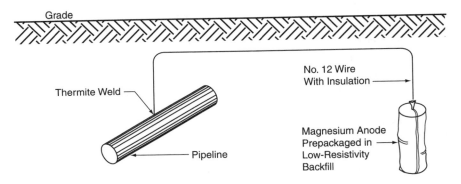

Grade

Thermite Weld

Pipeline

No. 12 Wire With Insulation

Magnesium Anode Prepackaged in Low-Resistivity Backfill

FIGURE 14-19 Cathodic protection using sacrificial anodes
Courtesy of Farwest Corrosion Control Company

Bimetallic corrosion

Another form of corrosion a water system operator may encounter is bimetallic corrosion. This type of corrosion occurs when two electrochemically dissimilar metals are directly connected to each other. The combination most often noticed in plumbing systems is the direct connection of a brass fitting to galvanized iron pipe. The two metals form a corrosion cell or, more exactly, a galvanic cell (Figure 14-21). This can result in loss of the anodic metal (in this case, the galvanized pipe) and protection of the cathodic metal (the brass fitting).

A practical galvanic series for common pipe and fitting materials is shown in Table 14-5. Each metal is anodic to, and may be corroded by, any metal below it. The greater the separation between any two metals in the series, the greater the potential for corrosion and (barring other factors) the more rapid the corrosion process.

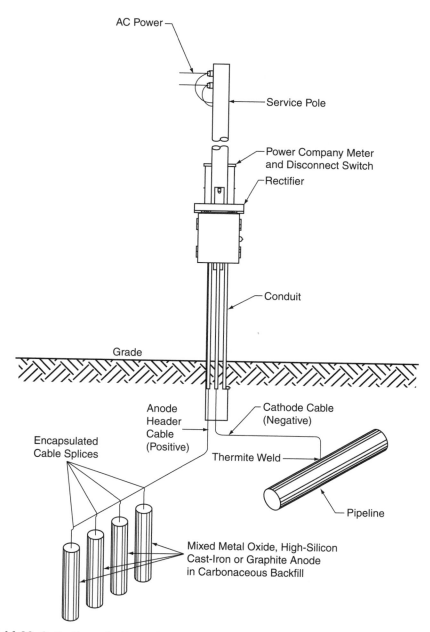

AC Power

Service Pole

Power Company Meter
and Disconnect Switch

Rectifier

Conduit

Grade

Anode
Header
Cable
(Positive)

Cathode Cable
(Negative)

Encapsulated
Cable Splices

Thermite Weld

Pipeline

Mixed Metal Oxide, High-Silicon
Cast-Iron or Graphite Anode
in Carbonaceous Backfill

FIGURE 14-20 Cathodic protection using impressed current
Courtesy of Dr. J.R. Myers, JRM Assoc., Franklin, Ohio 45005

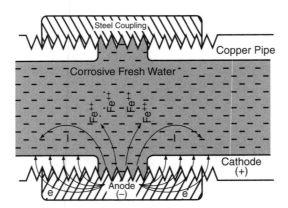

FIGURE 14-21 Small anode, large cathode

TABLE 14-5 Galvanic series for metals used in water systems

Corroded End (Anode)	Most Active
Magnesium	+
Magnesium alloys	
Zinc	
Aluminum	
Cadmium	
Mild steel	
Wrought (black) iron	Corrosion Potential
Cast iron	
Lead–tin solders	
Lead	
Tin	
Brass	
Copper	
Stainless steel	−
Protected End (Cathode)	Least Active

Corrosion control methods for bimetallic connections are designed to break the electrical circuit of the galvanic cell. Dielectric barriers, or insulators, are the method most commonly used to stop the flow of corrosion current through the metal. In addition, coating a buried bimetallic connection reduces the amount of corrosion current passing through the soil.

Stray-current corrosion

Stray-current corrosion is caused by DC current that leaves its intended circuit, collects on a pipeline, and discharges into the soil. The classic problem that affected many old water systems involved currents generated by electric trolley cars and light rail systems, as shown in Figure 14-22. The same effect occurs as a result of DC current from any number of sources, including cathodic protection being applied to other utilities or structures, as well as subways.

Corrosion caused by stray-current discharge often appears as deep pits concentrated in a relatively small area on the pipe. If a main or service has this appearance, particularly at the crossing of a cathodically protected pipeline or transit facility, electrical tests should be conducted to determine if interference is the cause of the problem. Since many liquid-fuel and natural-gas pipelines are required to have cathodic protection, these lines become potential sources of stray current.

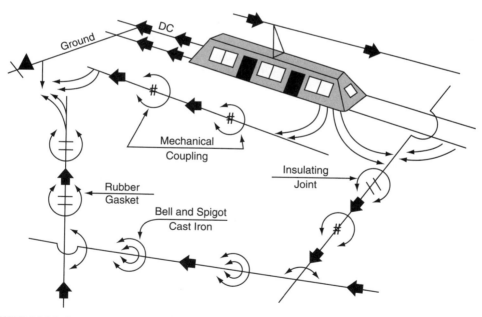

FIGURE 14-22 Stray current flow on underground pipelines

Like cathodic protection, stray-current corrosion is a very complicated process. Consultants and experts, available through engineering firms specializing in cathodic protection, may be required to help design a system that causes minimal interference with other metallic structures.

Operation and Maintenance Safety

The same safety precautions detailed in chapter 12 for water main installation apply to system operation and maintenance work. The one additional precaution that cannot be overemphasized is the need to work carefully under emergency conditions.

The old adage of "haste makes waste" is usually true. During an emergency repair, sufficient time must be taken to think through what must be done, how it is best done, and what tools and materials will be required. A few extra minutes taken in preparation can often save the extra time later required to get equipment that was forgotten. The urge to hurry repairs to save wasted water and minimize inconvenience to customers must not overshadow the safety of workers. A worker's life is worth a lot more than a few gallons of water.

BIBLIOGRAPHY

AWWA Research Foundation and DVGW-Forschungsstelle. 1996. *Internal Corrosion in Water Distribution Systems, 2nd ed.* Denver, Colo.: American Water Works Association Research Foundation.

AWWA Standard for Disinfecting Water Mains. ANSI/AWWA C651. Denver, Colo.: American Water Works Association (latest edition).

AWWA Standard for Distribution Systems Operation and Management. ANSI/AWWA G200. Denver, Colo.: American Water Works Association.

Cesario, L. 1995. *Modeling, Analysis, and Design of Water Distribution Systems.* Denver, Colo.: American Water Works Association.

Chadderton, R.A., G.L. Christensen, and P. Henry-Unrath. 1992. *Implementation and Optimization of Distribution Flushing Programs.* Denver, Colo.: American Water Works Association Research Foundation and American Water Works Association.

Clark, R.M., and W.M. Grayman. 1998. *Modeling Water Quality in Drinking Water Distribution Systems.* Denver, Colo.: American Water Works Association.

Deb, A.K., K.A. Momberger, Y.J. Hasit, and F.M. Grablutz. 2000. *Guidance for Management of Distribution System Operation and Maintenance.* Denver, Colo.: American Water Works Association Research Foundation and American Water Works Association.

Great Lakes–Upper Mississippi River Board of State Public Health and Environmental Managers. 2007. *Recommended Standards for Water Works: Policies for the Review and*

Approval of Plans and Specifications for Public Water Supplies, A Report of the Water Supply Committee of the Great Lakes–Upper Mississippi River Board of State and Provincial Public Health and Environmental Managers. Albany, NY.

Hydrant Flow Tests. 2010. Denver, Colo.: American Water Works Association. DVD.

Kirmeyer, G.J., M. Friedman, J. Clement, A. Sandvig, V. Snoeyink, W. Kriven, and A. Camper. 2000a. *Distribution System Water Quality Changes Following Implementation of Corrosion Control Strategies.* Denver, Colo.: American Water Works Association Research Foundation and American Water Works Association.

Kirmeyer, G.J., M. Friedman, J. Clement, A. Sandvig, P.F. Noran, K.D Martel, D. Smith, M. LeChevallier, C. Vlok, E. Antoun, D. Hiltebrand, J. Dyksen, and R. Cushing. 2000b. *Guidance Manual for Maintaining Distribution System Water Quality.* Denver, Colo.: AWWA Research Foundation and American Water Works Association.

Manual M3, Safety Practices for Water Utilities. 2002. Denver, Colo.: American Water Works Association.

Manual M20, Water Chlorination/Chloramination Practices and Principles. 2006. Denver, Colo.: American Water Works Association.

Manual M27, External Corrosion—Introduction to Chemistry and Control. 2004. Denver, Colo.: American Water Works Association.

Manual M28, Rehabilitation of Water Mains. 2001. Denver, Colo.: American Water Works Association.

Manual M36, Water Audits and Loss Control Programs. 2009. Denver, Colo.: American Water Works Association.

Mays, L.W. 1999. *Water Distribution Systems Handbook.* New York: McGraw-Hill.

National Fire Protection Association. 2007. *Standard on Water Supplies for Suburban and Rural Fire Fighting.* NFPA 1142. Quincy, Mass.: NFPA.

O'Connor, J.T., and B.J. Brazos. 1993. *The Effect of Lower Turbidity on Distribution System Water Quality.* Denver, Colo.: American Water Works Association and AWWA Research Foundation.

Orlando Utilities Commission and CH2M HILL, Inc. 1995. *Evaluation of the Effects of Electrical Grounding on Water Quality.* Denver, Colo.: American Water Works Association and AWWA Research Foundation.

Peet, J.R., S.J. Kippin, J.S. Marshall, and J.M. Marshall. 2000. *Water Quality Impacts From Blending Multiple Water Types.* Denver, Colo.: American Water Works Association and AWWA Research Foundation.

Pierson, G., K. Martel, A. Hill, G. Burlingame, and A. Godfree. 2001. *Practices to Prevent Microbiological Contamination of Water Mains.* Denver, Colo.: AWWA Research Foundation and American Water Works Association.

Reiber, S. 1993. *Chloramine Effects on Distribution System Materials.* Denver, Colo.: American Water Works Association and AWWA Research Foundation.

Tikkanen, M., J. Schroeter, L. Leong, and R. Ganesh. 2001. *Guidance Manual for Disposal of Chlorinated Water.* Denver, Colo.: AWWA Research Foundation and American Water Works Association.

Water Distribution Operator Training Handbook. 2005. Denver, Colo.: American Water Works Association.

Water Quality Modeling in Distribution Systems. 1991. Denver, Colo.: American Water Works Association and AWWA Research Foundation.

WSO: Flushing and Cleaning. 2006. DVD. Denver, Colo.: American Water Works Association.

CHAPTER 15

Water Services

Water service lines are the pipes and tubing that lead from a connection on the water main to a connection with the customer's plumbing. Service pipe size varies depending on the pressure at the main, the distance from the main to the building, the quantity of water required, and the residual pressure required by the customer. In most cases, each customer is served through an individual service line. However, there are some exceptions, such as multifamily housing units, which often have only a single service line.

METER LOCATIONS

Good water utility practice is to have a meter on every water service, including schools, churches, parks, public buildings, and the drinking fountain in the town square. The meter's purpose is not simply to keep track of what the customer must pay. A meter also reduces waste by identifying leaking plumbing. It also provides the best possible accounting of the water used on the system. The location of the meter on a service is, for the most part, governed by both climate and local custom.

Exposed Meters

In areas where there is no danger of freezing, meters can be placed almost anywhere. Some water utilities allow meters to be located in a garage or on the side of a building, as long as the meters are reasonably protected from damage and vandalism.

Meter Boxes

Many water systems install meters in a box, which is usually located in the street right-of-way. Depending on local custom and the size or depth of the box required, meter boxes may also be called meter pits or meter vaults. The following are some of the primary advantages and disadvantages of installing meters outside in boxes:

- In areas where ground frost is nominal, a meter box can be relatively small, inexpensive, and easy to install.
- If the boxes can always be readily located, meter boxes make meter reading easier because meter readers do not have to enter a building. On the other hand, if all meters are located inside buildings and are equipped with remote reading devices, the reading time is generally about the same.
- In areas where most buildings are built without basements, it is often difficult to find a satisfactory inside meter location. Placing all meters outside eliminates the need to find suitable locations.

- Placing meters outside eliminates the need to enter buildings to replace the meters, which is difficult in some inside locations.
- It is common to install meters in a box if there is a long distance between the main and buildings. An example is a rural water system that serves individual farmhouses that may be some distance from the main.
- In areas where there is deep frost, meter pits are much more expensive to construct. However, many water systems still require meter pits if there is no appropriate location for a meter in a building.
- Deeper pits often fill with water and must be pumped before the meter can be read. In addition, pit covers are subject to damage if driven over by a heavy vehicle.
- It is sometimes difficult to locate pits that are covered with snow. It may be necessary to remove snow before the meter can be read.

A typical water service with a shallow meter box is illustrated in Figure 15-1. Where meter pits are used in cold climates, the meter can still be raised to near the top of the pit to facilitate reading; meters will not freeze in these locations. If the pit bottom is below the frost line and the cover is tight, the pit will gather enough warmth from the ground to avoid freezing (Figure 15-2). Under extremely cold conditions, pits are sometimes constructed with a double cover to provide added insulation.

Meters Located in Buildings

In areas where most buildings have basements, it is common to locate meters in the basement, as illustrated in Figure 15-3. The service pipe is usually brought up through the floor adjacent to the foundation at a point nearest to the main. The meter should be installed at this point. It is not considered good practice to locate the meter at a distance from the entry point because the property owner might install an illegal connection ahead of the meter.

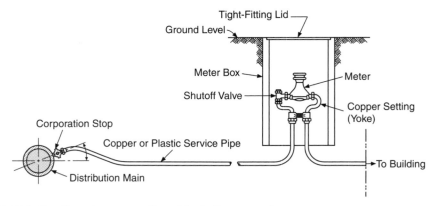

FIGURE 15-1 Small service connection with shallow meter box

Water utilities that install meters in basements may be faced with the problem of where to install meters if the buildings only have a crawl space or are constructed on a concrete slab. For these buildings, some utilities require meters to be installed in meter boxes. Other utilities will allow installation in a crawl space or utility closet as long as (1) the meter can easily be accessed by a meter reader or (2) the meter is equipped with a remote reading device.

SERVICE LINE SIZES, MATERIALS, AND EQUIPMENT

This section discusses the sizes and materials of service lines, as well as various additional pieces of equipment.

Service Line Size

The proper size service required to serve residences and other buildings depends primarily on the following conditions:

- The maximum water demand the customer will require
- The main pressure under peak demand conditions
- Any substantial elevation difference between the main and the highest portion of the building to be served
- The distance between the main and the connection to building plumbing

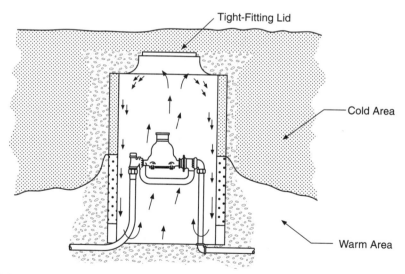

FIGURE 15-2 Cold-climate meter installation

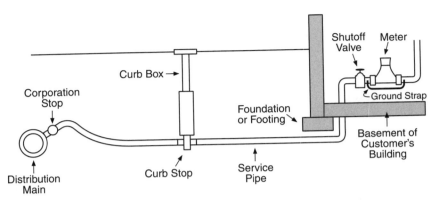

FIGURE 15-3 Small service connection with the meter located in a basement

Single-family residences are most commonly served with a ¾-in. (19-mm) service line. Larger residences, buildings with flushometer toilets, and buildings located far from the main connection should have a 1-in. (25-mm) or larger service. Apartment buildings, hospitals, schools, and industries all have different use patterns. The principal factor is how much water is likely to be required at one time. Additional details on proper sizing of service lines are included in American Water Works Association (AWWA) Manual M22, *Sizing Water Service Lines and Meters.*

Types of Service Line Pipe and Tubing

The materials that have been used for service line pipe and tubing include lead, galvanized iron, copper, and plastic. All materials used for service lines should comply with the applicable standards (AWWA Standard C800, ANSI/NSF Standard 61).

Lead pipe

Water services installed on older water systems were generally made of lead. Lead pipe was used as far back as the time of the Roman Empire. It was favored because it is relatively long-lasting and flexible. Joining and attaching fittings to lead pipe involved a molten lead process known as lead wiping, which required considerable expertise by professional plumbers.

Use of lead services gradually decreased as newer materials were found to be less expensive in terms of both material costs and installation. Most older water systems still have many old lead services in use, but these services are gradually being replaced. Most water utilities now have a policy that a leaking lead service cannot be repaired. Instead, it must be replaced with a new service made of another material. The principal reason for such a policy is that evidence points to health risks due to lead leached from plumbing materials. It is therefore advisable to eliminate lead from the system whenever reasonably possible.

Galvanized iron pipe

Many older systems allowed the use of galvanized iron pipe in place of lead because of the considerably lower cost. The pipe was usually connected to the corporation stop with a lead "gooseneck" a few feet long to provide flexibility at the connection.

In some soil, galvanized pipe works fairly well. However, it can fail within a year or so under corrosive soil conditions. In addition to general corrosion from contact with soil, galvanic corrosion often causes disintegration of the pipe at the connection with brass fittings unless dielectric fittings are used.

Newer materials have now been found to be much longer-lasting. They have essentially replaced galvanized pipe for new installations. Some older systems, though, may have many galvanized iron services still in use.

Copper tubing

Copper tubing became a popular replacement for lead and galvanized iron in service line installations because it is flexible, easy to install, corrosion resistant in most soils, and able to withstand high pressure. It is not sufficiently soluble in most water to be a health hazard, but corrosive water may dissolve enough copper to cause green stains on plumbing fixtures. Allowable copper concentration in drinking water at the tap is regulated by the Lead and Copper Rule. In some cases, corrosion inhibitors must be used to control the copper content. Brass and bronze valves and fittings can be directly connected to the pipe without causing appreciable galvanic corrosion. Copper water service tubing is usually connected by either flare or compression fittings. Although interior copper plumbing is usually connected with solder joints, this method is rarely used for buried service lines.

Plastic tubing

Of the many types of plastic tubing available, three are generally used for water services: polyvinyl chloride (PVC), polyethylene (PE), and polybutylene (PB). PB is not generally used for new water services since structural problems have occurred in older services. Plastic pipe has a very smooth interior surface, so it has lower friction loss than other types of pipe. It is also very lightweight, making installation easy. The types of tubing generally used for small water services are quite flexible, which allows considerable deflection as the pipe is laid. The pipe is also relatively resistant to both internal and external corrosion.

Plastic tubing must not be used at any location where gasoline, fuel oil, or industrial solvents have contaminated the soil or could do so in the future. Not only will these substances soften the plastic and cause it to eventually fail, but low concentrations will pass through the plastic walls by a process known as permeation. The chemical contamination of the water can cause taste and odor and it may also cause adverse health effects in persons drinking the water. Plastic, because it is nonmetallic, has other disadvantages. It cannot be located with an electronic pipe finder and it cannot be thawed by electric current.

All plastic tubing to be used for potable water must have the NSF International seal of approval printed along the exterior. This ensures that the pipe is made of a material that will not leach products that will lead to tastes, odors, or the formation of harmful chemicals. Plastic pipe is also covered in several AWWA standards in the C900 series. Additional information on specifications and approval for plumbing materials is provided in appendix A.

Adapters and Connectors

Flare fittings are available for connecting lengths of copper tubing and adapting to iron pipe thread. The same fittings can be used for plastic pipe of the same dimensions.

Compression fittings can be used with either copper or plastic tubing or iron or lead pipe. These fittings seal by means of a compressed beveled gasket. Compression fittings also have a gripping mechanism that positively secures the pipe or tubing. When they are used for PE or PB plastic tubing, it is also recommended that an insert stiffener be placed inside the tube to strengthen the joint. Where stray currents are a problem, dielectric couplings should be used to prevent corrosion.

Corporation Stops

The valve used to connect a small-diameter service line to a water main is called a corporation stop. The plug-type corporation stop has been used for connecting water services to mains for more than 100 years. The corporation stop is also referred to as the corporation cock, corporation tap, corp stop, corporation, or simply corp or stop.

Corporation stops are available in many different sizes and styles, including several combinations of threads and connections for the inlet and outlet ends. The inlet thread on the standard AWWA corporation stop is the one most commonly used. It is commonly known as the Mueller thread. The other thread in general use is the iron-pipe thread. Both threads have approximately the same number of threads per inch. The Mueller thread has a larger diameter and a steeper taper, which gives it greater strength (see AWWA Standard C800, most recent edition).

Corporation-stop outlet ends are available for connection to a variety of service line pipe and tubing materials, including flared copper service connections, iron-pipe threads, increasing iron-pipe threads, lead flange connections, and compression couplings for various materials. Some corporation-stop outlets also include an internal driving thread used for attaching an installation machine adapter. Three plug-type corporation stops are illustrated in Figure 15-4.

A relatively new development is a corporation stop that has a ball valve instead of a plug valve (Figure 15-5). This type of valve operates more easily. It is generally rated for higher pressures than a plug-type corporation valve.

Curb Stops and Boxes

When meters are located outside in a meter box, a shutoff valve for use in temporarily shutting off the service is installed in the box on the inlet side of the meter. When meters are located inside, a shutoff valve called a curb valve or curb stop is installed in the service line. The valve is usually located either between the street curb and the sidewalk or on the property line.

FIGURE 15-4 Plug-type corporation stops with different pipe connections
Courtesy of the Ford Meter Box Company, Inc.

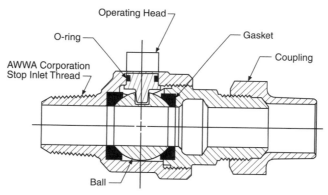

FIGURE 15-5 Principal parts of a ball-style corporation stop
Courtesy of A.Y. McDonald Mfg. Co., Dubuque, Iowa

A curb box is a pipe that extends from the curb stop to the ground surface. It allows the valve to be operated with a special key if an operator needs to temporarily discontinue water service to a building. Two styles of curb stops and boxes are in general use in the United States. Arch-pattern base boxes are intended to straddle a curb stop that has no threads on the top (Figure 15-6). The other type is the Minneapolis-style box, which is threaded at the bottom to screw onto threads provided on Minneapolis-style curb stops (Figure 15-7). Water systems generally use one style or the other as a standard.

FIGURE 15-6 Arch-pattern curb box and curb stop
Courtesy of the Ford Meter Box Company, Inc.

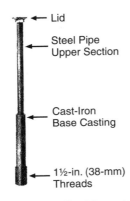

Lid

Steel Pipe
Upper Section

Cast-Iron
Base Casting

1½-in. (38-mm)
Threads

FIGURE 15-7 Minneapolis-style curb box and curb stop
Courtesy of A.Y. McDonald Mfg. Co., Dubuque, Iowa

Each style has some advantages and disadvantages. The arch-pattern base can get sand or soil worked up from the bottom. It can also shift so that it is difficult or impossible to engage the valve handle with a shutoff key. However, if the box is pulled from the ground by construction equipment, the service line won't normally be damaged. With Minneapolis-style curb stops and boxes, the valve key will almost always engage easily. However, if the box is pulled from the ground, it will probably bring the curb stop and service line up with it.

Curb stops are available with a variety of connections for different types of service pipe and tubing. They may have either plug-style or ball-style valves (Figure 15-8).

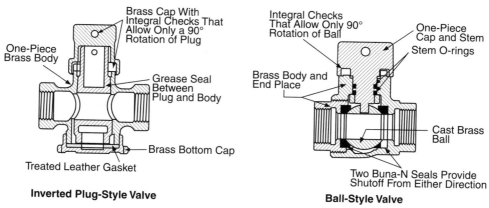

Inverted Plug-Style Valve

Ball-Style Valve

FIGURE 15-8 Principal styles of curb stops
Courtesy of A.Y. McDonald Mfg. Co., Dubuque, Iowa

WATER SERVICE TAPS

Connections for water services can be made either when the water main pipe is empty (dry taps) or when the pipe is filled with water under pressure (wet taps). Dry taps are usually made only when a new main is installed. Wet tapping is preferred when a service connection is being added to an existing main. It allows the connection to be made without turning off the water or interrupting service to existing customers. There is also less chance of contamination because the pressure in the main tends to expel any foreign matter.

When small taps are made on cast-iron or ductile-iron pipe, the corporation stop is usually screwed directly into a threaded hole in the pipe wall. This process is called direct tapping. Drilling-and-tapping machines make it possible to perform this operation without shutting down the main.

Direct tapping of 6-in. (150-mm) and larger asbestos–cement (A–C) and PVC pipe with ¾-in. (19-mm) and 1-in. (25.4-mm) corporation stops is possible, but it must be done very carefully. The pipe manufacturer's instructions should be followed explicitly. It is a good idea to make a practice tap on a piece of scrap pipe before attempting one on a pressurized main. Utility personnel should use a service saddle if they have any doubts about making the installation. AWWA withdrew all of its asbestos-cement standards.

Direct Insertion

Tapping a water main and inserting a corporation stop directly into the pipe wall requires a tapping machine. This machine actually performs three operations: drilling, tapping (threading), and inserting the corporation stop. The machine can also be used to remove a corporation stop. AWWA Standard C223, *Fabricated Steel & Stainless Steel Tapping Sleeves*, describes fabricated steel and stainless steel tapping sleeves use to provide outlets on pipe. The sleeves are intended for pipe sizes 4 in. (100 mm) through 48 in. (1,200 mm) with branch outlets through 36 in. (900 mm). This standard includes requirements for materials, dimensions, tolerances, finishes, and testing. This standard is not intended to apply to tapping sleeves welded to pipe. Fabricated tapping sleeves shall be manufactured from steel or stainless steel and are intended for use in systems conveying water. For outlets and main sizes greater than those specified, consult the manufacturer.

Small tapping machines (Figure 15-9) are used for direct installation of ½-in. through 2-in. (13-mm through 50-mm) corporation stops into a main. The hole is drilled into the pipe and the threads are cut by a combination drill and tap. The process of making a tap generally involves the following steps, as shown in Figure 15-10:

1. The water main pipe is excavated and cleaned in the area to be tapped. The drilling-and-tapping machine is clamped in place.
2. The first operation with the machine is to bore a hole into the pipe wall. The water pressure is contained within the sealed body of the machine. After the drill has fully penetrated the pipe, it is advanced further into the pipe to engage the tap. The tap then cuts threads for the corporation stop.
3. The boring bar is retracted. Operators contain the pressurized water from the drilled hole by closing the flapper valve.
4. The combined drill-and-tap tool is removed from the end of the boring bar and replaced with a corporation stop. The corporation stop must be in a closed position before it is inserted.

FIGURE 15-9 Drilling-and-tapping machine
Courtesy of Mueller Company, Decatur, Ill.

5. The bar is reinserted into the machine, and the corporation stop is screwed into the threaded hole.

6. The machine can then be removed. The corporation stop is ready to have a service line attached to it.

7. After the service line installation is completed, the corporation-stop valve is opened. The excavation is ready to be backfilled.

Service Saddles

For taps larger than 1 in. (25 mm) on A–C or PVC pipe or for smaller taps where conditions make it advisable to ensure a perfect tap, a service clamp should be used instead of a direct tap (Figure 15-11). Using a clamp eliminates the chances of the pipe splitting because of the tap and it avoids the problem of the corporation-stop threads not holding properly in the pipe material. Many water utilities have found it safest to use a clamp regularly for all taps on A–C pipe.

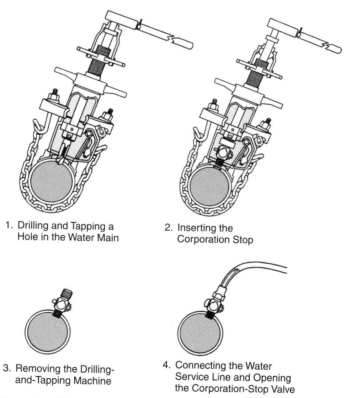

1. Drilling and Tapping a Hole in the Water Main

2. Inserting the Corporation Stop

3. Removing the Drilling-and-Tapping Machine

4. Connecting the Water Service Line and Opening the Corporation-Stop Valve

FIGURE 15-10 Steps in making a service tap
Courtesy of Mueller Company, Decatur, Ill.

FIGURE 15-11 Some styles of service saddles
Courtesy of the Ford Meter Box Company, Inc.

Small Drilling Machines

Small drilling machines may be used to connect service lines 2 in. (50 mm) and smaller, using a service clamp. A manual drilling machine is shown in Figure 15-12. As illustrated in Figure 15-13, the steps in completing a tap are as follows:

1. The pipe is cleaned all the way around and the service clamp is installed.
2. The opened corporation stop is screwed into the service clamp. The drilling machine is then attached to the corporation stop.
3. The drill bit of the drilling machine is extended through the open corporation stop and penetrates completely through the wall of the main.
4. The drill bit is then backed out until it is clear of the corporation-stop valve. The valve is then closed.
5. The drilling machine is removed, and the corporation stop is ready for use.

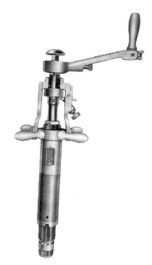

FIGURE 15-12 Manual drilling machine
Courtesy of Mueller Company, Decatur, Ill.

Self-Contained Taps

Special service connectors for PVC pipe are available that incorporate a built-in tapping tool. Figure 15-14 illustrates a connector of this type.

Tap Location

It is generally agreed that the best location for a tap on a main is at an angle of about 45° down from the top of the pipe. A tap located directly on the top of a main is more liable to draw air into the service, and a tap near the bottom could draw in sediment.

The tap should be on the same side of the main as the building to be served. The service pipe should then be supported in a wide S-curve down and into the ditch. Plenty of slack should be provided to allow for earth settlement and pipe expansion and contraction.

LEAKS AND BREAKS

Water utilities with old lead and galvanized iron services find that small leaks are relatively common. Some lead pipe will crystallize as it gets old and will crack if flexed. Leaks are most commonly located at the connections to curb stops. These leaks are probably caused by valve movement due to frost heaving or weight applied to the box at the surface. Lead is also prone to leaking after it has been disturbed by adjacent excavation, such as during the installation of a sewer or gas main. Galvanized pipe is more likely to develop a leak

1. With the service clamp attached to the main, the corporation stop is threaded into the clamp. The machine is then mounted on the corporation stop, and the stop is opened.

2. The drill penetrates the main without water escaping.

3. The drill bit is retracted, and the corporation stop is closed. The stop now controls the water.

4. The machine is removed, the service line connected, and the corporation stop reopened to activate water service.

FIGURE 15-13 Using a service clamp to install a corporation stop

Courtesy of Mueller Company, Decatur, Ill.

FIGURE 15-14 Service connector made for use with PVC pipe

Courtesy of Dresser Piping Specialties

after being disturbed. The best policy is not to repair lead or iron services but to replace the entire service, or at least the entire section with the leak.

Copper and plastic pipe rarely have small leaks. The most common problems they encounter are being pulled from a fitting by ground settlement or being cut by adjacent excavation. Copper can sometimes be pulled all the way to the surface by a backhoe without breaking. However, the pipe will probably be kinked in the process, so a new piece will have to be spliced in as a replacement. Plastic pipe is more easily broken or pulled from a connection.

THAWING

In addition to interrupting water service to customers, freezing can also rupture lines and damage meters. The best way to prevent freezing is to ensure that services are buried below the maximum frost line for the area.

Electrical Thawing

Electrical thawing of a service line is possible if the pipe is metallic. A current is run through the pipe, causing heat to be generated that will melt the ice. The current is usually supplied by a portable source of direct current, such as a welding unit.

Electrical thawing can be dangerous and can cause damage to the service line, the customer's plumbing, and electrical appliances. One major concern is that there will be poor conductivity between sections of the service line. Another is that the service line might be in direct contact with other metal pipes or conductors, such as gas pipelines. When these conditions exist, the path of the applied electrical current may be diverted from the service line. The current may then enter adjacent buildings.

Another danger involved with electrical thawing is that the current may damage O-rings, gaskets, and soldered joints on the service. This damage can result in joint failure and leakage. The dangers of stray current and of damage to the service can both be reduced if only low-voltage generators are used and if the voltage and amperage are closely monitored. Only experienced operators should attempt electrical thawing.

Before thawing begins, the property owner and tenant, if any, must be informed of the risks involved and be required to sign a waiver form. This waiver absolves the utility of liability in case of accident or damage. Thawing is a service performed for the customer on the customer's property. The utility must have written confirmation from its insurer stating that adequate liability insurance is in effect to cover possible consequences of the work.

Other Methods

Hot-water thawing (Figure 15-15) is becoming more common because it is less dangerous than electrical thawing. In addition, it can be used for plastic pipe. The technique typically involves pumping hot water through a small flexible tube that is fed into the frozen service line.

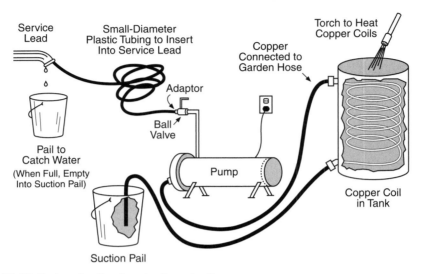

FIGURE 15-15 System for thawing plastic water line
Adapted from drawing by Randal W. Loeslie, manager, G.F.-Traill Water District, Thompson, N.D.

If a service line or meter is frozen only in the meter box, an effective method of thawing is to circulate warm air from a heat gun or hair dryer. Workers should be careful when using a propane torch for thawing because of the possibility of explosive gases that could accumulate in the vault. The intense, localized heat of a torch can also cause the pipe to expand too rapidly or it can create steam and rupture a meter or pipe.

After a frozen service has been thawed, it is usually necessary for the customer to leave a faucet running continuously to keep water flowing through the service until the frost in the ground recedes from around the pipe.

SERVICE LINE RESPONSIBILITY

Policies regarding who is responsible for maintaining water service lines vary for different water systems. Most systems have one of the following policies:

- The water utility assumes responsibility for the portion of the service in the street right-of-way (i.e., from the main to the lot line). The portion on private property is the responsibility of the property owner.

- The water utility maintains the service from the main to the curb stop or to the meter if it is in a vault. The remainder is the responsibility of the property owner.

- The entire service is the responsibility of the property owner.

Policies also vary on who does the work of installing the initial service. However, most systems allow the customer's contractor to install the entire service, as long as criteria provided by the water system are followed.

Most water utilities that own tapping equipment prefer to make the taps themselves. This ensures that the work is properly done and that the water system is not likely to be contaminated. The tap is not usually made until the service is installed and inspected. This practice provides a means of ensuring that the service is installed to the satisfaction of the water utility.

The cost of making the tap is then usually billed to the new customer. This bill is often combined with a "connection fee," which is also intended to defray other water system costs related to initiating a new service.

SERVICE LINE RECORDS

Water service information is often maintained on file cards or computer files, indexed by street address. Some important information that should be obtained and recorded before the pipe is covered in the trench includes

- measurements locating the water main tap,
- the type and size of pipe,
- tap size,
- burial depth,
- measurements to the curb stop or meter pit,
- location of the pipe at various points,
- location of the pipe entry into the building,
- date of installation, and
- address of the building served.

It is particularly important to keep good records of plastic pipe installations because there is no way of electronically locating them later. Additional information on suggested distribution system records is given in chapter 16.

BIBLIOGRAPHY

ANSI/NSF Standard 61: Drinking Water System Components—Health Effects. Ann Arbor, MI: NSF International.

AWWA Standard for Molecularly Oriented Polyvinyl Chloride (PVCO) Pressure Pipe, 4 In. Through 24 In. (100 mm Through 600 mm), for Water, Wastewater, and Reclaimed Water Service. ANSI/AWWA C909. Denver, Colo.: American Water Works Association (latest edition).

AWWA Standard for Polyethylene (PE) Pressure Pipe and Fittings, 4 In. Through 63 In., for Water Distribution and Transmission. ANSI/AWWA C906. Denver, Colo.: American Water Works Association (latest edition).

AWWA Standard for Polyethylene (PE) Pressure Pipe and Tubing, ½ In.(13 mm) Through 3 In. (76 mm), for Water Service. ANSI/AWWA C901. Denver, Colo.: American Water Works Association (latest edition).

AWWA Standard for Injection-Molded Polyvinyl Chloride (PVC) Pressure Fittings, 4 In. Through 12 In. (100 mm Through 300 mm) for Water Distribution. ANSI/AWWA C907. Denver, Colo.: American Water Works Association (latest edition).

AWWA Standard for Polyvinyl Chloride (PVC) Pressure Pipe and Fabricated Fittings, 4 In. Through 12 In. (100 mm Through 300 mm), for Water Transmission and Distribution. ANSI/AWWA C900. Denver, Colo.: American Water Works Association (latest edition).

AWWA Standard for Polyvinyl Chloride (PVC) Pressure Pipe and Fabricated Fittings, 14 In. Through 48 In. (350 mm Through 1,200 mm). ANSI/AWWA C905. Denver, Colo.: American Water Works Association (latest edition).

AWWA Standard for Underground Service Line Valves and Fittings. ANSI/AWWA C800. Denver, Colo.: American Water Works Association (latest edition).

Lead and Copper Rule Guidance Manual, Volume 2: Corrosion Control Treatment. Denver, Colo.: American Water Works Association.

Manual M3, Safety Practices for Water Utilities. 2002. Denver, Colo.: American Water Works Association (latest edition).

Manual M22, Sizing Water Service Lines and Meters. 2003. Denver, Colo.: American Water Works Association.

Manual M36, Water Audits and Loss Control Programs. 2009. Denver, Colo.: American Water Works Association.

Thompson, D.M., S.A. Weddle, and W.O. Maddaus. 1992. *Water Utility Experience With Plastic Service Lines.* Denver, Colo.: American Water Works Association Research Foundation and American Water Works Association.

Water Distribution Operator Training Handbook. 2005. Denver, Colo.: American Water Works Association.

CHAPTER 16

Information Management

Water utilities generate and use vast amounts of information. Previous chapters have discussed the many types of information—valve records, maintenance reports, customer meter readings, and residual chlorine measurements. Most water utility information can be grouped into the general categories of maps and records, maintenance management, source-of-supply and treatment plant process information, leakage control and emergency preparedness, laboratory information, and metering and customer information. Water utilities also generate business-related information such as personnel and finance, but this chapter discusses only operations-related information, in particular maps and records and maintenance management.

Information *systems* are grouped sets of data, related sets of data, or records. Most information systems of any size and significance are kept in computer databases, which allow personnel to easily retrieve and modify the systems. Some data sets are kept on cards, journals, or other handwritten records and stored in filing cabinets.

Information *management* is an important activity of water utilities. Many of the current state and federal water quality regulations require collecting and reporting information, such as levels of contaminants and types of service line materials. Efficient water utilities are usually adept at gathering, storing, and retrieving information. Efficient sharing of information among information systems—or among different departments in a utility—is also important, allowing operators and managers to make informed decisions.

COMPUTERS

Computers are commonly used in water distribution system operation and control (Figure 16-1). The personal computer (PC), which is commonly used in homes and businesses, is also used by water utilities. PC technology gives an individual user a powerful self-contained tool. Even those larger utilities with computer networks use the stand-alone PC for their individual terminals.

In addition to the PC, other types of computers used include (in order of increasing power) workstations, minicomputers and servers, and mainframe computers. The use of PC networks is the most common system for larger utilities. Because hardware and software are evolving and changing so quickly, the most current information cannot be provided here.

Water Utility Computer Uses

The different uses of computers in a water system can be divided into the following broad categories:

- Supervisory control
- Data retention and reports

FIGURE 16-1 SCADA control room

- System analysis
- Records and inventory control

These broad use categories will be discussed briefly before the specific information systems are presented in more detail.

Supervisory control

As discussed in chapter 9 on instrumentation and control, supervisory control is increasingly making use of digital technology in the form of computers and microprocessors. Chapter 9 also introduced supervisory control and data acquisition (SCADA) systems. The SCADA system has been described as the heart of all information systems, particularly those networked together with process control facilities, as shown in Figure 16-2. Figure 16-3 shows how SCADA interacts with other information systems, which are discussed in this chapter.

Data retention and reports

Shortly after electric typewriters became available, data-logging equipment was introduced. The intent was to replace most of the recording charts in a control system. Data such as pressure and flow are periodically transmitted to a central location, where the information is typed out to provide a permanent record. The system has the capability of defining allowable limits for each signal, such as high and low pressure, and the operator will be alerted and provided a special printout if a limit is exceeded.

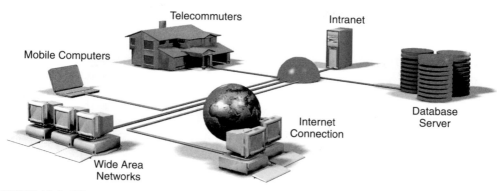

FIGURE 16-2 Client/server environment
Courtesy of Hansen Information Technologies

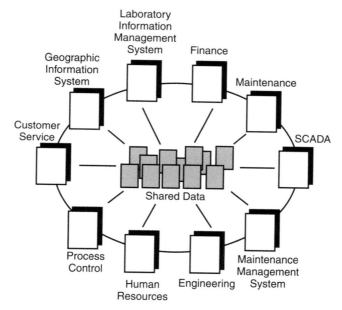

FIGURE 16-3 Integrated water control system

Although data logging created much more information for possible future analysis and reduced the number of circular and strip charts that had to be stored, it still created large quantities of paper. In addition, finding specific data and analyzing the data required an operator to review individual sheets.

The same type of data acquisition is done with a computer, but the equipment and processes are much more sophisticated. Data of every type can now be retained because of the computer's storage capability. Moreover, any part of the stored data can quickly be found and analyzed using a simple computer command. And when the data is determined to be obsolete, it can simply be erased from the computer memory.

Data collected and stored by a computer can be particularly helpful in providing reports of system operation. State and federal regulations are requiring increasingly detailed reports of system operation, particularly for plants using surface water sources. These reports can be generated quickly and easily by a computer. Quarterly or annual reports of water system operations can also be prepared quickly. Information such as the highest-use day and hour—which with manual records required picking through report sheets—can be provided in an instant by the computer.

System analysis

A number of computer programs are available that can analyze a specific water system function. One of the earliest applications was for water distribution system analysis. The original Hardy–Cross method of computing the flow in water mains was complicated and laborious to use, even with mechanical calculators. The computer version can now predict water flow and deficiencies in a fraction of the time. Some other examples of specific analysis that can be done on a computer are as follows:

- *Main failure analysis:* The program maintains a detailed history of water main breaks and identifies problem areas in the distribution system.
- *Financial planning:* The program provides a structured cash-flow analytical model and can be used for developing required water rates.
- *Corrosion control:* The program determines water stability and predicts changes in water quality with the addition of various chemicals.
- *Fire-flow simulations:* Hydraulic distribution models can be used to simulate fire-flow demand and allow operators to practice their response.

Records and inventory control

Water utilities store their system maps, drawings, and records on a computer. Likewise, information on treatment facility equipment can be stored on a computer to provide easy access. Some examples are as follows (they will be discussed in more detail in later sections):

- A meter inventory program stores all data on each meter such as make, model, date of last test, and installation location. Many different types of reports can be generated from the stored data, including lists by size and age and identification of those that should be tested or replaced. Figure 16-4 shows a sample computer screen by which an operator can access information about a specific piece of equipment. When the record number is entered into the computer, the entire listing for that particular piece of equipment is displayed on-screen.

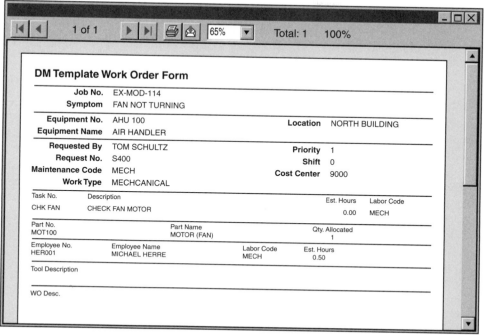

FIGURE 16-4 Equipment information screen
Courtesy of Eagle Technology, www.eaglecmms.com

- In a hydrant management program, all background data are maintained on each hydrant, including make, model, installation date, and location, as well as maintenance and test records. Complete data are instantly available for every hydrant, and all types of reports or analyses of the hydrants on the system can be generated.

- All background data and maintenance records on service lines can be maintained by computers. As with meters, many reports on the status of services can then be easily run from the stored data.

- Some or all water distribution system maps can be transferred to computer format, making the information much easier to find and update. Updating the maps is also easier if computer-assisted design (CAD) programs are used.

- As with distribution system maps, the drawings of treatment plant equipment and piping systems can be computerized, with the same advantages of providing quick access.

- Complete records on chemical feeders, pumps, mixers, and other plant equipment can be stored in a computer for quick and easy reference.

Computer Integration Trends

The computer has proven its value in water utility operations. This was facilitated by the development of industry standards that allow easy connections between hardware, software, and different systems.

MAPS AND RECORDS

Maps, drawings, and records provide a way to record the information necessary for the efficient operation and maintenance of a water distribution system. Since most of the distribution system is underground, maps and records are critical for finding system components. Maps, drawings, and records also provide vital information to other utilities that may be working belowground in a specific area. They aid in troubleshooting, planning, and systematically developing the distribution system. They also furnish utility management with an accurate, comprehensive record of the physical assets of the system.

Maps and records must be updated regularly to include system changes or new construction. Out-of-date maps and records not only hinder daily system operation but also pose problems during emergency situations.

Recognizing the importance of these records, a water utility that does not have the staff capability to prepare and maintain proper maps and drawings can hire a consulting engineer. The engineer can assist in initially bringing maps, technical drawings, and records up to date and then periodically provide corrections and enter new information.

Some or all water distribution system maps and records can be transferred to computer format through databases and a computer-assisted mapping program. A principal advantage of using computers is that information is much easier to find. Rather than leafing through rolls of plans or large plat books, an operator can find maps quickly and display them on-screen.

Distribution System Maps and Records

In addition to providing information on main, hydrant, and service locations, distribution system records provide a permanent record of the system's assets. Permanent records of physical facilities and appurtenances are essential to determine system development changes and to document the system's underground investments. Aboveground assets can easily be inventoried, but underground assets, which often constitute a major portion of the distribution system, can be inventoried only through maps and records.

Maintaining up-to-date maps, drawings, and records should not be considered a secondary function to construction or other work. Complete and accurate records help promote the efficient operation and maintenance of the distribution system. They allow personnel to determine which sections are undersupplied and why, whether hydrants are too close together or too far apart, and whether mains in a certain area are capable of handling a major fire. Maps, drawings, and records also aid in troubleshooting and systematically developing the system.

Some states have statutes or regulations requiring the maintenance of drawings and records so that the utility can respond to emergencies, personnel injuries, or damage to underground utility services during excavations. Distribution system maps, drawings, and records provide essential information to other utilities, such as gas, electric, or telephone utilities, whose services are also below ground.

A number of records are needed for the efficient operation and maintenance of a water distribution system. Traditional distribution system records can generally be divided into the following four groups:

1. Mapped records
2. Construction drawings
3. Card reports
4. Statistical records

Mapped records

Three principal types of mapped records are

1. Comprehensive maps outlining the entire system
2. Sectional maps providing a more detailed picture of the system on a larger scale
3. Valve and hydrant maps pinpointing valve and hydrant locations throughout the system

Large systems may also have arterial maps, valve closure maps, leak survey maps, and pressure zone maps.

Comprehensive maps. A comprehensive map (sometimes referred to as the wall map) provides a clear picture of the entire distribution system. It is primarily used by the system manager or engineering department to identify sections of the system that either need improvement or are not yet fully developed. The map allows engineers to identify places where short extensions will eliminate dead ends, where mains are inadequate to supply hydrants in case of large fires, or where additional valves or hydrants are needed. A comprehensive map also serves to document the financial value of physical assets within the distribution system. These costs are used to determine system development charges, which are used in setting water rates.

The comprehensive map should not be cluttered with distracting information. Important information to put on a comprehensive map includes

*   street names,
*   water mains,
*   sizes of mains,
*   fire hydrants,
*   valves,

- reservoirs and tanks,
- pump stations (discharging into the distribution system),
- water source,
- scale of the map,
- orientation arrow,
- date last corrected,
- pressure zone limits, and
- closed valves at pressure zone limits.

Because the primary purpose of the map is to provide an overall picture, the map's scale should be as large as possible. Typical comprehensive maps have scales ranging between 500 and 1,000 ft to 1 in. (6 and 12 m to 1 mm), with some smaller communities using scales less than 500 ft to 1 in. (6 m to 1 mm).

A scale that allows the map to be drawn on one sheet should be selected, although some larger systems may have to use two sheets attached together. Some systems also have an overlay with contour lines (topographic lines) for use with the comprehensive map.

Water suppliers should try to reference their comprehensive maps to the state's coordinate system or range and section lines. Referencing to such systems will facilitate easy additions whenever necessary. Ordinary commercial maps available in some cities must be checked carefully for accuracy prior to their use as a base map for a comprehensive map. Agencies such as highway departments and the US Geological Survey in the Department of the Interior are excellent sources for base maps. Once an accurate base map has been selected, photostatic enlargements or reductions must be done carefully so that the base map does not become distorted. The actual map should be drawn in ink on a durable material such as heavy tracing cloth. Prints can then easily be made from the master for regular use.

Copies of the comprehensive map are provided to the system manager and engineering department. Copies are also often provided to the mayor, city manager, city fire stations, and the water distribution maintenance department.

Comprehensive maps must be updated periodically. The normal practice is to update them either once or twice a year or immediately after the completion of major system extensions.

Sectional maps. A sectional map or plat (Figure 16-5) provides a detailed picture of a section of the distribution system. Sectional maps are generally used for day-to-day operations. They reveal the locations and valving of existing mains, the locations of fire hydrants, and the locations of active service lines. They also match account numbers with lots.

Because detail is important on a sectional map, the scale is generally much larger than that used on a comprehensive map. Therefore, several maps are required to cover the entire system. Typical scales range from 50 to 100 ft to 1 in. (15.2 to 30.4 m to 25.4 mm)

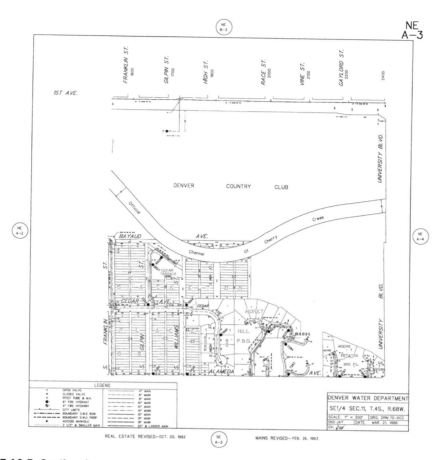

FIGURE 16-5 Sectional map
Courtesy of Denver Water

for most communities, to 200 ft to 1 in. (6.1 m to 25.4 mm) for larger systems. Information important to position on a sectional map include the following:

- Section designation or number
- Adjacent section numbers
- Street names
- Mains and sizes
- Materials of mains
- Date of main installation
- Distances from property line

- Fire hydrants and numbers
- Valves and numbers
- Valve sheet designation shown in margin
- Intersection numbers (if valve intersection plats are used)
- Block numbers
- Lot numbers
- House numbers
- Water account numbers
- Measurements to service lines
- Distances—main to curb box
- Distances to angle points
- Distances to fittings
- Dead ends and measurements
- Date last corrected
- Orientation or north arrow
- Scale
- Closed valves at pressure zone limits
- Pump stations (discharging into distribution system)
- Service limits
- Tanks and reservoirs
- Pressure zone limits

Sectional maps should not overlap each other. Instead they should butt up against one another to avoid confusion. Each section should be indexed (individually numbered) and should show matching points for adjacent sections. Indexing sectional maps is not difficult, and the comprehensive map should be used as a base map for indexing. North–south and east–west lines should be drawn on a copy of the comprehensive map to divide it into sections based on the appropriate scale, such as 50 ft (15 m) or 100 ft (30 m), for the sectional maps. These sections should then be numbered 1, 2, 3, 4, and so on from north to south and lettered A, B, C, D, and so on from east to west (Figure 16-6).

If there are not enough letters in the alphabet for east–west indexing, odd numbers can be used for north–south designations and even numbers for east–west designations (Figure 16-7A). Indexing should always begin some distance outside the present distribution system to allow for system growth.

Areas that experience growth in all directions may prefer a method in which the indexing starts in the middle of the mapped areas and is expanded outward by quadrant (for example, NW, SE). Columns would be lettered sequentially from south to north in the NE and NW quadrants, and from north to south in the SE and SW quadrants. Rows

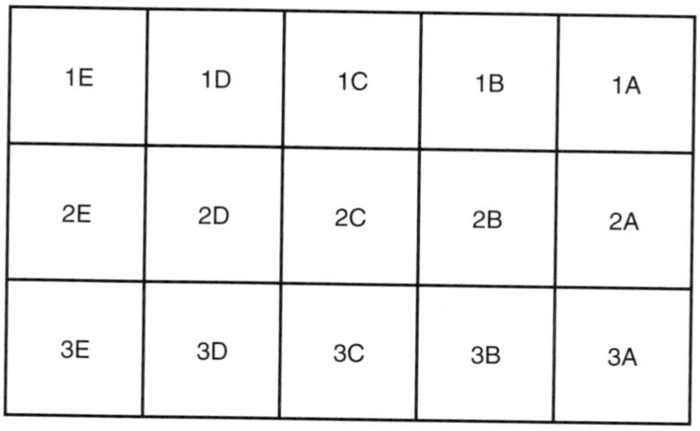

1E	1D	1C	1B	1A
2E	2D	2C	2B	2A
3E	3D	3C	3B	3A

FIGURE 16-6 Small comprehensive map divided into sections

NW   NE

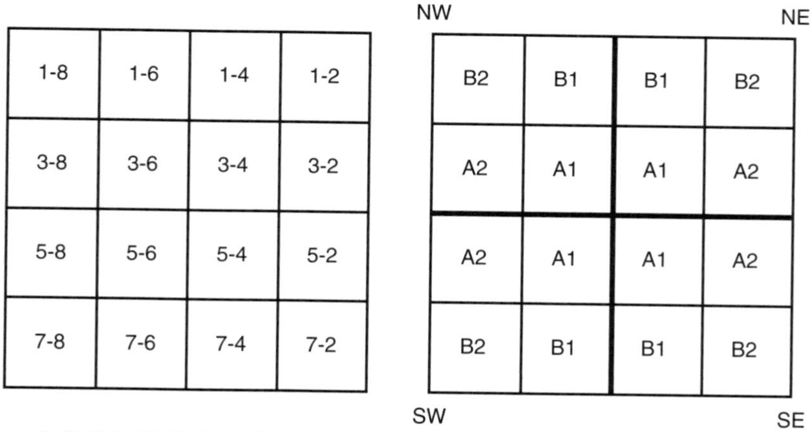

1-8	1-6	1-4	1-2
3-8	3-6	3-4	3-2
5-8	5-6	5-4	5-2
7-8	7-6	7-4	7-2

B2	B1	B1	B2
A2	A1	A1	A2
A2	A1	A1	A2
B2	B1	B1	B2

SW   SE

A. Odd for N–S, Even for E–W   B. Sequential by Quadrants

FIGURE 16-7 Large comprehensive map divided into sections A. Odd for N-S, Even for E-W B. Sequential by Quadrants

would be numbered sequentially from west to east in the NE and SE quadrants, and from east to west in the NW and SW quadrants (Figure 16-7B).

Tax assessment maps, insurance maps, subdivision maps, or city engineering maps may be used as sectional base maps provided they are checked carefully for accuracy. Photostatic enlargements or reductions can be used to arrive at a desired scale, but extreme care must be taken by the person doing the work so that there is no distortion. Many communities find that drawing an accurate base map at the desired scale is easier than searching for a suitable map and then checking its accuracy.

Sectional maps should be drawn in ink on either heavy tracing cloth (from which prints can be made) or another durable art material. Whether tracing cloth or another material is used, a set of duplicate sectional maps should be kept in a safe location, preferably a fireproof vault.

Sectional maps must be updated more frequently than comprehensive maps. Many systems update working copies of their sectional maps daily as field reports come back to the main office. Original sectional map prints are then updated periodically (monthly or quarterly) from the working copies, and new copies with the corrected sections are distributed as often as necessary. In situations where major system extensions are added or major changes are made, the original print should be updated as soon as possible. A copy of the corrected section should be issued for each copy in use.

Valve and hydrant maps. Valve and hydrant maps pinpoint valve and hydrant locations throughout the distribution system. Valve and hydrant maps are primarily used by field crews. They enable crews to locate all valves and hydrants quickly for general maintenance work or emergency situations when a section of main must be repaired. To be of value, valve and hydrant maps should be bound in books and then passed out to appropriate crews for field use.

Valve and hydrant maps should provide measurements from permanent reference points to each valve within the system and they should pinpoint hydrant locations. Communities with flush-type hydrants at the surface level will need especially good measurements from permanent reference points. Valve and hydrant maps should also provide appropriate information such as direction to open, number of turns to open, model, type, and date installed for hydrants and valves. Some maps also include the last date tested or repaired.

There are a number of methods for setting up valve and hydrant maps. The two most common are the plat-and-list method and the intersection method. In the plat-and-list method, the plat is the map or drawing that shows street names, mains, main sizes, numbered valves, and numbered hydrants. Following the plat is a list providing appropriate information on the numbered valves and hydrants, such as reference measurements, size and make, direction of operation and number of turns, date installed, date tested, and any remarks. Plats generally have a scale of around 500 ft to 1 in. (15.2 m to 25.4 mm).

Plats can be indexed in a manner similar or identical to that of sectional maps. Valves and hydrants on each plat should be assigned a number or designation. A reduced print of the comprehensive map or city map showing the coordinates for each plat may be used as a field index for valve and hydrant maps. These indexes need not be updated (unless extensions to the original system are added) because they are used only to show the general area each plat covers.

In the intersection method, plats (maps) of intersections (Figure 16-8) are drawn on a very large scale, such as 20 to 30 ft to 1 in. (6.1 to 9.1 m to 25.4 mm), showing property or street lines, mains, hydrants, valves, and other features. This large scale permits measurements from permanent reference points (property or street lines, for example) to be entered on the intersection plat. Valves and hydrants that are located between intersections cannot be shown to scale but they can be located based on their measurements from intersecting streets.

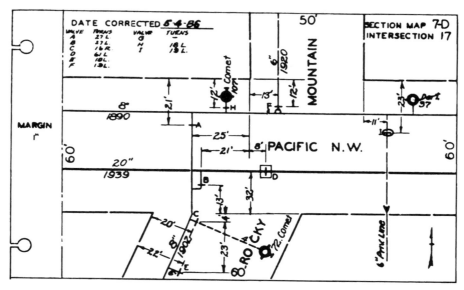

FIGURE 16-8 Valve intersection plat

Indexing intersections involves giving each intersection a number. The intersection number can then be attached to the end of the sectional plat number, which pinpoints the intersection's location. Valves and hydrants at or near an intersection can then be given a letter to pinpoint their location.

If the size of the community is not extremely large, intersections are sometimes sequentially numbered. A cross index of streets in alphabetical order (Figure 16-9) or a print of the comprehensive map showing sectional map areas and intersection designations (Figure 16-10) can be used as a field index. If appropriate, intersection maps, like plat-and-list maps, can have factual information concerning the valves and hydrants. Such information can be listed in a corner or suitable open space on the intersection drawing. For more complex systems, a book or a series of overlay maps and indexes can catalog this information (for example, there could be a map book and a valve book). These books can then be divided into geographic areas for field use.

In very large systems, intersection drawings can become quite bulky, sometimes making it unfeasible to equip field crews with complete records. Some systems, therefore, keep the intersection maps in the main office. However, this defeats one of the purposes for developing the maps—field use.

Valve-and-hydrant maps must be updated for efficient system operation and maintenance. Working valve-and-hydrant map copies should be corrected daily in large systems, and original prints should be updated monthly or quarterly. After the original prints are corrected, new copies of the corrected prints should be issued to replace those copies in use.

5th St. So. &

Strodman Avenue NE Corner	432
Wyatt Avenue NE Corner	434
2041 - 5th Street S. - Property Line	436
2251 - 5th Street S. - North Property Line	438
Pepper Avenue NE Corner	439
Clyde Avenue NE Corner	561
Glennwood Heights NE Corner	563
3021 - 5th Street S. -	
Property Line South of Shady Lane	564

Shady Lane &

2940 Shady Lane	565
Grove Avenue SW Corner	567

Webb Avenue &

711 Webb Avenue - West Property Line	574

Airport Avenue &

Angle Drive	569
East End of Angle Drive	570

6th Street &

Daly Avenue SW Corner	440
Taylor Avenue	442
Miller Avenue in Front of 1941	444
Goodnow Avenue	446
Pepper Avenue NW Corner	448
Cook Avenue	571

7th Street South &

1820 - 7th Street South, East Side of Street	450
1980 - 7th Street South, Eash Side of Street	451
2131 - 7th Street South, South Lot Line	452
Pepper Avenue NE Corner	453

FIGURE 16-9 Cross index of streets
Courtesy of Water Works and Lighting Commission, Wisconsin Rapids, Wis.

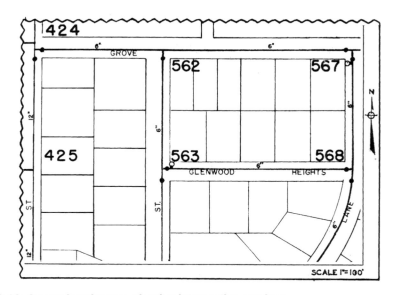

FIGURE 16-10 Comprehensive map showing intersection numbers
Courtesy of Water Works and Lighting Commission, Wisconsin Rapids, Wis.

An alternative to using a comprehensive map, a sectional map, and a valve-and-hydrant map is to incorporate the valve-and-hydrant map into the other two. In fact, this approach is often more practical for small systems. Hydrant and valve locations and operating positions are generally shown on the comprehensive map, with more detailed information shown on the sectional map. This may require the sectional map to be of a larger scale than would be necessary if a valve-and-hydrant map were used. However, the advantage of this type of system is that one fewer map needs to be updated. Reduced-scale reproductions can then be made for easier handling in the field, as long as the information is still readable.

Plan and profile drawings

Plan and profile drawings, which are used in the construction of parts of the system, are engineering records showing pipe depth, pipe location, and other information as originally proposed. These drawings should be marked as construction takes place to indicate any variations from the original plan. The revised drawings should then be carefully kept in a file for future reference.

Occasionally, water distribution system operators will deal with plan and profile drawings. These engineering drawings should be referenced to section lines or property lines. Along with the other installation information, the drawings show the correct distance from a starting reference point (stationing). Stationing is normally specified in an engineering form; for example, 60 + 40 means 6,040 ft from the starting point.

Common abbreviations found on plan and profile drawings include

- POT—point on tangent
- POC—point on curve
- BC (or PC)—beginning of curve
- EC (or PT)—end of curve
- PI—point of intersection
- EL—elevation

Supplemental mapped records

In addition to a comprehensive map, sectional maps, and valve-and-hydrant maps, some systems maintain supplemental mapped records. For example, large systems may keep an arterial map and a comprehensive map at a scale of 2,000 to 4,000 ft to 1 in. (610 to 1,220 m to 25.4 mm) to show primary distribution mains 8 in. (200 mm) or larger. These maps are primarily used for system analysis.

Systems that extend over a wide variation of ground elevations usually maintain pressure-zone maps (water-gradient contour maps). These maps record simultaneous pressure readings taken at different points throughout the system, which are then plotted on a print of the comprehensive map. Contour lines can then be drawn to identify zones of equal pressure. From this map, low-pressure or high-pressure zones can be identified and controlled. Some systems maintain pressure zones directly on the comprehensive map.

Other systems maintain leak frequency maps. On these, different-colored push pins are used to identify different types of leaks on a print of the comprehensive map. Points where soil conditions or electrolysis is bad or where faulty installation techniques or poor materials were used show up very clearly after a few years. Leak survey maps are used by many systems. When regular leak survey work is conducted, these maps show the valves to be closed and the areas to be isolated.

Statistical records

Statistical records are used for reporting construction progress and growth. The types generally used are as follows:

- Statistical records for mains summarize the total length for each size of main laid or retired, plus the total length of pipe in use in the system.

- Valve and hydrant statistical records summarize the number of valves and hydrants installed for given sizes.

- Service statistical records document the total length of each size of service pipe laid, the number and size of corporation stops installed, and the number and size of curb stops installed. If services are installed and owned by the customer, service statistical records need only summarize the number added or abandoned.

- Meter statistical records report the number of each size and type of meter installed in the system.

Many water utilities must maintain statistical records for use in preparing annual reports, audits, reports to stockholders, or tax forms. Statistical records are also used to report construction growth or measure the efficiency of the distribution workforce. Statistical records (Figure 16-11) may be compiled monthly, semiannually, or annually, depending on reporting requirements.

Statistical records lend themselves well to graphing because they deal primarily with numbers. Graphing such records enables water supply personnel to compare past and present figures at a glance and identify system trends. If the records are maintained on a computer, graphing can be done very easily if the data are fed into the proper computer program.

Record requirements for small and large systems

Distribution system records are just as important for small systems as they are for large systems. The only real difference between the two should be the number of records kept. Smaller systems, which are generally less complicated, must still keep the same records, but they can often consolidate certain records. Generally, most small systems can consolidate their mapped records on one detailed map similar to a larger system's sectional map.

					Daily Filter Record Filter Plant No. 2					

Filter No. _____   Prev. Run _____ Hours   Date: ___ / ___ / 20 _____

Oper.	Time	Rate of Flow, mgd	Loss of Head, ft	Surface Wash, min.	Rate Water Wash, mgd	Water Wash, min	Polymer Feed On	Polymer Feed Off	Wash Water Used	Water Filtered, mil gal
	12 mid									
	1 a.m.									
	2									
	3									
	4									
	5									
	6									
	7									
	8									
	9									
	10									
	11									
	Noon									
	1									
	2									
	3									
	4									
	5									
	6									
	7									
	8									
	9									
	10									
	11									
	12 mid									

FIGURE 16-11 Typical daily filter record

Although record keeping at smaller water distribution systems may be more consolidated than at larger ones, the problem of maintaining records is often more difficult. Small system operators frequently have additional duties beyond operating the water distribution system. These duties compound an already busy workload and leave less time for record maintenance. In such cases, it is important that the small system operator set aside time to maintain and update distribution system records so that they will provide accurate reference information for system operation and maintenance.

Map symbols

Every water distribution system should adopt a set of standardized map symbols to denote different items on its mapped records. Generally used symbols are shown in Table 16-1. These symbols were selected because of their simplicity, clarity, and acceptance in current practice.

Water distribution system employees involved in drawing or using mapped records should become familiar with their system's symbols. This will prevent mistakes in drawing or interpreting system maps and promote more efficient map use. Many systems include a copy of their symbols in all their record books for easy reference.

Automated Mapping/Facility Management/Geographic Information Systems

An automated mapping/facility management/geographic information system (AM/FM/GIS) collects, stores, manipulates, and analyzes water system components for which geographic location is an important characteristic. Information about public and private properties; road networks; topography; distribution and transmission lines; and facilities such as tanks, pump stations, pipes, wells, and treatment plants are all tied to geographic location.

In an AM/FM/GIS, geographic data are represented as points, lines, and areas on a computerized map. The computerized map has many layers, each layer corresponding to related data, such as topography, streets, water distribution, or electric power system. The concept is similar to a map with many transparent overlays but much more precise and detailed. In addition to maps, the data may be presented in the form of tables or even as text descriptions. The AM/FM/GIS may have hundreds of factors associated with each feature or location.

Basic elements of an AM/FM/GIS

The basic elements of an AM/FM/GIS to handle geographic information include data entry (collection, management, and storage), data manipulation (modeling and analysis), and data output (retrieval, management, display, and storage). Table 16-2 lists the attributes of some data management systems.

Data entry requires converting data from an existing form into one that can be used by an AM/FM/GIS. As with any engineering system, the results are only as accurate as the initial data. The initial data entry is a complicated and labor-intensive procedure and thus is expensive.

The data manipulation function includes modeling and analysis. An AM/FM/GIS can automate certain activities within the organization. It will also change the way the organization works. For example, an AM/FM/GIS can quickly generate more alternatives for a project, helping with decision making. An AM/FM/GIS is not merely a high-tech way to produce system maps.

TABLE 16-1 Typical map symbols

Item	Job Sketches	Sectional Plats	Valve Record Intersection Sheets	Comprehensive Map and Valve Plats
3-in. (80-mm) and smaller mains				
4-in. (100-mm) mains				
6-in. (150-mm) mains				
8-in. (200-mm) mains				
Larger mains	Size Noted	Size Noted	12 in. (300 mm) / 24 in. (600 mm) / 36 in. (900 mm)	12 in. (300 mm) / 24 in. (600 mm) / 36 in. (900 mm)
Valve				
Valve, closed				
Valve, partly closed				
Valve in vault				
Tapping valve and sleeve				
Check valve (flow →)				
Regulator	®️	®️	®️	®️
Recording gauge	G	G	G	G
Hydrant (2½-in. [65-mm]) hose nozzles				
Hydrant with hose and steamer nozzles				
Crossover (option 1)				
Crossover (option 2)				
Tee and cross	BSB BSBB			
Plug, cap, and dead end	Plug Cap			
Reducer	BS BS	12 in. (300 mm) 8 in. (200 mm)		
Bends, horizontal	Deg. Noted	Deg. Noted	Deg. Noted	
Bends, vertical	Up Down	No Symbol	No Symbol	No Symbol
Sleeve				
Joint, bell and spigot	Bell Spigot			
Joint, dresser-type				
Joint, flanged				
Joint, screwed				

① Open circle: hydrant on 4-in. (100-mm) branch
② Closed circle: hydrant on 6-in. (150-mm) branch
B = bell
S = spigot

TABLE 16-2 Attributes of data management systems

GIS	AM/FM	OSS/DSS/BSS
Macro view	Micro view	Date intensive
Small scale; 1:2,000 typical	Large scale; 1:50, 1:100, 1:200 typical	Tabular
Spatial analysis	Network analysis	Aspatial analysis
Spatial topology required	Network topology required	No topology
Not time sensitive	Time sensitive	Mixed time sensitivity
More static	Dynamic	Historical
Not mission critical	Often mission critical	Often mission critical
Few users	Many users	Many users

OSS = operational support systems, DSS = decision support systems, BSS = business support systems

The output functions, determined by the user's requirements, include the retrieval, management, display, and storage functions. Output will be in the form of printed tables and reports, plotted maps and drawings, or digital files.

System software

The software used depends on the size of the database, whether data are to be shared among multiple users, and what functions will be performed. The size of the database dictates the level of sophistication required for the software to locate the requested information quickly. The need to share common data requires the software to distribute data to multiple users accurately and quickly. The types of functions to be performed determine the database management architecture, as well as the data structure and the functionality required. Figure 16-12 shows a complex software configuration.

Water utility applications

The objective of an AM/FM/GIS is to provide a fully integrated database to support management, operation, and maintenance functions at all levels. The applications can range from the daily transactions of preparing work orders to repairing meters to providing analyses of alternative design plans for new water main locations. Table 16-3 enumerates the typical water utility data sets, and Figure 16-13 shows the layered database categories that are commonly used. Table 16-4 lists some of the applications the AM/FM/GIS would provide for a water utility.

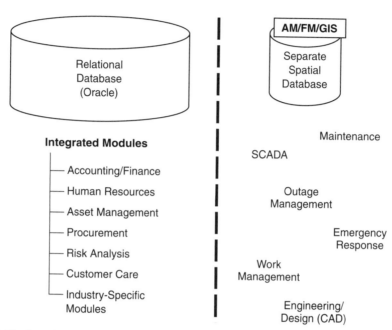

FIGURE 16-12 Complex software configuration for a water utility

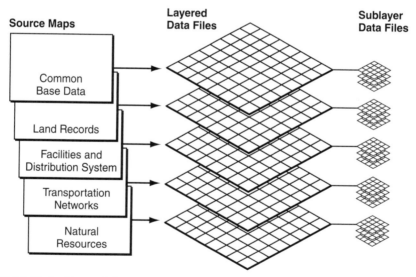

FIGURE 16-13 Typical layered database categories

TABLE 16-3 Typical water utility data sets

Data Category	Example Map Layers
Base data	Control information
	Planimetric features
	Hydrology features
Facilities and distribution systems data	Water piping
	Water valves and utility holes
	Service areas
	Water plant facilities (wells, pump stations, tanks, treatment plants)
	Other utilities
Land records data	Property boundaries
	Easements
	Right-of-ways
Natural resources data	Groundwater data
	Drainage data
	Soils data
	Flood plain boundaries
	Topographical features
	Vegetation information
Transportation network data	Road center lines
	Pavement locations
	Road intersections
	Bridges

Some specific applications are described in the following paragraphs.

Water resource management. The AM/FM/GIS can create distribution system maps showing water quality data (e.g., bacteria counts, heavy metal concentrations) overlaid and related to the location of water production (e.g., wells, treatment plants). This graphical representation is used by the operator to control pumping and treatment facilities based on location within the community and water network, so that high-quality water is

TABLE 16-4 Typical water utility applications

Type of Application	Example
Facility management	Update, display, and analyze facility data; support the planning, design, operations, and maintenance to water facilities
Emergency response	Provide display of emergency vehicle routing; analyze frequency and location of emergency event
Area mapping and reporting	Analyze and display maps; produce maps and reports
Facility inspection	Schedule and track inspection; perform safety violations inspections; log and process complaints
Permitting and licensing	Process and track information
Land-use and environmental planning	Display and analyze land-use and environmental data such as water quality information by source
Facility siting	Select optimal locations for new utility facilities
Code enforcement	Schedule and track code violations; display and analyze enforcement data
Customer service	Analyze complaints, billing errors, accounts receivable, water usage by location and demographics

available to the customer. Planning personnel also use water quality trends, again overlaid onto the graphical model of the network, to analyze any degradation of water sources in relation to other variables such as consumption, season, or weather.

Customer services. Information on billing data and meter numbers can be graphically represented and located on distribution system maps. Analyzing such things as accounts receivable information, consumption, and complaints with respect to street address, sector in the community, and location in the water network provides important information about the effectiveness of service, system reliability, and customer satisfaction.

Operations. An AM/FM/GIS connected with a SCADA system is an effective tool for utility management. For example, suppose a control center operator receives a trouble call from a customer. The AM/FM/GIS links the customer's address or meter number to the collection or distribution system piping schematic (Figure 16-14). The pump station status is available on a SCADA display. An informed decision, based on current data, is made by the central operator, and crews are dispatched quickly with adequate information to resolve the problem.

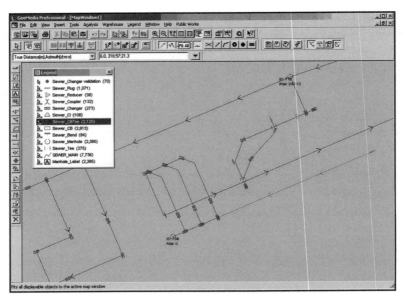

FIGURE 16-14 Customer records linked with mapping information
Courtesy of Integraph Mapping & Geospatial Solutions

Water demand forecasting. Utilities can graphically represent the utility system on an AM/FM/GIS for short-term demand forecasting. For example, water demand is highly dependent on rainfall, because customers tend to decrease or stop irrigation when it rains. Plotting the relationships between rainfall measurements, map location, and demand provides useful historical information when a weather-sensitive, short-term demand forecast needs to be made. The utility can then use the demand forecast to help establish a pumping schedule based on economics or reliability.

Water system modeling. An accurate and precise AM/FM/GIS distribution system network database can be accessed automatically by the planning model, resulting in more precise modeling. Overlaying accurate data with data from the SCADA system also provides a powerful tool for water system analysis.

MAINTENANCE MANAGEMENT

In addition to the basic records discussed previously, certain other types of records are important. Work orders are used to communicate between the office and the field, and technical information on installed or available equipment must be available for maintenance and design.

Work Orders

Distribution system records are only as accurate as the information recorded. Therefore, every system must have an established procedure by which field crews complete and submit accurate information on the work that they do. A good working relationship between field and office personnel will help ensure a reliable record system.

Many water distribution systems use a work order to instruct field crews on specific work to be done. In addition, work orders are also used for other purposes, such as ordering materials, organizing field crews, and obtaining easements.

Blanket work orders, which cover general repairs that are expected to fall within a specific time frame or dollar amount (as opposed to costly or unplanned repairs), are often used for work such as grounds keeping, changing light bulbs, painting, safety meetings, and special projects. They track maintenance hours used for indirect and nonspecific maintenance activities. This enables managers to effectively plan, schedule staff, and track estimated-versus-actual time usage, permitting more accurate scheduling in the future. Blanket work requests are usually initiated by maintenance supervision or management.

When the job is completed, the field crew completes the form to show that the job was completed. A work order usually contains information such as

- the location and description of the job to be performed,
- a listing of all the materials needed and used,
- information on the time and human resources required,
- a list of equipment used, and
- the total cost of the work performed.

A work order system should be designed to be functional rather than just additional paperwork. A sample work order is shown in Figure 16-15.

Technical Information

Water distribution systems are often swamped with technical information—information accompanying valves, hydrants, pumps, pipe, tapping machines, backhoes, and other distribution system components and equipment. Today's competitive sales market further adds to the volume of material by providing technical information through the mail. Technical information should not be discarded simply because of its bulk. Many technical bulletins and pamphlets are extremely important because they provide specifications for installation, operating tips, or maintenance procedures.

Technical information should be evaluated and maintained on the computer for easy reference. Each major item within the file system should have its own section. For example, a section on hydrants could be established and subdivided by manufacturer. Files on each hydrant manufacturer would specify where important information pertaining to the hydrant is stored. They would also contain any documents, such as warranties, letters, or

WORK ORDER

Work Order Number: [WorkOrderID]

In Service: _____ Out of Service: _____

WO Initiated Date: [InitiateDate]	Supervisor:	[Supervisor]	Team Members:	[EstLaborName]	Shop: [Shop]

Activity Performed: [Description]

Address: [Location]	Key Map: [MapPage]

Crew Type: WR SR SS FH MR VS VR EV PM **Location of Worksite:** Front Rear Side Arrival: _____ Departure: _____

Size Main: $^{3}/_{4}$" 1" 1$^{1}/_{4}$" 1$^{1}/_{2}$" 2" 4" 6" 8" 10" 12" Other _____ Size Service Connection: $^{3}/_{4}$" 1" 1$^{1}/_{4}$" 1$^{1}/_{2}$" 2" 4" 6" 8" Other _____

Pipe Type: A.C. C.I. Clay Concrete PVC Steel Other _____ Material (quick codes) used: _____

Further Work Pending: Yes No Landscape Area: _____ Concrete Cut: _____ Asphalt Cut: _____ Fence: _____ wood chain link
wrought iron brick other _____

Vehicle (quick code): _____ Equipment (quick code): _____ Time Updated to C.O.S.: _____ Work Order Status: [Status]
Update Map: [UpdateMap]

Instructions:	[Instructions]

Comments:	[Comments]

Associated Service Requests

RequestID	Date/Time Initialized	Priority	Description	Problem Address
[SRequestID]	[SRDateTimeInit]	[SRPriority]	[SRDescription]	[ProbAddress]
[associatedsr]				

Tasks

Order	Name	Start	Status	Proceed	Assigned To	Request Code	Comments
[TaskSeqID]	[TaskName]	[StartDate]	[TaskStatus]	[Proceed OK]	[AssignedTo Name]	[TaskCode]	[TaskComments]
[tasktable]							

Employee #s _____ _____ _____ _____ _____ Planner/Scheduler Sign Out: _____

FIGURE 16-15 Sample work order form
Courtesy of Department of Public Works and Engineering, City of Houston

manufacturers' phone numbers. A filing system might also be established on valves, subdivided by valve type and manufacturer.

File systems can be arranged in any number of ways. The important consideration is that every water supplier should establish an efficient means of organizing important technical information for future reference.

The date received should be stamped or written on new catalogs or pieces of literature as it is received. Manufacturers often do not date their material, and when several different catalogs end up in a file, there is no way of telling which is most current unless they have been dated.

Computerized Maintenance Management

A computerized maintenance management system (MMS) is an organized way for a utility to keep track of its maintenance needs. An MMS has a wide range of features, including work order management, inventory management, purchase requisitioning, labor planning and time management, and analytic or predictive reporting. Other application systems that are not part of an MMS but can, with customization, be integrated include general ledger, accounts payable, accounts receivable, payroll, geographic information, automated mapping and facility management, and engineering sciences or modeling programs. Other, less typical applications are also possible.

Maintenance generates vast amounts of information and a great deal of data that must be managed in some way. All of these conditions make the maintenance department an ideal candidate for computerized information and work management systems. For example, a computerized tracking mechanism can estimate the time and materials needed for repairs based on such factors as reported symptoms, historic data on similar repair types, and previous work management information. Data listings show how much work is waiting to be done, how many jobs are currently in progress, how much work is completed and is waiting for final check or approval, and how many repairs are waiting for parts to arrive.

An effective maintenance system generates several kinds of work orders: preventive, corrective, emergency, and blanket. It allows for a routine scheduling of a known work backlog and the interrupt scheduling of nonroutine work that has a higher urgency.

Managers need an MMS that allows them to prioritize all work so that it can be completed in a timely and commonsense manner. When prioritizing work, managers should consider (1) whether the task is critical to system operation and (2) whether the task is critical to the organization. Many tasks, although not critical to continuous operation, may still be time-sensitive or highly valuable to the organization (e.g., housekeeping or safety tasks). The planning and scheduling procedures of a maintenance organization are largely dependent on the priorities of the tasks that need to be completed.

Work can be scheduled in a number of ways. The system can monitor work requests with higher priorities and insert tasks into schedules automatically or the maintenance manager can adjust schedules based on predetermined priority codes as he or she sees fit.

Formal requests for emergency work are usually entered into the system after the fact. Data regarding the nature of the repair, labor effort expended, and materials used are entered into the system. Including emergency repair information allows for a complete historical picture of equipment maintenance.

Other important elements of an MMS are described in the following paragraphs.

Reporting

The maintenance system must keep track of labor and material use. This data can help determine correct staffing levels or identify frequently used materials or parts. Reports should cover user-specified time periods and reflect data such as labor hours and cost (both regular and overtime), types of repairs, parts and materials used, and the uses and costs of outside contractors. These reports can be customized to provide executive reporting, which provides specific indicators tailored for overview by management.

Preventive maintenance

Preventive maintenance (PM) is a significant function of automated maintenance systems. Effective PM means less time spent on corrective maintenance, resulting in better planning and increased productivity. The reliability of equipment can also be substantially improved.

The consistent, timely completion of regular PM tasks involves orderly procedures, schedules, and controls. The respective tasks to be performed on each piece of equipment should be identified and entered into the system. The usual sources of information are operational and maintenance manuals, equipment manufacturers' recommendations, and the experience of the plant personnel.

Tasks are categorized according to frequency of performance (e.g., daily, weekly, monthly, quarterly, per run-time hours, or by specific calendar date). A detailed procedure is defined for each task. This approach assists with new employee training. It also ensures uniformity (a standard operating procedure) in the performance of each task.

Equipment histories

A maintenance history records all corrective and PM tasks performed on a particular piece of equipment for a specified period of time (1 to 5 years). It is vital for making repair-versus-replacement decisions.

Historical reports also include information vital to the analysis of equipment effectiveness. Data such as parts and materials used, personnel who performed the work, duration of tasks, reasons for equipment failure, outside contractors used, or environmental conditions can be important when trends or statistics need to be determined.

System integration

Maintenance management systems should not be installed to run independently. System integration is important.

Connection points between maintenance systems can be divided into two categories: those that solely provide input to maintenance and those that also require output from maintenance. Repair notification can come from any interfacing departments. Any existing SCADA system can be interfaced to provide equipment run times for maintenance. This information, in turn, can automatically trigger appropriate PM tasks.

The operations department can provide information to the maintenance department regarding the intended operating plans. As a result, major repairs and PM tasks can be performed during scheduled shutdown periods.

The accounting department can provide budgetary information so that dollar amounts and limits can be established and monitored by account number.

The personnel department can establish codes for specialized skills and training of individual employees. It can update the system when new training and certification are acquired by each individual. This will allow for the scheduling of staff who possess the proper experience to do the job. The department can also provide labor rates and update accordingly so that the cost of each work order can be calculated.

The purchasing department can provide valuable vendor information regarding suppliers. It can monitor vendor performance, take advantage of cost-saving opportunities, and obtain material in a timely fashion. Purchase order status should be accessible to the maintenance department so that the expected delivery date of parts can be incorporated into the planning process. Maintenance also needs to be notified when parts are received.

Maintenance locators require information and maps from the geographic information system to mark infrastructure repair sites accurately. The engineering department can provide current, as-built drawings of equipment, piping, and wiring to assist with many repair tasks. Maintenance can monitor the progress of construction projects being conducted by the engineering and planning departments. This way, maintenance can be prepared to assume its activities when the projects are completed. New equipment and parts descriptions should be transferred to the maintenance database from project files before the equipment is put into use. This allows appropriate PM routines and procedures to be identified and incorporated.

Stores (warehouse) inventory provides the on-hand balance of parts for work order planning. This balance permits scheduling of only those work orders that have all of the necessary parts to complete the task. Inventory balances should automatically update the status of work orders. When all reserved parts are in stock, the work order can be released for scheduling and execution. A history of parts usage and an analysis of that usage can be determined from the information within the stores' inventory records.

Labor costs, material costs, and work order details are available to the accounting department. This department will incorporate the information into its budgeting process and possibly charge fees to user departments, other agencies, or customers. Labor hours charged to work orders can be input to payroll for time keeping and check processing. The purchase of special items required by the maintenance department can be controlled by the purchasing department and validated by the system via a purchase requisition system that is closely integrated with the maintenance department.

Work order status on infrastructure-mapping tasks is accessible through the GIS for use by any user department. Work order status on infrastructure and meter repairs is also available to the customer information system (CIS), so that customer service representatives can respond to inquiries from the public. In addition, the work order status, labor and material costs, and other data needed for managerial reports are available to the approved employees. Work order status can also be linked to equipment in the laboratory information management system (LIMS), so that the lab can plan when equipment will be used and when routine or corrective repairs will be completed.

All of these interface points provide the basis for a complete and comprehensive system integration plan. They allow for the accountability of each department's unique data while providing for a cooperative, shared information environment.

OTHER MAJOR INFORMATION SYSTEMS

In addition to SCADA, AM/FM/GIS, and MMS, other major information systems are used by a water utility to manage the water treatment and distribution network. All of these systems can be connected together (i.e., interfaced) to share common data and applications. This section discusses the other major computer information system interfaces, along with the corresponding shared data and applications.

Source-of-Supply Systems and Treatment Plant Process Control Systems

The source-of-supply system is concerned with the quantity and quality of raw water. The treatment plant process control system is concerned with the quantity and quality of water being treated. Maintaining the quality and quantity of water is a function of the demand. Computer systems can handle the large amounts of data from historical records and from real-time, or current, network sensors.

How these two systems work with the other systems depends on how the system network as a whole is managed. If the network is automated but does not include demand forecasts, the impact of consumption on the network is passed on to the treatment facility, which in turn passes this impact on to the supply source. Network demand variations are partially smoothed out by the use of treated-water reservoirs, which reduce the need for flow variations at the treatment plant.

If the network is driven by an optimized system that takes into account demand forecasts, these forecasts are passed directly on to both treatment facilities and supply sources. These forecasts improve the utility's ability to keep a constant flow at the treatment facilities.

Laboratory Information Management System

A laboratory information management system collects, stores, and processes water quality information. It also gives feedback to enhance water treatment and generates reports.

Information exchanges with an LIMS are limited to network quality management data. The control center receives real-time quality data from the LIMS. These data are used as they are received, especially in the form of alarms, but they are also stored in a water quality database, as are other analyses coming from the laboratory. These data can then be used to simulate water quality changes in the network, in connection with data on pump operation and valve status. Conversely, when the laboratory detects an abnormal analysis, the data are transmitted to the network manager so that corrective measures can be taken immediately.

Leakage Control and Emergency Response

The way leakage is traditionally detected is described in detail in chapter 13. Leak detection software can also be used. For instance, the software can receive readings of pressures, reservoir levels, and flow measurements in order to send input to a hydraulic model. Alternatively, the software can receive meter readings from the distribution network to calculate consumption per district. Unusual readings can indicate a leak.

Emergency response information systems can also benefit from the integration of computerized systems. Figure 16-16 illustrates how shared data from inside the water utility, as well as from outside sources including emergency services and other utilities, are available in an emergency situation. Having all of the required information readily available allows a water utility to react much more quickly in an emergency.

Customer Information System

The CIS contains data such as consumption and the customer's category and address. This information is used to calculate the meshed network for demand forecasts and to calculate network yields. The remote or automatic meter-reading methods described in chapter 10 provide data that can be used for demand forecasting as well as for leak detection.

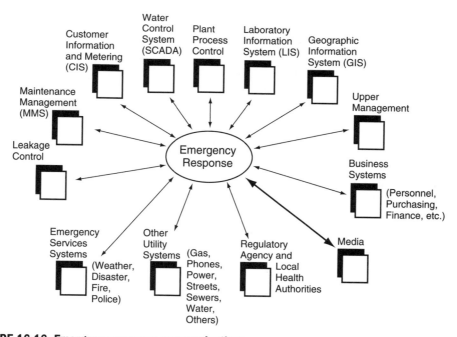

FIGURE 16-16 Emergency response communications

BIBLIOGRAPHY

Bein, R. 1998. *A Complete Integration of Water Distribution Computer Applications: Tomorrow's Water Utility*. AWWA Information Management & Technology Conference Proceedings, Reno-Sparks, Nev.

Casey, R. 2001. *Integrating Modelling, SCADA, GIS and Customer Systems to Improve Network Management*. AWWA Information Management & Technology Conference Proceedings, Atlanta, Ga.

Jordan, J.K. 2000. *Maintenance Management for Water Utilities*. Denver, Colo.: American Water Works Association.

Keiji, G., J.K. Jacobs, S. Hosoda, and R.L. Gerstberger, eds. 1994. *Instrumentation and Computer Integration of Water Utility Operations*. Denver, Colo.: AWWA Research Foundation and Japan Water Works Association.

Lindley, T.R., and S.G. Buchberger. 2001. *Managing Water Distribution Systems for Potential Intrusions Using Integrated Modeling and GIS*. AWWA Information Management & Technology Conference Proceedings, Atlanta, Ga.

Manual M2, Instrumentation & Control. 2001. Denver, Colo.: American Water Works Association.

Shamsi, U. 2002. *GIS Tools for Water, Wastewater, and Stormwater Systems*. New York, NY: ASCE Press, American Society of Civil Engineers.

Shamsi, U. 2005. *GIS Applications for Water, Wastewater, and Stormwater Systems*. Boca Raton, Fla.: CRC Press.

Water Distribution Operator Training Handbook. 2005. Denver, Colo.: American Water Works Association.

Water Environment Federation. 2005. *GIS Implementation for Water and Wastewater Treatment Facilities*. WEF Manual of Practice No. 26. New York, NY: WEF Press and McGraw-Hill.

WSO: SCADA and Instrumentation. 2005. DVD. Denver, Colo.: American Water Works Association.

CHAPTER 17

Systems Security and Emergency Response

WATER SUPPLY SYSTEM THREATS

The types of threats that would affect water utilities include both natural and accidental disasters and intentional disasters, which are discussed below. Natural disasters include floods, windstorms, ice storms, snowstorms, fires, droughts, and earthquakes. Accidental disasters include chemical spills, fires, transportation accidents, and explosions. Intentional disasters include the terrorist activities that are covered in this chapter. Three main types of consequences may occur: (1) complete interruption of supply, (2) sufficient quantity, but compromised quality, and (3) sufficient quality, but insufficient quantity.

Water distribution systems are extensive with many components and subcomponents. The components and subcomponents are relatively unprotected and accessible and are often isolated. The physical destruction of a water distribution system's components and subcomponents or the disruption of water supply could be more likely than the introduction of contaminants to a system. The actual probability of a terrorist threat to drinking water is probably very low; however, the consequences could be extremely severe for exposed populations.

Four major types of intentional threats to drinking water systems (Mays 2004) are discussed here.

Cyber Threats

- Physical disruption of a supervisory control and data acquisition (SCADA) network
- Attacks on central control system to create simultaneous failures
- Software attacks using worms/viruses
- Network flooding
- Jamming
- Disguise data to neutralize chlorine and add no disinfectant allowing addition of microbes

Physical Threats

- Physical destruction of system's assets or disruption of water supply could be more likely than contamination. A single terrorist or a small group of terrorists could easily cripple an entire city by destroying a critical component of the water system.

- Loss of water pressure compromises fire-fighting capabilities and could lead to possible bacterial buildup in the system.
- Potential for creating a water hammer effect by opening and closing major control valves and turning pumps off and on too quickly, resulting in simultaneous main breaks.

Chemical Threats

There are numerous chemical warfare agents and industrial chemical poisons. Some of the chemical warfare agents include hydrogen cyanide, tabun, sarin, VX, lewisite (arsenic fraction), sulfur mustard, 3-quinucli dinyl benzilate, and lysergic acid diethylamide. Some of the industrial chemical poisons include cyanides, arsenic fluoride, cadmium, mercury, dieldrin, sodium flouroacetate, and parathion.

Biological Threats

Several pathogens and biotoxins exist that have been weaponized, are potentially resistant to disinfection by chlorination, and are stable for relatively long periods in water. The pathogens include *Clostridium perfringens*, plague, and others. Biotoxins include botulium, aflatoxin, ricin, and others. Water does provide dilution potential; however a neutrally buoyant particle of any size could be used to disperse pathogens into drinking water systems. Water storage and distribution systems can facilitate the delivery of an effective dose of toxicant to a potentially large population. A more extensive discussion of microbiological contaminants and threats of concern is presented by Abbaszadegan and Alum (2004).

WATER SUPPLY SYSTEM VULNERABILITIES

Vulnerability of Water Supply Systems

Water supply systems are designed to operate under pressure and supply most of the water for fire-fighting purposes. Either a loss of water or a loss of pressure could interrupt service, disable fire-fighting ability, and disrupt public confidence. Such loss of pressure and/or water could result from sabotaging pumps that maintain flow and pressure or from disabling electric power sources, causing long-term disruption and taking months to replace custom-designed equipment.

Locations of vulnerability in these systems include the following (Clark and Deininger 2000):

- Raw water source (surface or groundwater)
- Raw water channels and pipelines
- Raw water reservoirs
- Treatment facilities

- Connections to the distribution system
- Pump stations and valves
- Finished water tanks and reservoirs

Figure 17-1 shows a range of potential contamination scenarios.

Vulnerabilities of Computer System Infrastructure

The computer infrastructure of a medium to large water utility has numerous systems including the financial, human resources, laboratory information management, maintenance management, supervisory control and data acquisition (SCADA), and others. The term SCADA is often used to include both in-plant computer-based process control systems and computer-based systems providing monitoring and control of geographically distributed raw water production and treated water distribution systems. Figure 17-2 simplifies a utility's network architecture into blocks showing the composite of a network's vulnerabilities.

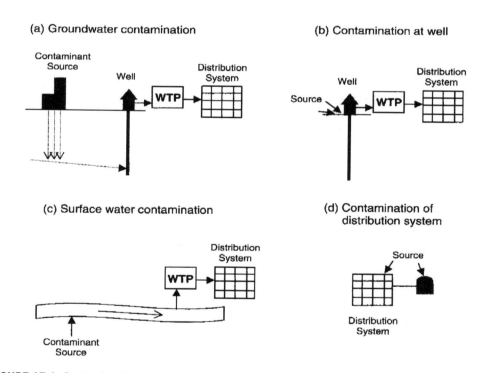

FIGURE 17-1   Contamination scenarios (Grayman et al. 2004a)

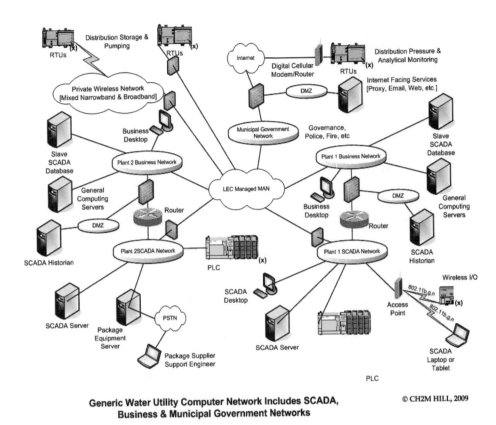

Generic Water Utility Computer Network Includes SCADA,
Business & Municipal Government Networks

© CH2M HILL, 2009

FIGURE 17-2 Typical mid-to-large water utility information flow requirements (W.R. Phillips Jr. CH2MHill 2009)

The list of vulnerabilities determined from a USEPA-mandated assessment of SCADA systems of a number of large utilities in the United States are as follows (Panguluri et al. 2004):

- Operator station logged on all the time even when the operator is not present at the workstation, thereby rendering the authentication process useless.
- Physical access to the SCADA equipment is relatively easy.
- Unprotected SCADA network access from remote locations via Digital Subscriber. Lines (DSL) and/or dial-up modem lines.
- Insecure wireless access points on the network.
- Most of SCADA networks directly or indirectly connected to the Internet.

- No firewall installed or the firewall configuration is weak or unverified.
- System event logs not monitored.
- Intrusion detection systems not used.
- Operating and SCADA system software patches not routinely applied.
- Network and/or router configuration insecure; passwords not changed from manufacturer's default.

VULNERABILITY ASSESSMENTS

A vulnerability assessment is the process of identifying, quantifying, and prioritizing (or ranking) the vulnerabilities in a system. Vulnerability assessments help water utilities to evaluate their susceptibility to potential threats and identify corrective actions to reduce or mitigate the risk of serious consequences from vandalism, insider sabotage, or terrorist attack. As required under the Bioterrorism Act (Public Health, Security, and Bioterrorism Preparedness and Response Act, PL 107-188, June 2002), a drinking water utility serving more than 3,300 persons must do the following:

- Conduct a vulnerability assessment, certify to USEPA that the assessment has been completed, and submit a copy of the assessment to USEPA.
- Show that the system has updated or completed an emergency response plan outlining response measures if an incident occurs.
- Certify to USEPA Administrator, upon completing the vulnerability assessment, that the system has completed or updated their emergency response plan.

Common elements of vulnerability assessments for a water system are (USEPA 2002):

- Characterization of the water system including its mission and objectives
- Identification and prioritization of adverse consequences to avoid
- Determination of critical assets that might be subject to malevolent acts that could result in undesired consequences
- Assessment of the likelihood (qualitative probability) of such malevolent acts from adversaries
- Evaluation of existing countermeasures
- Analysis of current risk and development of a prioritized plan for risk analysis

The complexity of vulnerability assessments ranges based upon the design and operation of the system.

SCADA Systems

A 21-step guide to improve cyber security of SCADA networks was developed by the US Department of Energy (2002) listed below:

1. Identify all connections to SCADA networks.
2. Remove unnecessary connections to the SCADA network.
3. Evaluate and strengthen the security of any remaining connections to the SCADA network.
4. Harden SCADA networks by removing or disabling unnecessary services.
5. Do not rely on proprietary protocols to protect the system.
6. Implement security features provided by device and system vendors.
7. Establish strong controls over any medium that is used as a backdoor into the SCADA network.
8. Implement internal and external intrusion detection systems and establish 24-hour-a-day incident monitoring.
9. Perform technical audits of SCADA devices and networks, and any other connected networks, to identify security concerns.
10. Conduct physical security surveys and assess all remote sites connected to the SCADA network to evaluate their security.
11. Establish SCADA "Red Teams" to identify and evaluate possible attack scenarios.
12. Clearly define cyber security roles, responsibilities, and authorities for managers, system administrators, and users.
13. Document network architecture and identify systems that serve critical functions or contain sensitive information that require additional levels of protection.
14. Establish a rigorous, ongoing risk management process.
15. Establish a network protection strategy based on the principle of defense in depth.
16. Clearly identify cyber security requirements.
17. Establish effective configuration management processes.
18. Conduct routine self-assessments.
19. Establish system backups and disaster recovery plans.
20. Senior organizational leadership establish expectations for cyber security performance and hold individuals accountable for their performance.
21. Establish policies and conduct training to minimize the likelihood that organizational personnel will inadvertently disclose sensitive information regarding the SCADA system design, operations, or security controls.

EARLY WARNING SYSTEMS

Early warning systems (EWS) for both source and finish (distributed) water are intended to reliably identify low-probability/high impact contamination events. Figure 17-3 illustrates

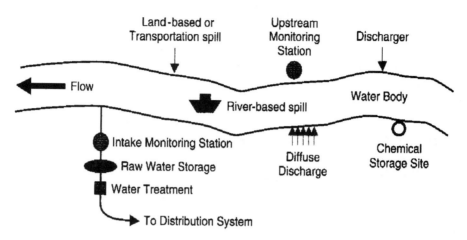

FIGURE 17-3 Elements of an early warning system (Grayman et al. 2004b)

the elements of a source early warning system. The International Life Sciences Institute (ILSI 1999) identified the following objectives for hazardous events in water:

*   Provide warning in sufficient time to respond to a contamination event and prevent exposure of the public to the contaminant
*   Have the capability to detect all potential contamination threats
*   Can be operated remotely
*   Can identify the point at which the contaminant was introduced
*   Have a low rate of false positive and false negative results
*   Provide continuous, year-round surveillance
*   Produce results with acceptable accuracy and precision
*   Require low skill and training
*   Is affordable to the majority of public water systems

Keeping in mind that a contamination event in a source or in a distributed water system must be identified in time to allow for an appropriate response that mitigates or eliminates the adverse impact, the following are features of an ideal EWS (Clark et al. 2004):

*   Rapid response time
*   Fully automated
*   Screens for a range of contaminants
*   Specific for the contaminants of concern
*   Sufficient sensitivity

- Low occurrence of false positives and false negatives
- High rate of sampling
- Reliable and rugged
- Requires minimal skill and training
- Affordable cost
- Maintain online model ready to use for source tracing after a contamination event

SECURITY HARDWARE AND SURVEILLANCE SYSTEMS

Intrusion Detection Systems

An intrusion detection system (IDS) has the purpose of detecting an intruder approaching a site, facility, or area as early as possible. These systems may provide the first and only indication that someone or something is trying to enter a facility. Key issues to consider in designing a IDS are the following (Booth et al. 2004):

- What is the facility, space, or area to be protected?
- Who is the perceived threat?
- What are known vulnerabilities of the area or space?
- What are the key assets or targets at the space?
- How will system monitoring take place?
- What power and communications methods exist?

The design of perimeter IDS needs to follow the following concepts with the goal to achieve the best possible performance (Booth et al. 2004):

- No gaps in coverage—a continuous line of detection around the perimeter area or interior space.
- Suitability for physical and environmental conditions—the sensor must be appropriate for the area being monitored (temperature, humidity, rain, fog, wind, pollution).
- Layers of protection—a fundamental security concept is that multiple, layered detection systems are much more effective than single systems. If one system is bypassed or defeated, the remaining systems are still in place to detect the intruder.

There are several intrusion detection sensor categories, including the following:

- *Exterior intrusion sensors* are used to sense an intrusion crossing an outdoor perimeter boundary. There are several types of exterior intrusion detection sensors including: buried line sensors, pressure/seismic sensors, magnetic field sensors, ported-coaxial buried cable systems, and fiber-optic buried cable systems.

- *Fence-mounted cabling sensors* include electromechanical vibrating sensing, coaxial strain sensitive cable, fiber-optic strain sensitive cable, taut-wire systems, and fence-mounted electric field sensors.
- *Free-standing exterior sensors* include active infrared sensors, passive infrared sensors, microwave, and dual technology (passive infrared [PIR] and microwave).
- *Interior sensors* types include interior volumetric sensors, ultrasonic and microwave sensors, passive infrared, and dual technology sensors.
- *Interior boundary penetration sensors* include glass break sensor, door switch, and linear beam (photoelectric beam).

Digital motion detection analyzes video streams of closed-circuit television cameras and compares those video streams to a still image in the unit's memory. The cameras must be in a fixed position and not pan/tilt type units. Wireless intrusion detection sensors include door switches, PIR, and dual technology (microwave plus PIR) sensors. Wireless sensors can provide cost savings over hard-wired sensors.

General intrusion detection recommendations for water supply systems are (Booth et al. 2004):

Reservoirs and elevated tanks
- Perimeter detection
- Monitoring ladder
- Monitoring vaults
- Monitoring hatches

Pump stations
- Perimeter detection
- Entrance detection
- Volumetric detection

Water treatment stations
- Perimeter detection
- Entrance detection
- Volumetric detection
- Monitoring clearwells
- Monitoring SCADA control rooms
- Monitoring chemical storage and dispensing areas

Raw water intake stations
- Perimeter detection
- Entrance detection
- Volumetric detection

Closed Circuit Television (CCTV)

Closed circuit television (CCTV) has been used for decades as an integral part of comprehensive security systems. The evolution in digital hardware has made CCTV smarter, more reliable, more efficient, and more effective for premise security in all types of applications (Booth et al. 2004). CCTV system components include cameras, the switcher or multiplexer, the transport media, and the wireless transmission.

Access Control Systems

Access control systems permit only authorized personnel to enter and exit a restricted area. Electronic access control systems are used to control entry into a perimeter, area, or interior space. Layered security systems for a water supply system may include four or five security access control levels (Booth et al. 2004):

- Public zone (level 1)
- Clear zone (level 2)
- Building lobby area (level 3)
- Internal circulation area (level 4)
- High value areas, if needed (level 5): for example, SCADA rooms, security equipment rooms, laboratory areas, chemical storage, etc.

Locking systems include key locks, mechanical or electrical keypads, electrified locking systems, fail-safe locking systems, and fail-secure locking systems. Card reader systems provide the most reliable, flexible method of controlling access to a facility. The access control for the interior circulation area could be a card access, and the access control to the high value areas could be a card reader plus a PIN. Figure 17-4 illustrates a typical single door card reader installation, and Figure 17-5 illustrates a typical access card system block diagram (with video surveillance). A guard tour system requires designated security staff conduct a security tour of a facility at specified frequencies and durations.

Role of SCADA Systems for Security

SCADA systems can be made a central part of security efforts so that security measures are coordinated with operations. Linking SCADA systems to perimeter monitoring devices provides constant monitoring and reduces the need for manned patrols. Security systems and equipment can be interfaced directly to the SCADA system or through a remote terminal unit (RTU). The SCADA system can react to conditions and perform control actions automatically. These actions could include the starting and stopping of pumps, the opening and closing of valves, emergency shutdowns, and others. Portions of a water distribution system can be isolated by stopping pumps and closing valves. The SCADA system can also include alarm management. SCADA systems can coordinate

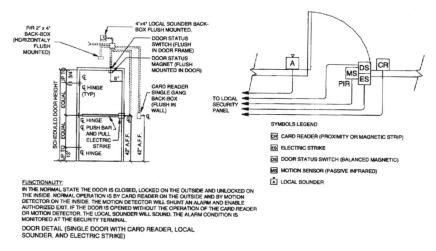

FIGURE 17-4 Typical single door card reader installation
Courtesy of Booth et al. 2004

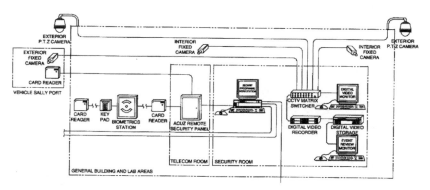

FIGURE 17-5 Typical access card system block diagram (with video surveillance)
Courtesy of Booth et al. 2004

security measures with process operations; reduce or eliminate manned patrols; provide constant monitoring, system-wide; and record alarms and events. SCADA systems can be expanded by using additional I/O points, RTU devices, and network links. SCADA systems can be enhanced even further if they incorporate advanced capabilities such as hydraulic modeling and simulation (network analysis).

EMERGENCY RESPONSE PLANS

An Emergency Response Plan (ERP) provides a step-by-step response to, and recovery from, incidents related to situations of emergency. The ability of water utility staff to respond rapidly in an emergency will help prevent unnecessary complications and protect consumers' health and safety. Proper preparedness is the key to achieving emergency response success.

System components could become less susceptible to harm by taking mitigation measures, which are actions that are taken to eliminate or reduce the harmful effects of water systems emergencies. The following items provide a guide to mitigation measures for water distribution, which should be described in an ERP (Manitoba Office of Drinking Water 2009):

Mitigation at the raw water source includes the following:

- Having access to an alternate raw water source, if situation allows
- Restricting access of unauthorized persons by fence and gate
- Facilitating access to the water source by utility staff (access by boat, road)
- Maintaining wells and surface water intakes; applying setback distance to wellhead
- Having a source water protection plan and wellhead protection plan

Mitigation of water distribution system failures consists of the following:

- Having spare parts available (valves, pipes, repair kits)
- Maintaining networks by replacing old, damaged, and poorly built distribution system components; regular flushing, valve and hydrant exercising
- Having redundancy by close-looping of networks and installing sufficient check valves, other control valves, etc.
- Preparing/updating distribution network mapping

Components of an ERP

The components of an ERP could include (Manitoba Office of Drinking Water 2009):

- A detailed map of the distribution system, detailed locations of each valve in the system, including references that will aid in locating these valves, and a map of well locations and surface water intakes, as applicable.
- A detailed map of electrical diagrams clearly showing generator and power source change-over.
- A contact list of emergency services, regulators, suppliers, contractors, water users with critical needs, media, phone companies, and water utilities.

- A statement of amounts budgeted for emergency use, along with a statement showing who may authorize expenditures for such purpose and under what conditions.

- A determination of not less than nine most likely emergencies that may affect the water system and procedures to be followed and actions necessary to provide service during emergencies.

- A determination of who would operate the system if all operators are off (i.e., pandemic flu).

- A description of ways to obtain and transport water from an alternate source, should it become necessary. It is advisable to have arrangements for obtaining water from at least two alternative sources that are not likely to be affected by the same hazards at the same time.

- A description of how often the plan should be revised (at least every two years), who has copies of the plan, etc.

- A description of methods of notification of water users that an emergency is under way.

An example ERP and an example emergency action chart are shown in Figures 17-6 and 17-7, respectively. (See the following pages.)

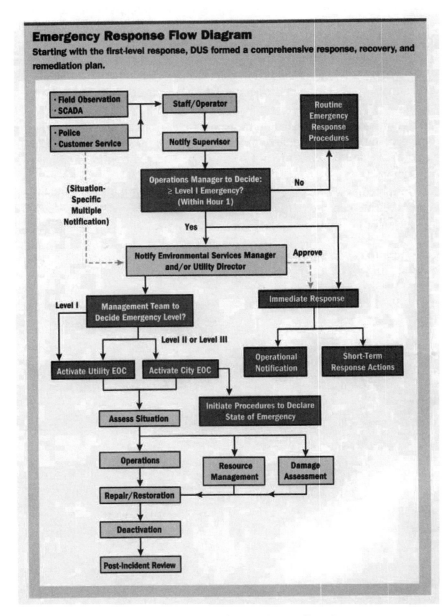

Emergency Response Flow Diagram

Starting with the first-level response, DUS formed a comprehensive response, recovery, and remediation plan.

FIGURE 17-6 Emergency response flow diagram for Henderson, Nevada

Courtesy of Johnson and Gabriel 2009

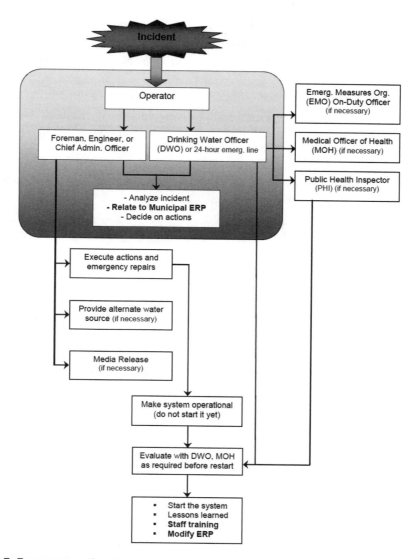

FIGURE 17-7 Emergency action chart
Courtesy of Manitoba Office of Drinking Water 2009

BIBLIOGRAPHY

Abbaszadegan, M. and A. Alum. 2004. Microbiological Contaminants and Threats of Concern. In *Water Supply Systems Security*, ed. L.W. Mays. New York: McGraw-Hill.

AWWA Standard for Security Practices for Operation and Management. ANSI/AWWA G430. Denver, Colo.: American Water Works Association (latest edition).

Booth, R., A. Bowman, F. Gist, and J. Ringold. 2004. Security Hardware and Surveillance Systems for Water Supply Systems. In *Water Supply Systems Security*, ed. L.W. Mays. New York: McGraw-Hill.

Clark, R.M. and R.A. Deininger. 2001. Minimizing the Vulnerability of Water Supplies to Natural and Terrorist Threats. In *Proceedings of the American Water Works Association IMTech Conference*, Atlanta, GA.

Clark, R.M., Panguluri, S., and Haught, R.C. 2004. Remote Monitoring and Network Models: Their Potential for Protecting US Water Supplies. In *Water Supply Systems Security*, ed. L.W. Mays. New York: McGraw-Hill.

Finnan, K. Water Security: The Role of the SCADA System. Bristol Babcock, http://www.automation.com/resources-tools/articles-white-papers/hmi-and-scada-software-technologies/water-security-the-role-of-the-scada-system.

Grayman, W.M., R.M. Clark, B.L. Harding, M. Maslia, and J. Aramini. 2004a. Reconstructing Historical Contamination Events. In *Water Supply Systems Security*, ed. L.W. Mays. New York: McGraw-Hill.

Grayman, W.M., R.A. Deininger, R.M. Males, and R.W. Gullick. 2004b. Source Water Early Warning Systems. In *Water Supply Systems Security*, ed. L.W. Mays. New York: McGraw-Hill.

International Life Sciences Institute, Risk Science Institute. 1999. *Early Warning Monitoring to Detect Hazardous Events in Water Supplies.* Washington, D.C.: ILSI Press.

Johnson, A., and R. Grabriel. 2009. Preparedness Reduces Risk. *Opflow*, 35(9):14–17.

Manitoba Office of Drinking Water. 2009. Emergency Planning for Water Utilities in Manitoba.

Mays, L.W., ed. 2004. *Water Supply Systems Security.* New York: McGraw-Hill.

Panguluri, S., W.R. Phillips Jr., R.M Clark. 2004. Cyber Threats and IT/SCADA System Vulnerability. In *Water Supply Systems Security*, ed. L.W. Mays. New York: McGraw-Hill.

Phillips Jr., W.R. 2003. Solving the Puzzle of Providing Appropriate Cyber Security While Maintaining Operations Effectiveness and Efficiency. Paper presented at the Florida Water Resources conference.

US Department of Energy, President's CIP Board. 2002. 21 Steps to Improve Cyber Security of SCADA Networks. Available at www.counterterrorismtraining.gov/upgates_102002.html.

US Environmental Protection Agency. 2002. *Vulnerability Assessment Fact Sheet.* Office of Water (4601M), EPA 816-F-02-025, available at www.epa.gov/ogwdw/security/index.html or www.epa.gov/safewater/watersecurity/pubs/va_fact_sheet_12-19.pdf.

CHAPTER **18**

Public Relations

It is important to maintain the public's confidence in the quality of drinking water and the services provided by a utility. Satisfied customers will pay their bills promptly and will provide political support for necessary rate increases or bond issues. They will also be less likely to turn to bottled water or home treatment devices of questionable quality.

The public relations activities of a utility are directed at maintaining public confidence and customer satisfaction. Water distribution personnel generally have the greatest exposure to the public and can do more to ensure public confidence than all the formal media campaigns and large-scale public relations projects put together. Conversely, management's formal public relations campaigns can be undermined by an employee's poor attitude or unwillingness to be of service beyond the immediate requirements of the job.

THE ROLE OF PUBLIC RELATIONS

The role of public relations is to help create and maintain public confidence in the utility's product and organization. The active cooperation of all utility personnel is essential. Every customer contact should be viewed as an opportunity to improve communications and build goodwill.

Utilities depend on customer support for new budgets and for the implementation of special projects that require additional charges. It can be goodwill that tips the balance in favor of a badly needed bond issue. Customers who see their utility in a favorable light will generally vote in a like manner. Customers who are favorably impressed with water system operations are most likely to support increased service charges when they are required. In addition, where there are good public relations, customers will also be more tolerant of problems such as temporary tastes and odors, voluntary conservation measures, or repairs that require disruption of service.

Conversely, a utility that fosters an "us versus them" attitude can expect little customer cooperation. In fact, this approach may prompt outright antagonism from customers who already envision the utility as an unfriendly institution with the power to withhold a product vital to life. Everybody suffers in such cases, including water distribution personnel, who may bear the brunt in terms of layoffs, lower salaries, or budget reductions affecting the system.

Customers must never be taken for granted. Although a utility rarely finds itself in the situation of a retail store, with competition offering lower prices or better service on the next corner, the same "best value for the dollar" approach to customer service should be applied. Higher prices must be justified by better products and services. Customer satisfaction must be a top priority.

THE ROLE OF WATER DISTRIBUTION PERSONNEL

As noted earlier, water distribution personnel are often in much closer contact with customers than anyone else in the utility. They may be the only contact between the utility and some customers. It may seem unfair, but a customer will remember the meter reader who took time at the doorway to wipe mud from his or her shoes much longer than today's favorable newspaper article about the utility. Likewise, the customer will recall the repair crew that heckled the family's cocker spaniel more vividly than the last rate hike.

Field personnel can add or detract from a utility's public image. On-the-job behavior can tell customers that the utility values their business or simply does not care. Once in the field, the utility employee *is* the utility. If utility employees do not respect the customer, in the customer's mind, the utility itself does not respect the customer.

Having effective public relations with customers requires three basic ingredients:

1. *Good communications* means really listening to what the customer has to say, then explaining policy, answering questions, or pointing out how the customer can save water or money.
2. *Caring* involves employees taking pride in themselves, their appearance, and the well-being of the customer.
3. *Courtesy* requires field personnel to follow the commonsense rules of polite behavior.

Meter Readers

Meter readers find themselves in the unenviable position of being associated with the water bill. Whether the customer blames a high bill on a misreading of the meter or simply feels that rates are too high, the meter reader is the most available target for complaint. The meter reader must also deal with unfriendly dogs, bad weather, and the general reluctance of homeowners to allow a stranger into their homes.

Irate customers are generally the exception, but an overdose of complaints can make any job difficult. However, the meter reader should remain informative and polite. Meter readers are not expected to be superhuman, but a cheerful and helpful outlook goes a long way toward effective public relations and makes the job much easier.

Here are some basic behavioral guidelines that a meter reader can follow in performing day-to-day tasks:

- For a good first impression, maintain a neat appearance (Figure 18-1). Most water utilities now feel it is well worth the cost to provide uniforms so that the employee is easily identifiable. The employee also has no excuse not to wear appropriate clothes. The clothes should be clean and well pressed.
- Meter readers should display name tags and carry credentials in the event they are asked to verify their employment with the water system. Do not make customers take your word that you are who you say you are. Many customers are wary about admitting strangers into their homes. Anyone can claim to be a meter reader. Be prepared to prove it.

FIGURE 18-1 Meter reader appearance is important
Courtesy of Colorado Springs Utilities

- Meter readers must be polite. They don't have to answer detailed questions but they can inform customers about inquiry or complaint procedures. Many customers merely want sympathy. A person who listens well and is courteous can often calm an irate customer.

- Short, succinct answers should be given whenever possible. A long, drawn-out explanation is confusing—and it also takes up time when the meter reader is supposed to be reading meters. If the utility furnishes informative brochures, a few of them can be carried to hand out when appropriate.

- Any leaks found on the premises should be reported. Make sure customers know that the water utility does not want them to pay for unnecessary water use.

- If customers read their own meters, time should be taken to explain the procedure. This helps prevent errors or future bill adjustments.

- Meter readers should show enthusiasm and keep a smile in their voice.

- Customers should be addressed properly. Use *Miss, Sir,* or *Mrs.*, rather than *Lady,* or *Hey You!*

- Good judgment must be used. For example, wipe off muddy shoes; don't kick the dog; don't smoke in homes, and preferably not in public; don't swear; don't chat with a customer for long periods of time; walk on sidewalks, not gardens or lawns; obey all driving and parking rules. In short, the meter reader should always try to be helpful and polite and let customers know that he or she and the utility are on the customer's side.

Maintenance and Repair Crews

Like meter readers, maintenance and repair crews are highly visible to the public. Field personnel who are conducting routine inspections or performing minor repairs should stay as well groomed as possible. In most cases, maintenance or repair crews will have little face-to-face contact with customers. Personnel should answer questions politely and to the best of their ability or refer the customer to a supervisor or the appropriate customer relations representative.

A common questions is, What are you doing? The inclination by workers who are busy and don't want to be bothered is to give a curt reply, or worse, an inaccurate reply. It must be remembered that most people who ask questions really are interested and want a correct answer. Although the worker can't take the time to provide a long dissertation, a reasonable response should be provided.

As a matter of routine, customers should be notified when service is to be temporarily discontinued. Shutoffs should be scheduled to coincide with low-water-use hours. Figure 18-2 shows a sample doorknob card that is often used for this purpose. Be sure the card isn't hung on a door that residents rarely or never use. In most cases, it is best to hang such cards on the back door. Other water systems print letters on letterhead paper and have workers hand them to homeowners or slip them under the door if nobody is at home.

If water must be shut off immediately, customers should be given at least a few minutes' notice to finish a shower or collect a pitcher of drinking water. At the same time, customers should be warned that water may be cloudy for a short time after it is turned back on. They should be informed that this presents no health threat, but that clothes should not be washed until the water clears.

Property should be respected and the work site kept as clean as possible. Utilities usually have policies governing lunch and breaks, but under no circumstances should field crews litter an area with wastepaper or soda cans. In general, customers don't like having workers lounging on their lawn while eating lunch. Smokers should not use the customer's lawn as an ashtray.

Damage to lawns, sidewalks, gardens, or streets should be kept to a minimum, and customers should be forewarned if property damage is expected to occur. Repairs should be completed and property restored to its original state before workers leave the site.

Vehicles should be clean and parked out of the way. They should not block driveways or alleys, nor should they hamper the flow of traffic in an area. Workers should never nap in a utility vehicle—this gives the impression that they are permitted to sleep on the job.

If streets are to be dug up, the neighborhood should be warned well in advance. The utility should suggest that those people affected move their vehicles. Good safety habits also promote public relations. Road barriers and warning signs communicate the utility's concern about customer and employee safety.

During the course of the job, workers should maintain a friendly attitude and a genuine desire to accomplish their work quickly and in the most inconspicuous manner possible.

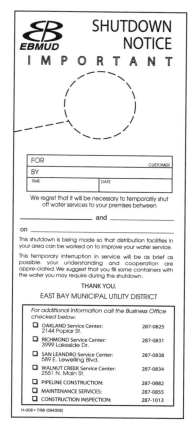

FIGURE 18-2 Example of a customer service card
Courtesy of EBMUD

Public Relations Behind the Wheel

At some time, most water distribution personnel will use a company vehicle. Public image is enhanced by good driving habits. If the driver of a utility vehicle drives over the speed limit or tailgates the driver in front, it will not be long before a customer telephones the utility to complain. Careless driving is dangerous, makes people angry, and gives the utility a bad image.

Driving rules should always be obeyed. Vehicles should be parked so as not to impede the flow of traffic. Parked vehicles should not block driveways, intersections, or alleys unless absolutely necessary. If blocking cannot be avoided, warning signals should be used on the vehicle. Where work must be performed in the street, appropriate cones, barricades, and other work-area protection methods should be employed to channel traffic smoothly around the jobsite.

Utility personnel must use good judgment regarding parking vehicles in front of coffee and doughnut shops. Although workers may be taking a legitimate coffee break, customers who see trucks regularly parked there get the impression that workers are spending most of their working day there.

Some water systems have developed a program with the local police department whereby every few years, all service personnel take a 4-hour safe driving course. Workers don't usually like having to go through the training, but it is still worth the time in terms of reduced accidents and improved public relations.

Public Relations and the Media

Although water distribution personnel may never be confronted with news reporters, a situation could arise—a main break, for instance, or several months of water conservation restrictions—that draws the attention of local newspapers and/or television stations. The general rule for talking to reporters is, *Don't!*

Large utilities maintain a public relations department whose job it is to prepare news releases and meet with the media. In other utilities, a manager or someone else is designated to take that responsibility. If approached by a reporter, distribution system operators should courteously but firmly state that they are not qualified to answer questions. Comments to the press will do little to help the situation. They may, on the contrary, do a great deal of damage.

THE ROLE OF INFORMED EMPLOYEES IN PUBLIC RELATIONS

When employees do not know or understand utility procedures, they may give customers the wrong information or anger them by refusing to answer questions. Communication between the utility and its customers requires well-informed employees.

Employee Training

Employee orientation programs should place a high priority on familiarizing new personnel with policies, regulations, and procedures. Information that relates to specific job assignments should be further explained during the initial period of employment. If an employee orientation program is not undertaken by the utility, new personnel should acquaint themselves with the utility's procedures. Some suggestions for doing this include the following:

*   If standard operating procedures (SOPs) exist, read them.
*   Keep a notebook and write down policies and procedures as you learn about them.
*   Do not be afraid to say "I don't know." Find out the answer, inform the customer, and keep track of the answer for future reference.
*   Ask questions.

All personnel should keep abreast of new developments. After initial employment, there may be no formal effort to update personnel. Workers should read bulletin boards and memos. Attending staff meetings can provide a forum for the exchange of ideas and information. Employees should take advantage of any training that is offered.

Procedural Manuals As a Public Relations Tool

Procedural manuals explain the governing policies of the utility, its services, turn-on and shutoff practices, and forms. They may double as employee handbooks, providing information on sick leave, benefits, and employee procedures.

Utilities that use procedural manuals usually have devised them to assist the customer service department. Traditionally, customer service representatives spend the bulk of their time on the telephone answering questions or handling complaints. When field personnel refer customers to the utility, their calls may end up in that department. A responsive utility will be open to procedural changes and will periodically review its operating practices.

Written guidelines keep employees informed and promote a well-run organization. Procedures should be clearly stated so that employees do not need to create their own interpretations. If a clearly worded policy is quoted to a customer, no argument as to the policy's meaning is likely to occur. Customers will be far more antagonistic toward a procedure that is unclear or sounds as if it has been made up on the spur of the moment.

FORMAL PUBLIC RELATIONS PROGRAMS

Although water distribution personnel play an integral part in creating a favorable image for the utility, there are many other facets of public relations that contribute to the total picture. The size of a water utility may influence the formality or scope of a public relations program, but not its importance.

Large utilities may confront political and environmental issues that affect a cross section of community interest. They may face a high volume of maintenance and repairs, billing, or collection problems; rate-allocation difficulties; or growth patterns that strain current treatment and personnel levels. A larger utility, therefore, will usually maintain a full-time public relations staff whose primary concern is to promote the company's image within the community. This staff will dispense information to the public and work closely with the media.

Smaller utilities are not immune to political, environmental, or community issues, but by necessity they may have to operate with a one- or two-person customer relations staff. In addition, there will likely be a manager who deals with local government, community representatives, and the public in general. Some functions of a larger utility's public relations efforts are discussed in the following sections.

Customer Service

Customer service representatives answer customer questions and handle complaints. They respond to telephone calls and complete the resulting paperwork. Some problems can be solved within minutes, whereas others are more complicated or involve intricate billing problems. Complex or particularly sensitive problems may be referred to a supervisor. In general, customer service representatives are well versed in telephone etiquette, active listening techniques, utility procedures, and persuasive speaking.

Public Information

Public relations specialists dispense information to the community. Their primary goal is to project a favorable image of the organization to the public. Public speaking engagements, participation in civic or professional clubs, and the creation of public service or special school projects consume much of the public information expert's time. Utilities that are conscious of their community image will often support local television, particularly educational or public service channels.

The public information specialist may also create and distribute literature explaining utility operations and policies or brochures giving useful water-related information to consumers.

Media Relations

Public relations personnel generally handle most communications with newspapers, magazines, radio, and television. An ongoing program highlights utility-sponsored projects, upper-management personnel changes, conservation efforts, or any newsworthy information. Bond issues, rate increases, special projects and how they affect homeowners and businesses, or emergency situations are explained to the public through the news media.

A press conference is an interview held for reporters. It is conducted by a utility spokesperson, usually a public relations or media relations expert, the general manager, or the manager's designate. A press conference is usually called to explain an emergency that has some impact on the community as a whole or to make some type of announcement. Media relations experts know what to say and how to say it in order to put the utility in the best possible light.

BIBLIOGRAPHY

Bloetscher, Frederick. 2009. *Water Basics for Decision Makers: Local Officials' Guide to Water & Wastewater Systems.* Denver, Colo.: American Water Works Association.

CH2M-HILL, Inc. *Public Involvement...Making It Work.* 2001. Denver, Colo.: American Water Works Association and AWWA Research Foundation.

Customer Service for Water Utilities. 2005. DVD. Denver, Colo.: American Water Works Association.

Miller, W.H. 1992. *Miller on Managing: Straight Talk on the Ups and Downs, the Do's and Don'ts of Managing a Water Utility.* Denver, Colo.: American Water Works Association.

Olstein, M.A., M.J. Stanford, and C.E. Day. 2001. *Best Practices for a Continually Improving Customer Responsive Organization.* Denver, Colo.: American Water Works Association and AWWA Research Foundation.

Symons, J.M. 2010. *Plain Talk About Drinking Water: Answers to Your Questions About the Water You Drink,* 5th ed. Denver, Colo.: American Water Works Association.

Specifications and Approval of Treatment Chemicals and System Components

DRINKING WATER ADDITIVES

Many chemicals are added to water during the various water treatment processes. Chemical additives are commonly used in the following processes:

- Coagulation and flocculation
- Control of corrosion and scale
- Chemical softening
- Sequestering of iron and manganese
- Chemical precipitation
- pH adjustment
- Disinfection and oxidation
- Algae and aquatic plant control

It is important that the chemicals added to water during treatment do not add any objectionable tastes, odors, or color to the water; otherwise, customers will likely complain. It is of even greater importance that the chemicals, or side effects that they cause, do not create any danger to public health.

COATINGS AND EQUIPMENT IN CONTACT WITH WATER

In addition to chemicals that are added to water for treatment, many products come in direct contact with potable water and could, under some circumstances, cause objectionable contamination. The general categories of products that could cause an adverse effect because of their contact with water are

- pipes, faucets, and other plumbing materials;
- protective materials such as paints, coatings, and linings;
- joining and sealing materials;

- lubricants;
- mechanical devices that contact drinking water; and
- process media, such as filter and ion exchange media.

Objectionable effects that can be caused by construction and maintenance materials include unpleasant tastes and odors, support of microbiological growth, and the liberation of toxic organic and inorganic chemicals.

NSF INTERNATIONAL STANDARDS AND APPROVAL

For many years, the water industry relied on American Water Works Association (AWWA) standards and on approval by the US Environmental Protection Agency (USEPA) of individual products to ensure that harmful chemicals were not unknowingly added to potable water. However, there was no actual testing of the products, and the water treatment industry had to rely on the manufacturers' word that their products did not contain toxic materials.

In recent years, increasing numbers of products have been offered for public water supply use, many of them made with new manufactured chemicals that have not been thoroughly tested. In addition, recent toxicological research has revealed potential adverse health effects due to rather low levels of continuous exposure to many chemicals and substances previously considered safe.

In the early 1980s, it became evident that more exacting standards were needed, as well as more definite assurance that products positively meet safety standards. In view of the growing complexity of testing and approving water treatment chemicals and components, the USEPA awarded a grant in 1985 for the development of private-sector standards and a certification program. The grant was awarded to a consortium of partners. National Science Foundation (NSF) International was designated the responsible lead. Other cooperating organizations were the Association of State Drinking Water Administrators (ASDWA), AWWA, the Water Research Foundation (formerly AwwaRF), and the Conference of State Health and Environmental Managers (COSHEM).

Two standards were developed by these organizations, along with the help of many volunteers from the water supply and manufacturing industries who served on development committees. These standards have now been adopted by the American National Standards Institute (ANSI), so they bear an ANSI designation in addition to the NSF reference.

ANSI/NSF Standard 60 essentially covers treatment chemicals for drinking water. The standard sets up testing procedures for each type of chemical, provides limits on the percentage of the chemical that can safely be added to potable water, and places limitations on any harmful substances that might be present as impurities in the chemical.

ANSI/NSF Standard 61 covers materials that are in contact with water, such as coatings, construction materials, and components used in processing and distributing potable water. The standard sets up testing procedures for each type of product to ensure that they do not unduly contribute to microbiological growth, leach harmful chemicals into the water, or otherwise cause problems or adverse effects on public health.

Manufacturers of chemicals and other products that are sold for the purpose of being added to water or that will be in contact with potable water must now submit samples of their product to NSF International or another qualifying laboratory for testing based on Standards 60 and 61. If a product qualifies, it is then "listed." There are also provisions for periodic retesting and inspection of the manufacturer's processes by the testing laboratory.

The listings of certified products provided by NSF International are used in particular by three groups in the water supply industry.

1. In the design of water treatment facilities, engineers must specify that pipes, paints, caulks, liners, and other products that will be used in construction, as well as the chemicals to be used in the treatment process, are listed under one of the standards. This ensures in a very simple manner that only appropriate materials will be used. It also provides contractors with specific information on what materials qualify for use, without limiting competition.

2. Individual states and local agencies have the right to impose more stringent requirements or to allow the use of products based on other criteria, but most states have basically agreed to accept NSF International standards. When state authorities approve plans and specifications for the construction of new water systems or improvements to older systems, they will generally specify that all additives, coatings, and components be listed as having been tested for compliance with the standards.

3. Water system operators can best protect themselves and their water systems from customer complaints or possibly even lawsuits by insisting that only listed products be used for everything that is added to or in contact with potable water. Whether it concerns purchasing paint for plant maintenance or taking bids for supplies of chemicals, the manufacturer or representative should be asked to provide proof that the exact product has been tested and is listed.

Copies of the current listing of products approved based on Standards 60 and 61 should be available at state drinking water program offices. A copy of the current listing can also be obtained by contacting the nearest NSF International regional office or the following address:

NSF International
P.O. Box 130140
789 N. Dixboro Road
Ann Arbor, MI 48113-0140
(734) 769-8010
www.nsf.org

AWWA STANDARDS

Since 1908, AWWA has developed and maintained a series of voluntary consensus standards for products and procedures used in the water supply community. These standards cover products such as pipe, valves, and water treatment chemicals. They also cover procedures such as disinfection of storage tanks and design of pipe. As of 2010, AWWA had approximately 156 standards in existence and many new standards under development. A list of standards currently available may be obtained from AWWA at any time.

AWWA offers and encourages use of its standards by anyone on a voluntary basis. AWWA has no authority to require the use of its standards by any water utility, manufacturer, or other person. Many individuals in the water supply community, however, choose to use AWWA standards. Manufacturers often produce products complying with the provisions of AWWA standards. Water utilities and consulting engineers frequently include the provisions of AWWA standards in their specifications for projects or purchase of products. Regulatory agencies require compliance with AWWA standards as part of their public water supply regulations. All of these uses of AWWA standards can establish a mandatory relationship, for example, between a buyer and a seller, but AWWA is not part of that relationship.

AWWA recognizes that others use its standards extensively in mandatory relationships and takes great care to avoid provisions in the standards that could give one party a disadvantage relative to another. Proprietary products are avoided whenever possible in favor of generic descriptions of functionality or construction. AWWA standards are not intended to describe the highest level of quality available, but rather describe minimum levels of quality and performance expected to provide long and useful service in the water supply community.

While maintaining product and procedural standards, AWWA does not endorse, test, approve, or certify any product. No product is or ever has been AWWA approved. Compliance with AWWA standards is encouraged, and demonstration of such compliance is entirely between the buyer and seller, with no involvement by AWWA.

AWWA standards are developed by balanced committees of persons from the water supply community who serve on a voluntary basis. Product users and producers, as well as those with general interest, are all involved in AWWA committees. Persons from water utilities, manufacturing companies, consulting engineering firms, regulatory agencies, universities, and others gather to provide their expertise in developing the content of the standards. Agreement of such a group is intended to provide standards, and thereby products, that serve the water supply community well.

Other Sources of Information

Many trade associations, publishers, and other groups have information available that can be useful to water system operators. Depending on the organization, they may have product literature, association standards, installation manuals, slide presentations, and video programs available. Some groups will also provide technical advice on their area of expertise.

The following are names and addresses of some of these organizations. Note that this list is not exhaustive. The local state water agency may be contacted for more sources.

American Concrete Pressure Pipe Association
3900 University Dr., Ste. 110
Fairfax, VA 22030-2513
(703) 273-PCCP (7227)
www.acppa.org

American Public Works Association
2345 Grand Blvd., Ste. 700
Kansas City, MO 64108-2625
(816) 472-6100
www.apwa.net

American Water Works Association
6666 W. Quincy Ave.
Denver, CO 80235-3098
(303) 794-7711
www.awwa.org

Association of State Drinking Water Administrators
1401 Wilson Blvd., Ste. 1225
Arlington, VA 22209
(703) 812-9505
www.asdwa.org

The Chlorine Institute Inc.
1300 Wilson Blvd.
Arlington, VA 22209
(703) 741-5760
www.chlorineinstitute.org

Ductile Iron Pipe Research Association
245 Riverchase Parkway East, Ste. O
Birmingham, AL 35244
(205) 402-8700
www.dipra.org

Hydraulic Institute
6 Campus Dr., First Floor NorthParsippany,
NJ 07054-4406
(973) 267-9700
www.pumps.org

International Desalination Association
P.O. Box 387
94 Central St., Ste. 200
Topsfield, MA 01983
(978) 887-0410
www.idadesal.org

International Ozone Association
Pan American Group
P.O. Box 28873
Scottsdale, AZ 85255
(480) 529-3787
www.io3a .org

National Association of Water Companies
2001 L Street NW, Ste. 850
Washington, DC 20036
(202) 833-8383
www.nawc.org

National Fire Protection Association
1 Batterymarch Park
Quincy, MA 02169-7471
(617) 770-3000
www.nfpa.org

National Ground Water Association
601 Dempsey Rd.
Westerville, OH 43081
(800) 551-7379
www.ngwa.org

National Lime Association
200 N. Glebe Rd., Ste. 800
Arlington, VA 22203
(703) 243-5463
www.lime.org

National Rural Water Association
2915 S. 13th St.
Duncan, OK 73533
(580) 252-0629
www.nrwa.org

National Small Flows Clearinghouse
West Virginia University/NRCCE
P.O. Box 6064
Morgantown, WV 26506-6064
(304) 293-4191
www.nesc.wvu.edu

North American Society for Trenchless Technology
1655 N. Ft. Myer Dr., Ste. 700
Arlington, VA 22209
(703) 351-5252
www.nastt.org

NSF International
P.O. Box 130140
789 N. Dixboro Rd.
Ann Arbor, MI 48113-0140
(734) 769-8010
www.nsf.org

Plastics Pipe Institute
1825 Connecticut Ave., NW, Ste. 680
Washington, DC 20009
(202) 462-9607
www.plasticpipe.org

Salt Institute
700 N. Fairfax St., Ste. 600
Alexandria, VA 22314-2040
(703) 549-4648
www.saltinstitute.org

Steel Tube Institute of North America
2516 Waukegan Rd., Ste. 172
Glenview, IL 60025
(847) 461-1701
www.steeltubeinstitute.org

Trench Shoring & Shielding Association
A Product Specific Group of the Association of Equipment Manufacturers
6737 W. Washington St., Ste. 2400
Milwaukee, WI 53214-5647
(414) 272-0943
www.aem.org/CBC/ ProdSpec/TSSA/

Uni-Bell PVC Pipe Association
2711 LBJ Freeway, Ste. 1000Dallas, TX 75234
(972) 243-3902
www.uni-bell.org

Utilimetrics
1400 E. Touhy Avenue, Ste. 258
Des Plaines, IL 60018
(847) 480-9628

Valve Manufacturers Association of America
1050 17th St., NW, Ste. 280
Washington, DC 20036-5521
(202) 331-8105
www.vma.org

Water Quality Association
4151 Naperville Rd.
Lisle, IL 60532-3696
(630) 505-0160
www.wqa.org

The following are some published sources of current information on water system equipment:

* *American Water Works Association Sourcebook*, sourcebook.awwa.org (covering the field of water supply and treatment). Published annually by the American Water Works Association, 6666 W. Quincy Ave., Denver, CO 80235.
* *Pollution Equipment News Buyer's Guide* (covering water supply and pollution fields). Published annually by Rimbach Publishing Inc., 8650 Babcock Blvd., Pittsburgh, PA 15237-5821
* *Public Works Manual* (covering the fields of general operations, streets and highways, water supply and treatment, and water pollution control). Hanley Wood Business Media, 8725 W. Higgins Rd., Suite 600,Chicago, IL 60631.

Glossary

AC   See *alternating current*.

A–C   Asbestos–cement.

across-the-line controller   See *full-voltage controller*.

actual cross-connection   Any arrangement of pipes, fittings, or devices that connects a potable water supply directly to a nonpotable source at all times. Also known as a *direct cross-connection*.

actuator   A device, usually electrically or pneumatically powered, that is used to operate valves.

air-and-vacuum relief valve   A dual-function air valve that (1) permits entrance of air into a pipe being emptied, thus preventing a vacuum, and (2) allows air to escape in a pipe while being filled or under pressure.

air binding   The condition in which air has collected in the high points of distribution mains, reducing the capacity of the mains.

air gap   In plumbing, the unobstructed vertical distance through the free atmosphere between (1) the lowest opening from any pipe or outlet supplying water to a tank, plumbing fixture, or other container and (2) the overflow rim of that container.

air purging   A procedure to clean mains less than 4 in. (100 mm) in diameter, in which air from a compressor is mixed with the water and flushed through the main.

air-relief valve   An air valve placed at a high point in a pipeline to release air automatically, thereby preventing air binding and pressure buildup.

alternating current (AC)   Electric current that flows first in one direction and then in the other. The sequence of one rise and fall in current strength in each direction is called a *cycle*.

altitude valve   A valve that automatically shuts off water flow when the water level in an elevated tank reaches a preset elevation, then opens again when the pressure on the system side is less than that on the tank side.

ambient   Relating to the prevailing environmental conditions in a given area.

AM/FM/GIS   See *automated mapping/facility management/geographic information system*.

ammeter   An instrument for measuring amperes.

ampere (amp or A)   The unit of measure for electric current. One volt will send a current of 1 ampere through a resistance of 1 ohm.

anaerobic   Absent of air or free oxygen.

analog   Continuously variable, as applied to signals, instruments, or controls. Compare *digital*.

angle of repose   The maximum angle or slope from the horizontal that a given loose or granular material, such as sand, can maintain without caving in or sliding. Can vary considerably with changes in moisture content.

appurtenances   Auxiliary equipment, such as valves and hydrants, attached to the distribution system to enable it to function properly.

arbor press   A special tool used to force a press-fitted impeller and bearings off of the pump shaft without damaging the parts.

armature   See *rotor*.

arterial-loop system   A distribution system layout involving a complete loop of arterial mains (sometimes called trunk mains or feeders) around the area being served, with branch mains projecting inward. Such a system minimizes dead ends.

arterial map   A comprehensive map showing primary distribution mains 8 in. (203 mm) or larger. Generally a supplemental mapped record that is used in system analysis.

as-built plans   Plans showing how the distribution system was actually constructed, including all modifications that were made during construction.

atmospheric vacuum breaker   A mechanical device consisting of a float check valve and an air-inlet port designed to prevent backsiphonage.

automated mapping/facility management/geographic information system (AM/FM/GIS)   A computerized system for collecting, storing, and analyzing water system components for which geographic location is an important characteristic.

automatic control   A system in which equipment is controlled entirely by machines or computers, without human intervention, under normal conditions.

automatic meter reading (AMR)   Any of several methods of obtaining readings from customer meters by a remote method. Methods that have been used include transmitting the reading through the telephone system, through the electric power network, through water lines (via sound transmission), through cable TV wiring, and by radio.

auxiliary supply   Any water source or system, other than the potable water supply, that may be available in the building or premises.

axial-flow pump   A pump in which a propeller-like impeller forces water out in a direction parallel to the shaft. Also called a *propeller pump*. Compare *mixed-flow pump, radial-flow pump*.

backfill   (1) The operation of refilling an excavation, such as a trench, after the pipeline or other structure has been placed into the excavation. (2) The material used to fill the excavation in the process of backfilling.

backflow   A hydraulic condition, caused by a difference in pressures, in which nonpotable water or other fluids flow into a potable water system.

backpressure   A condition in which a pump, boiler, or other equipment produces a pressure greater than the water supply pressure.

backsiphonage   A condition in which the pressure in the distribution system is less than atmospheric pressure, which allows contamination to enter a water system through a cross-connection.

ball valve   A valve consisting of a ball resting in a cylindrical seat. A hole is bored through the ball to allow water to flow when the valve is open. When the ball is rotated 90°, the valve is closed.

barrel   The body of a fire hydrant.

base   The inlet structure of a fire hydrant. An elbow-shaped piece that is usually constructed as a gray cast-iron casting. Also known as the *shoe, inlet, elbow,* or *foot piece*.

base map   A map used to prepare comprehensive, sectional, or other mapped records. Serves as the background for the mapped records. State agencies, such as highway departments, are a good source for base maps.

bearing   Antifriction device used to support and guide pump and motor shafts.

bedding   A select type of soil used to support a pipe or other conduit in a trench.

bellows sensor   A simple, accordion-like mechanical device for sensing changes in pressure.

bimetallic   Made of two different types of metal.

binary code   A method of representing numerical values using only two signal levels—on and off (or high and low). Used extensively in digital computer electronics.

blowoff valve   A valve installed in a low point or depression on a pipeline to allow drainage of the line. Also called a *washout valve*.

body   The major part of a valve, which houses the remainder of the valve assembly.

bonnet   The top cover or closure on the hydrant upper section. It is removable for the purpose of repairing or replacing the internal parts of the hydrant.

booster disinfection   The practice of adding additional disinfectant in the distribution system.

Bourdon tube   A semicircular tube of elliptical cross section, used to sense pressure changes.

brace   See *trench brace.*

breakaway hydrant   A two-part, dry-barrel post hydrant with a coupling or other device joining the upper and lower sections. The coupling and barrel are designed to break cleanly when the hydrant is struck by a vehicle, preventing water loss and allowing easy repair.

bronze seat ring   A machined ring, mounted in the body of a hydrant or valve, against which the moving disc of the valve closes.

brushes   Graphite connectors that rub against the spinning commutator in an electric motor or generator, connecting the rotor windings to the external circuit.

bubbler tube   A level-sensing device that forces a constant volume of air into the liquid for which the level is being measured.

butterfly valve   A valve in which a disc rotates on a shaft as the valve opens or closes. In the fully open position, the disc is parallel to the axis of the pipe.

bypass   (1) An arrangement of pipes, conduits, gates, or valves by which the flow may be passed around an appurtenance or treatment process. (2) In cross-connection control, any pipe arrangement that passes water around a protective device, causing the device to be ineffective.

bypass valve   A small valve installed in parallel with a larger valve. Used to equalize the pressure on both sides of the disc of the larger valve before the larger valve is opened.

cap nut   Piece of equipment that connects a standard compression hydrant valve assembly to the hydrant main rod.

carriers (of disease) Humans or animals who carry disease and can pass the germs to others without getting the disease themselves.

casing The enclosure surrounding a pump impeller, into which the suction and discharge ports are machined.

cathodic protection An electrical system for preventing corrosion to metals, particularly metallic pipe and tanks.

cavitation A condition that can occur when pumps are run too fast or water is forced to change direction quickly. During cavitation, a partial vacuum forms near the pipe wall or impeller blade, causing potentially rapid pitting of the metal.

centrifugal pump A pump consisting of an impeller on a rotating shaft enclosed by a casing that has suction and discharge connections. The spinning impeller throws water outward at high velocity, and the casing shape converts this high velocity to a high pressure.

channel See *transmission channel*.

check valve A valve designed to open in the direction of normal flow and close with reversal of flow. An approved check valve has substantial construction and suitable materials, is positive in closing, and permits no leakage in a direction opposite to normal flow.

CIS See *customer information system*.

close-coupled Relating to pump assembly for which the impeller is mounted on the shaft of the motor that drives the pump. Compare *frame-mounted*.

closed-loop control A form of computerized control that automatically adjusts for changing conditions to produce the correct output, so that no operator intervention is needed.

combination starter A motor starter in which the safety switch (to protect the motor against short circuits) is combined with the starter.

commutator A device that is part of the rotor of certain designs of motors and generators. The motor unit's brushes rub against the surface of the spinning commutator, allowing current to be transferred between the rotor and the external circuits.

compound meter A water meter consisting of two single meters of different capacities and a regulating valve that automatically diverts all or part of the flow from one meter to the other. The valve senses flow rate and shifts the flow to the meter that can most accurately measure it.

comprehensive map   A map that provides a clear picture of the entire distribution system. It usually indicates the locations of water mains, fire hydrants, valves, reservoirs and tanks, pump stations, pressure zone limits, and closed valves at pressure zone limits.

cone valve   A valve in which the movable internal part is a cone-shaped rotating plug. The valve is opened when the plug is turned through an angle of 90°, so fluid can pass through a port machined through the plug.

continuous feed method   A method of disinfecting new or repaired mains in which chlorine is continuously added to the water being used to fill the pipe, so that a constant concentration can be maintained.

contour lines   A line on a map that joins points having equal elevations.

control equipment   Mechanical, electrical, and hydraulic devices used to turn other machines on or off or to change their operating characteristics.

control relay   A device that allows low-power electrical signals to operate the on–off switch for high-power equipment. Also know as *power relay.*

coordinate system   Any standard system for determining the location of a particular point on the earth's surface.

Corey hydrant   A type of dry-barrel hydrant in which the main valve closes horizontally and the barrel extends well below the connection to the pipe. Also called *Iowa hydrant.*

corporation cock   See *corporation stop.*

corporation stop   A valve for joining a service line to a street water main. Cannot be operated from the surface. Also called *corporation cock.*

corrosion   The gradual deterioration or destruction of a substance or material by chemical action, frequently induced by electrochemical processes. The action proceeds inward from the surface.

coupling   A device that connects the pump shaft to the motor shaft.

coupon   In tapping, the section of the main cut out by the drilling machine.

cross-connection   Any arrangement of pipes, fittings, fixtures, or devices that connects a nonpotable system to a potable water system.

crushing strength   A measure of the ability of a pipe or other material to withstand external loads, such as backfill and traffic.

curb box   A cylinder placed around the curb stop and extending to the ground surface to allow access to the valve.

curb cock   See *curb stop.*

curb stop   A shutoff valve attached to a water service line from a water main to a customer's premises, usually placed near the customer's property line. It may be operated by a valve key to start or stop flow to the water supply line. Also called *curb valve.*

current   (1) The flow rate of electricity, measured in amperes. (2) In telemetry, a signal whose amperage varies as the parameter being measured varies.

current meter   A device for determining flow rate by measuring the velocity of moving water. Turbine meters, propeller meters, and multijet meters are common types. Compare *positive-displacement meter.*

customer information system (CIS)   A computerized system for collecting, storing, and reporting water consumption data by customer name, category, and address.

customer service   A division of a utility that is responsible for the direct contact with customers regarding service or billing inquiries. The division is staffed with customer service representatives.

cut-in valve   A specially designed valve used with a sleeve that allows the valve to be placed in an existing main.

C value   The Hazen–Williams roughness coefficient. A number used in the Hazen–Williams formula, which is used to determine flow capacities of pipelines. The C value depends on the condition of the inside surface of the pipe. The smoother the surface of the pipe wall, the larger the C value and the greater the carrying capacity of the pipeline.

D'Arsonval meter   An electrical measuring device, consisting of an indicator needle attached to a coil of wire, placed within the field of a permanent magnet. The needle moves when an electric current is passed through the coil.

DC   See *direct current.*

demand charge   An amount charged by electrical utilities in addition to the normal power rate to take into account the extra power capacity needed to supply an electric motor's increased current demand when the motor is starting.

detector-check meter   A meter that measures daily flow but allows emergency flow to bypass the meter. Consists of a weight-loaded check valve in the main line that remains closed under normal usage and a bypass around the valve containing a positive-displacement meter.

diaphragm element   A mechanical sensor used to determine liquid levels. Uses a diaphragm and an enclosed volume of air.

diffuser bowl   The segment of a turbine pump that houses one impeller stage.

diffuser vanes   Vanes installed within the pump casing of diffuser centrifugal pumps to change velocity head to pressure head.

digital   Varying in precise steps, as applied to signals or instrumentation and control devices. Compare *analog*.

direct cross-connection   See *actual cross-connection*.

direct current (DC)   Type of electrical current, such as that produced by a battery, for which the same polarity is maintained at all times.

direct-wire control   A system for controlling equipment at a site by running wires from the equipment to the onsite control panel.

directional flushing   A systematic approach to direct the flow from a clean source to the area to be flushed.

disc   A circular piece of metal used in many valves as the movable element that regulates the flow of water as the valve is operated.

dispatcher station   The central location from which an operator controls equipment at one or more remote sites.

distribution load-control center   A central site with equipment for one or more operators to monitor and operate the entire distribution system.

distribution main   Any pipe in the distribution system other than a service line.

distribution storage   A tank or reservoir connected with the distribution system of a water supply. Used primarily to accommodate changes in demand that occur over short periods (several hours to several days) and also to provide local storage for use during emergencies, such as a break in a main supply line or failure of a pumping plant.

Doppler effect   The apparent change in frequency (pitch) of sound waves due to the relative velocity between the source of the sound waves and the observer.

double-suction pump   A centrifugal pump in which the water enters from both sides of the impeller. Also called a *split-case pump*.

dry-barrel hydrant   A hydrant for which the main valve is located in the base. The barrel is pressurized with water only when the main valve is opened. When the main valve is closed, the barrel drains. This type of hydrant is especially appropriate for use in areas where freezing weather occurs.

dry tap   A connection made to a main that is empty. Compare *wet tap*.

dry-top hydrant   A dry-barrel hydrant in which the threaded end of the main rod and the revolving or operating nut is sealed from water in the barrel when the main valve of the hydrant is in use.

dysentery   A disease (sometimes waterborne) caused by pathogenic microorganisms and characterized by severe diarrhea with passage of mucus and blood.

eddy hydrant   A type of dry-barrel hydrant in which the main valve closes against pressure (downward) and the barrel extends slightly below the connection to the pipe. Compare *standard compression hydrant*.

elbow   See *base*.

electrodynamic meter   A device used to measure electrical power (in watts or kilowatts).

electromotive force (EMF)   The pressure that forces electrical current through a circuit, measured in volts.

elevated storage   In any distribution system, storage of water in a tank supported on a tower above the surface of the ground.

elevated tank   A water distribution storage tank that is raised above the ground and supported by posts or columns.

emergency storage   Storage volume reserved for catastrophic situations, such as a supply-line break or pump-station failure.

EMF   See *electromotive force*.

external load   Any load placed on the outside of the pipe from backfill, traffic, or other sources. Also known as *superimposed load*.

feedwater   Water that is added to a commercial or industrial system and subsequently used by the system, such as water that is fed to a boiler to produce steam.

fire demand   The required fire flow and the duration for which it is needed, usually expressed in gallons (or liters) per minute for a certain number of hours. Also used to denote the total quantity of water needed to deliver the required fire flow for a specified number of hours.

fire flow   The rate of flow, usually measured in gallons per minute (gpm) or liters per minute (L/min), that can be delivered from a water distribution system at a specified residual pressure for fire fighting. When delivery is to fire department pumpers, the specified residual pressure is generally 20 psi (140 kPa).

fire hydrant   A device connected to a water main and provided with the necessary valves and outlet nozzles to which a fire hose may be attached. The primary purpose of a fire hydrant is to fight fires, but it is also used for washing down streets, filling water-tank trucks, and flushing out water mains. Sometimes called a *fire plug*.

fire plug   See *fire hydrant*.

flexural strength   The ability of a material to bend (flex) without breaking.

float mechanism   A simple, mechanical device used to sense fluid level.

flood level   See *overflow level*.

flood rim   See *overflow rim*.

floor stand   A device for operating a gate valve (by hand) and indicating the extent of opening.

flow coefficient   See *C value*.

flow rate   The volume of water passing by a point per unit time. Flow rates are either instantaneous or average.

flush hydrant   A fire hydrant with the entire barrel and head below ground elevation. The head, with operating nut and outlet nozzles, is encased in a box with a cover that is flush with the ground line. Usually a dry-barrel hydrant.

foot piece   See *base*.

foot valve   A check valve placed in the bottom of the suction pipe of a pump, which opens to allow water to enter the suction pipe but closes to prevent water from passing out of it at the bottom end.

frame-mounted   Relating to centrifugal pumps in which the pump shaft is connected to the motor shaft with a coupling. Compare *close-coupled*.

full duplex   Capable of sending and receiving data at the same time. Compare *half duplex, simplex*.

full-voltage controller   An electric motor controller that uses the full line voltage to start the motor. Also called an *across-the-line controller*.

galvanic cell   A corrosion condition created when two different metals are connected and immersed in an electrolyte, such as water.

galvanic corrosion   A form of localized corrosion caused by the connection of two different metals in an electrolyte, such as water.

gastroenteritis   An intestinal disorder caused by pathogenic microorganisms, which involves inflammation of the stomach and intestines. Symptoms are diarrhea, pain, and nausea.

gate hydrant   A dry-barrel hydrant in which the main valve is a simple gate valve, similar to one side of an ordinary rubber-faced gate valve.

gate valve   A valve in which the closing element consists of a disc that slides across an opening to stop the flow of water.

globe valve   A valve having a round, ball-like shell and horizontal disc.

gooseneck   A flexible coupling, usually consisting of a short piece of lead on copper pipe shaped like the letter S.

grid system   A distribution system layout in which all ends of the mains are connected to eliminate dead ends.

ground-level tank   In a distribution system, storage of water in a tank whose bottom is at or below the surface of the ground.

groundwater supply system   A water system using wells, springs, or infiltration galleries as its source of supply.

half duplex   Capable of sending or receiving data but not both at the same time. Compare *full duplex, simplex*.

head (of a hydrant)   See *upper section*.

head (pressure)   (1) A measure of the energy possessed by water at a given location in the water system, expressed in feet (or meters). (2) A measure of the pressure (force) exerted by water, expressed in feet (or meters).

helical sensor   A spiral tube used to sense pressure changes.

hepatitis   An inflammation of the liver caused by a pathogenic virus. Symptoms are jaundice (yellowing of the skin), general weakness, nausea, and presence of dark urine.

horn   See *yoke*.

hose bibb   A faucet to which a hose may be attached. Also called a *sill cock*.

hydrant   See *fire hydrant*.

hydrant map   A mapped record that pinpoints the location of fire hydrants throughout the distribution system. Generally of the plat-and-list or intersection type.

hydraulic detention time   The time the water is in the system facility or system component (such as a storage tank).

hydraulically applied concrete   Concrete that is placed under pressure by a pneumatic gun.

hydropneumatic system   A system using an airtight tank in which air is compressed over water (separated from the air by a flexible diaphragm). The air imparts pressure to water in the tank and the attached distribution pipelines.

hydrostatic pressure   The pressure exerted by water at rest (for example, in a nonflowing pipeline).

impeller   The rotating set of vanes that forces water through a pump.

incubation period   The time period that elapses between the time a person is exposed to some disease and the time that person shows the first sign or symptom of the disease.

indexing   A system in which water system maps are individually numbered for ease of reference.

indicator   The part of an instrument that displays information about a system being monitored. Generally either an analog or digital display.

indirect cross-connection   See *potential cross-connection*.

inlet   See *base*.

in-plant piping system The network of pipes in a particular facility, such as a water treatment plant, that carry the water or wastes for that facility.

inrush current See *locked-rotor current*.

inserting valve A shutoff valve that can be inserted by special apparatus into a pipeline while the line is in service under pressure.

instrument Any device used to measure and display or record the conditions and changes in a system.

instrumentation See *instrument*.

integrator See *totalizer*.

interference fit A method of joining the pump impeller to the shaft by warming the impeller, then allowing it to cool and shrink around the shaft to provide a tight fit. Also called a *shrink fit*. (A slot and key are generally used to prevent rotational slippage of the installed impeller.)

internal backflow In a pump, leakage around the impeller from the discharge to the suction side.

internal load The load or force exerted by the water pressure on the inside of the pipe.

intersection method A method of preparing valve and hydrant maps. Maps of intersections are drawn on a very large scale, permitting valves, hydrants, and mains to be drawn to scale. Each intersection is then given a designation for easy reference.

Iowa hydrant See *Corey hydrant*.

isolation valve A valve installed in a pipeline to shut off flow in a portion of the pipe, for the purpose of inspection or repair. Such valves are usually installed in the mainlines.

jet pump A device that pumps fluid by converting the energy of a high-pressure fluid into that of a high-velocity fluid.

key (1) A small, rectangular piece of metal used to prevent a pump impeller or coupling from rotating relative to the shaft. (2) A metal rod with a long shank used for operating valves that are located below ground level.

kinetic pump See *velocity pump*.

laboratory information management system (LIMS)   A computerized system for collecting, storing, processing, and reporting water quality information and for enhancing the treatment process.

lantern ring   A perforated ring placed around the pump shaft in the stuffing box. Water from the pump discharge is piped to the lantern ring so that it will form a liquid seal around the shaft and lubricate the packing.

lateral   Smaller-diameter pipe that conveys water from the mains to points of use.

leak frequency map   A map showing different types of leaks throughout the distribution system. Colored pushpins are generally used to identify different leaks on a print of a comprehensive map.

leak survey map   A modified sectional or valve map showing valves to be closed and areas to be isolated during a leak survey.

LIMS   See *laboratory information management system.*

line of sight   An open air space, unbroken by buildings or other obstructions that would prevent high-frequency radio waves (and visible light) from passing between a transmitter and a receiver.

list   The text portion of the plat-and-list map. Provides information as to the size, make, direction to open, number of turns to open, date installed, date tested, and reference measurements for numbered valves and hydrants.

local manual control   A simple system of controlling equipment by direct human operation of switches and levers mounted on the equipment being operated.

locked-rotor current   The current drawn by the motor the instant the power is turned on while the rotor is still at rest. Also called the *motor-starting current* or *inrush current.*

loessial   Referring to loess, which is a fine-grained fertile soil deposited mainly by the wind.

loop system   See *arterial-loop system.*

lower barrel   The section of a hydrant that carries the water flow between the base and the upper section. Usually buried in the ground with the connection to the upper section approximately 2 in. (50 mm) above ground line.

lower main rod   The lower part of a standard compression hydrant rod. Attaches to the main valve assembly and is equipped with a spring to ensure positive closure.

lower section   The part of a dry-barrel hydrant that includes the lower barrel, the main valve assembly, and the base.

lower valve plate   The portion of the main valve assembly in a standard compression hydrant that connects the valve to the lower main rod.

magnetic meter   A flow-measuring device in which the movement of water induces an electrical current proportional to the rate of flow.

main rod   A rod, made of two sections, that connects the standard compression hydrant valve to the operating nut.

main valve   In a dry-barrel hydrant, the valve in the hydrant's base that is used to pressurize the hydrant barrel, allowing water to flow from any open outlet nozzle.

main valve assembly   A standard compression hydrant subassembly including the lower main rod, upper valve plate, resilient hydrant valve, lower valve plate, cap nut, and bronze seat ring. Screws into a bronze subseat or directly into threads cut into the base.

measuring chamber   A chamber of known size in a positive-displacement meter, used to determine the amount of water flowing through the meter.

mechanical joint   A type of joint for ductile-iron pipe. Uses bolts, flanges, and a special gasket.

mechanical seal   A seal placed on the pump shaft to prevent water from leaking from the pump along the shaft. Also prevents air from entering the pump. Mechanical seals are an alternative to packing rings.

media relations   The methods and activities used to promote an informed public and a favorable image of an organization through good relations with print and broadcast media.

meter box   A pit-like enclosure that protects water meters installed outside of buildings and allows access for reading the meter. Also known as *meter pit*.

metering flume   A flow-measuring device such as a Parshall flume that is used to measure flow in an open channel.

meter pit   See *meter box*.

mixed-flow pump   A pump that moves water partly by centrifugal force and partly by the lift of vanes on the liquid. With this type of pump, the flow enters the impeller axially and leaves axially and radially. Compare *axial-flow pump, radial-flow pump*.

mode   Any of several logical schemes for determining the interaction between instruments and controls.

motor-starting current   See *locked-rotor current*.

multijet meter   A type of current meter in which a vertically mounted turbine wheel is spun by jets of water from several ports around the wheel.

multiple protection barriers   A series of system components each of which provides a barrier to contaminants entering the water supply.

multiplexing   The use of a single wire or channel to carry the information for several instruments or controls.

multiplier   A number noted on the meter face, such as 10× or 100×. The reading from the meter must be multiplied by that number to provide the correct volume of water.

nonrising-stem valve   A gate valve in which the valve stem does not move up or down as it is rotated.

nozzle section   See *upper section*.

nutating-disk meter   A type of positive-displacement meter that uses a hard rubber disc that wobbles (rotates) in proportion to the volume of water flowing through the meter. Also called a *wobble meter*.

ohm   A measure of the ability of a path to resist or impede the flow of electric current. One volt will send a current of 1 ampere through a resistance of 1 ohm.

ohmmeter   An instrument for measuring the resistance of a circuit (in ohms). Usually combined with a voltmeter in test equipment.

on–off differential control   A mode of controlling equipment in which the equipment is turned fully on when a measured parameter reaches a preset value, then turned fully off when it returns to another preset value.

open-loop control   A computer-based form of control that advises the operator of system status, then sequences the operation of pumps and valves in response to the operator's commands.

operating nut   A nut, usually pentagonal or square, rotated with a wrench to open or close a valve or hydrant valve. May be a single component or it may be combined with a weather shield.

operating storage A tank supplying a given area and capable of storing water during hours of low demand, for use when demands exceed the pumps' capacity to deliver water to the district.

orifice meter A type of flowmeter consisting of a section of pipe blocked by a disc pierced with a small hole or orifice. The entire flow passes through the orifice, creating a pressure drop proportional to the flow rate.

outlet nozzle A threaded bronze outlet on the upper section of a fire hydrant, providing a point of hookup for hose lines or suction hose from hydrant to pumper truck.

outlet-nozzle cap Cast-iron cover that screws on to the outlet nozzle of a fire hydrant, protecting it from damage and unauthorized use.

outlying station A remote location of unattended distribution system equipment, such as pumps and valves, controlled by the central station.

overflow level The maximum height that water or liquid will rise in a receptacle before it flows over the overflow rim. Also known as *flood level*.

overflow rim The top edge of an open receptacle over which water will flow. Also known as *flood rim*.

packing Rings of graphite-impregnated cotton, flax, or synthetic materials, used to control leakage along a valve stem or a pump shaft.

packing gland A follower ring that compresses the packing in the stuffing box.

packing ring See *packing*.

parameter A measurable physical characteristic, such as pressure, water level, or voltage.

Parshall flume A calibrated channel for measuring the flow of liquid in an open conduit.

PDM See *pulse-duration modulation*.

peak-hour demand The greatest volume of water in an hour that must be supplied by a water system during any particular time period, such as a year, to meet customer demand.

pig Bullet-shaped polyurethane foam plug, often with a tough, abrasive external coating, used to clean pipelines. Forced through the pipeline by water pressure.

pilot valve The control mechanism on an automatic altitude or pressure-regulating valve.

piston meter   A water meter of the positive-displacement type, generally used for pipe-line sizes of 2 in. (50 mm) or less, in which the flow is registered by the action of an oscillating piston.

pitometer   A device operating on the principle of a Pitot tube, principally used for determining velocity of flowing fluids at various points in a water distribution system.

Pitot gauge (Pitot tube)   A device for measuring the velocity head of the stream as an indicator of velocity. Consists of a small tube pointed upstream, connected to a gauge on which the velocity head may be measured.

plan and profile drawings   Engineering drawings showing depth of pipe, pipe location (both horizontal and vertical displacements), and the distance from a reference point.

plat   A map showing street names, mains, main sizes, numbered valves, and numbered hydrants for the plat-and-list method of setting up valve and hydrant maps.

plat-and-list method   A method of preparing valve and hydrant maps. *Plat* is the map position, showing mains, valves, and hydrants. *List* is the text portion, which provides appropriate information for items on the plat.

plug valve   A valve in which the movable element is a cylindrical or conical plug.

pneumatic   Operated by air pressure.

polling   A technique of monitoring several instruments over a single communications channel with a receiver that periodically asks each instrument to send current status.

positive-displacement meter   A meter that measures the quantity of flow by recording the number of times a known volume is filled and empties. Primarily used for low flows. There are two common styles: the disc type and the piston type.

positive-displacement pump   A pump that delivers a precise volume of liquid for each stroke of the piston or rotation of the shaft.

post hydrant   A fire hydrant with an upper section that extends at least 24 in. (600 mm) above the ground.

potable   Relating to water that does not contain objectionable pollution, contamination, minerals, or infective agents and is considered satisfactory for domestic consumption.

potential cross-connection   Any arrangement of pipes, fittings, or devices that indirectly connects a potable water supply to a nonpotable source. This connection may not be present at all times but it is always there potentially. Also known as an *indirect cross-connection.*

power   A measure of the amount of work done per unit time by an electrical circuit, expressed in watts.

power factor   The ratio of useful (real) power to the apparent power in an alternating current (AC) circuit.

power relay   See *control relay.*

pressure differential   The difference in pressure between two points in a hydraulic device or system.

pressure-reducing valve   A valve with a horizontal disk for automatically reducing water pressures in a main to a preset value.

pressure-relief valve   A valve that opens automatically when the water pressure reaches a preset limit to relieve the stress on a pipeline.

pressure vacuum breaker   A device designed to prevent backsiphonage, consisting of one or two independently operating, spring-loaded check valves and an independently operating, spring-loaded air-inlet valve.

pressure zone map   A map showing zones of equal pressure. Pressure zone maps are sometimes called *water-gradient contour maps* and may be prepared on a print of a comprehensive map or maintained as a supplemental mapped record.

prestressed concrete   Reinforced concrete placed in compression by highly stressed, closely spaced, helically wound wire. The prestressing permits the concrete to withstand tension forces.

primary instrumentation   Sensors that are used to measure process variables such as flow, pressure, level, and temperature.

prime mover   A source of power, such as an internal combustion engine or an electric motor, designed to supply force and motion to drive machinery, such as a pump.

priming   The action of starting the flow in a pump or siphon. With a centrifugal pump, this involves filling the pump casing and suction pipe with water.

propeller meter   A meter that measures flow rate by measuring the speed at which a propeller spins as an indication of the velocity at which the water is moving through a conduit of known cross-sectional area.

propeller pump   See *axial-flow pump*.

proportional control   A mode of automatic control in which a valve or motor is activated slightly to respond to small variations in the system, but activated at a greater rate to respond to larger variations.

proportional meter   Any flowmeter that diverts a small portion of the main flow and measures the flow rate of that portion as an indication of the rate of the main flow. The rate of the diverted flow is proportional to the rate of the main flow.

public information   Information that is disseminated through various communications media to attract public notice or to educate the public at large.

public relations   The methods and activities employed to promote a favorable relationship with the public.

pulse-duration modulation (PDM)   An analog type of telemetry-signaling protocol in which the time that a signal pulse remains on varies with the value of the parameter being measured.

pumper outlet nozzle   A large fire-hydrant outlet, usually 4½ in. (114 mm) in diameter, used to supply the suction hose for fire department pumpers. Sometimes called a *steamer outlet nozzle* because it was originally used to supply steam-driven fire engines.

purchased water system   A water system that purchases water from another water system and so generally provides only distribution and minimal treatment.

push-on joint   The joint commonly used for ductile-iron, asbestos–cement, and polyvinyl chloride piping systems. One side of the joint has a bell with a specially designed recess to accept a rubber ring gasket. The other side has a beveled-end spigot.

racking   A condition in which a pump is subjected to frequent start–stop operations because of pressure surges affecting the pump controller. Can also result from a malfunctioning controller.

radial-flow pump   A pump that moves water by centrifugal force, spinning the water radially outward from the center of the impeller. Compare *axial-flow pump*, *mixed-flow pump*.

receiver   (1) The part of a meter that converts the signal from the sensor into a form that can be read by the operator; also called the *receiver–indicator*. (2) In a telemetry system, the device that converts the signal from the transmission channel into a form that the indicator can respond to.

receiver–indicator   An instrument component that combines the features of a receiver and an indicator.

reciprocating pump   A type of positive-displacement pump consisting of a closed cylinder containing a piston or plunger to draw liquid into the cylinder through an inlet valve and force it out through an outlet valve. When the piston acts on the liquid in one end of the cylinder, the pump is termed single-action; when the piston acts in both ends, the pump is termed double-action.

recorder   Any instrument that makes a permanent record of the signal being monitored.

reduced-pressure backflow preventer (RPBP)   A mechanical device consisting of two independently operating, spring-loaded check valves with a reduced-pressure zone between the check valves. Designed to protect against both backpressure and backsiphonage.

reduced pressure zone (RPZ) backflow preventer   Another term for *reduced-pressure backflow preventer*.

reduced-voltage controller   An electric controller that uses less than the line voltage to start the motor. Used when full line voltage may overload or damage the electrical system.

register   That part of the meter that displays the volume of water that has flowed through the meter. Meter registers are generally either of the straight or circular type.

regular operating delivery pressure   The water pressure in the distribution system during normal operation.

relay   An electrical device in which an input signal, usually of low power, is used to operate a switch that controls another circuit, often of higher power.

relay station   A combination receiver and transmitter, placed between the originator of a telemetry transmission and the final receiver, that boosts the power of the signal so that the signal can travel a greater distance.

remote-manual control   A type of system control in which personnel in a central location manually operate the switches and levers to control equipment at a distant site.

remote meter-reading system   A means of obtaining meter readings through a register that is installed and can be used at a location some distance from where the meter is located.

remote terminal unit (RTU)   A computer terminal used to monitor the status of control elements, monitor and transmit inputs from instruments, and respond to data requests and commands from the master station.

reservoir   (1) Any tank or basin used for the storage of water. (2) A ground-level storage tank for which the diameter is greater than the height.

residual pressure   The pressure remaining in the mains of a water distribution system when a specified rate of flow, such as that needed for fire-fighting purposes, is being withdrawn from the system.

resilient   Able to return to its original shape after being compressed (a common characteristic of many synthetic rubbers and plastics).

resilient hydrant valve   A fire-hydrant valve made of resilient materials to ensure effective shutoff.

resilient-seated gate valve   A gate valve with a disc that has a resilient material attached to it to allow leak-tight shutoff at high pressure.

resistance   A characteristic of an electrical circuit that tends to restrict the flow of current, similar to friction in a pipeline. Measured in ohms.

riser   The vertical supply pipe to an elevated tank.

rotary pump   A type of positive-displacement pump consisting of elements resembling gears that rotate in a close-fitting pump case. The rotation of these elements alternately draws in and discharges the water being pumped. Such pumps act with neither suction nor discharge valves, operate at almost any speed, and do not depend on centrifugal forces to lift the water.

rotor   The rotating part of an electric generator or motor.

roughness coefficient   See *C value.*

RPBP   See *reduced-pressure backflow preventer.*

RPZ   See *reduced pressure zone backflow preventer.*

RTU   See *remote terminal unit.*

rubber seat   A valve seat made of rubber.

rural water system   A water system that has been established to serve widely spaced homes and communities in areas having no available groundwater or water of very poor quality.

saddle   A device attached around a main to hold the corporation stop. Used with mains that have thinner walls to prevent leakage. Also called a *service clamp*.

salmonellosis   A disease that affects the intestinal tract and is caused by pathogenic bacteria. The primary symptom is severe diarrhea.

Saybolt standard units (SSU)   Standard measure of the viscosity of oil and grease used for lubricating bearings.

SCADA   See *supervisory control and data acquisition*.

scanning   A technique of checking the value of each of several instruments, one after another. Used to monitor more than one instrument over a single channel.

seat   The portion of a valve that the disk compresses against to achieve valve shutoff.

secondary instrumentation   Instruments that respond to and display information from primary instrumentation.

sectional map   A map that provides a detailed picture of a portion (section) of the distribution system. Reveals the locations and valving of existing mains, locations of fire hydrants, and locations of active service lines.

semiautomatic control   A form of system control equipment in which many actions are taken automatically but some situations require human intervention.

sensor   The part of an instrument that responds directly to changes in whatever is being measured, then sends a signal to the receiver and indicator or recorder.

service   See *service line*.

service clamp   See *saddle*.

service connection   The portion of the service line from the utility's water main to the curb stop at or adjacent to the street line or the customer's property line.

service line   The pipe (and all appurtenances) that runs between the utility's water main and the customer's place of use, including fire lines.

service valve   A valve, such as a corporation stop or curb stop, that is used to shut off water to individual customers.

shaft   (1) The bearing-supported rod in a pump, turned by the motor, on which the impeller is mounted. (2) The portion of a butterfly valve attached to the disc and a valve actuator. The shaft opens and closes the disc as the actuator is operated.

shaft bearing   Corrosion-resistant bearing that fits around the shaft on a butterfly valve to reduce friction when the shaft turns.

shielding   A method to protect workers against cave-ins through the use of a steel box open at the top, bottom, and ends. Allows the workers to work inside the box while installing water mains.

shoe   See *base*.

shoring   A framework of wood and/or metal constructed against the walls of a trench to prevent cave-in of the earth walls.

shrink fit   See *interference fit*.

sill cock   See *hose bibb*.

silt stop   A device placed at the outlet of water storage tanks to prevent silt or sediment from reaching the consumer.

simplex   Related to a telemetry or data transmission system that can move data through a single channel in only one direction. Compare *half duplex*, *full duplex*.

single-phase power   Alternating current (AC) power in which the current flow reaches a peak in each direction only once per cycle.

single-phasing   A phase-imbalance condition that occurs in a three-phase electrical circuit when power to one leg of the circuit is interrupted.

single-suction pump   A centrifugal pump in which the water enters from only one side of the impeller. Also called *end-suction pump*.

siphon   A bent tube, hose, or pipe that uses atmospheric pressure on the surface of a liquid to carry the liquid out over the top edge of a container. One end of the tube is placed in the liquid and the other end is placed outside the container at a point below the surface level of the liquid. The tube must be filled with the liquid (primed) before flow will start.

slip (1) In a pump, the percentage of water taken into the suction end that is not discharged because of clearances in the moving unit. (2) In a motor, the difference between the speed of the rotating magnetic field produced by the stator and the speed of the rotor.

sloping A method of preventing cave-ins that involves excavating the sides of the trench at an angle (the angle of repose) so that the sides will be stable.

slug method A method of disinfecting new or repaired water mains in which a high dosage of chlorine is added to a portion of the water used to fill the pipe. This slug of water is allowed to pass through the entire length of pipe being disinfected.

sluice gate A single, movable gate mounted in a frame, used in open channels or conduits to regulate flow.

smoothness coefficient See *C value.*

solenoid An electrical device that consists of a coil of wire wrapped around a movable iron core. When a current is passed through the coil, the core moves, activating mechanical levers or switches.

solvent A liquid that can dissolve other substances. Water is an excellent solvent.

sonic meter A meter that sends sound pulses alternately in opposite diagonal directions across the pipe. The difference between the frequency of the sound signal traveling with the flow of water and the signal against the flow of water is an accurate indication of the water's velocity.

SOPs See *standard operating procedures.*

split-phase motor See *squirrel-cage induction motor.*

split-tee fitting A special sleeve that is bolted around a main to allow a wet tap to be made. Also called a *tapping sleeve.*

spring line The horizontal center line of a pipe.

squirrel-cage induction motor The most common type of induction electric motor. The rotor consists of a series of aluminum or copper bars parallel to the shaft, resembling a squirrel cage. Also known as a *split-phase motor.*

SSU See *Saybolt standard units.*

stability A measure of a water's tendency to corrode pipes or deposit scale in pipes.

standard compression hydrant   A type of dry-barrel hydrant in which the main valve closes upward with the water pressure, creating a positive seal. Compare *eddy hydrant*.

standard operating procedures (SOPs)   Written procedures that set forth policy in the form of routine practices.

standpipe   A ground-level water storage tank for which the height is greater than the diameter.

starter   A motor-control device that uses a small push-button switch to activate a control relay, which sends electrical current to the motor.

statistical record   A summary record containing information such as the number of valves and hydrants installed, the total length of water mains laid, the amount of service pipe installed, or the number of meters installed.

stator   The stationary member of an electric generator or motor.

steamer outlet nozzle   See *pumper outlet nozzle*.

steam turbine   A prime mover in which the pressure or motion of steam against vanes is used for the generation of mechanical power.

strain gauge   A type of pressure sensor that is commonly used in modern instrumentation systems, consisting of a thin, flexible sheet with imbedded electrical conducting elements.

stringer   The horizontal member of a shoring system, running parallel to the trench, to which the trench braces are attached. The stringers hold the uprights against the soil.

stringing (hydrants)   The practice of dropping a weighted string down the barrel of a hydrant to check if the barrel has fully drained.

stuffing box   A portion of the pump casing through which the shaft extends and in which packing or a mechanical seal is placed to prevent leakage.

submersible pump   A vertical turbine pump with the motor placed below the impellers. The motor is designed to be submersed in water.

suction lift   The condition existing when the source of water supply is below the center line of the pump.

superimposed load   See *external load*.

supervisory control and data acquisition (SCADA)   A methodology involving equipment that both acquires data on an operation and provides limited to total control of equipment in response to the data.

supervisory control equipment   Centrally located controls that transmit commands over a telemetry system to remote equipment.

surface water system   A water system using water from a lake or stream for its supply.

surge arrester   An electrical protective device installed to protect electrical motors from high-voltage surges in the power lines. Also called *surge suppressor*.

surge pressure   A momentary increase of water pressure in a pipeline due to a sudden change in water velocity or direction of flow.

surge suppressor   See *surge arrester*.

surge tank   In cross-connection control, the receiving, nonpressurized storage vessel immediately downstream of an air gap. Nonpotable sources can be connected to the tank without threatening the potable supply.

swab   Polyurethane foam plug, similar to a pig but more flexible and less durable.

synchronous motor   An electric motor in which the rotor turns at the same speed as the rotating magnetic field produced by the stator. This type of motor has no slip.

synchronous speed   The rotational speed, in revolutions per minute, of the magnetic field produced by an electric motor's stator.

tablet method   A method of disinfecting new or repaired water mains in which calcium hypochlorite tablets are placed in a section of pipe. As the water fills the pipe, the tablets dissolve, producing a chlorine concentration in the water.

tachometer generator   A sensor for measuring the rotational speed of a shaft.

tank   A structure used in a water system to contain large volumes of water or other liquids.

tank farm   A large facility for the storage of chemical or petroleum products in aboveground storage tanks.

tapping   The process of connecting laterals and service lines to mains and/or other laterals.

tapping sleeve   See *split-tee fitting*.

tapping valve   A special shutoff valve used with a tapping sleeve.

telemetry A system of sending data over long distances, consisting of a transmitter, a transmission channel (wire, radio, or microwave), and a receiver. Used for remote instrumentation and control.

tensile strength A measure of the ability of pipe or other material to resist breakage when it is pulled lengthwise.

thermistor A semiconductor type of sensor that measures temperature.

thermocouple A sensor, made of two wires of dissimilar metals, that measures temperature.

thickness classes The standard wall thicknesses in which ductile-iron pipe is available.

thin-plate orifice meter See *orifice meter*.

three-phase power Alternating current (AC) power in which the current flow reaches three peaks in each direction during each cycle.

throttling The act of opening or closing a valve to control the rate of flow. Usually used to describe closing the valve.

thrust (1) A force resulting from water under pressure and in motion. Thrust pushes against fittings, valves, and hydrants; it can cause couplings to leak or to pull apart entirely. (2) In general, any pushing force.

thrust anchor A block of concrete, often a roughly shaped cube, cast in place below a fitting to be anchored against vertical thrust, and tied to the fitting with anchor rods.

thrust bearing A bearing designed to resist thrust in line with a turning shaft (as well as the usual force at right angles to the shaft). Often a cone-shaped roller bearing, but specially designed ball bearings are also used.

thrust block A mass of concrete cast in place between a fitting to be anchored against thrust and the undisturbed soil at the side or bottom of the pipe trench.

till A type of soil consisting of a mix of clay, sand, and gravel.

tone-frequency multiplexing A method of sending several signals simultaneously over a single channel by converting the signals into sounds (tones) and assigning a specific tone to each signal.

totalizer A device for indicating the total quantity of flow through a flowmeter. Also called an *integrator*.

traffic load The load placed on a buried pipe by the traffic traveling over it.

transducer See *transmitter.*

transmission channel In a telemetry system, the wire, radio wave, fiber-optic line, or microwave beam that carries the data from the transmitter to the receiver.

transmission line The pipeline or aqueduct used for water transmission, i.e., movement of water from the source to the treatment plant and from the plant to the distribution system.

transmitter In telemetry or remote instrumentation, the device that converts the signal generated by the sensor into a signal that can be sent to the receiver–indicator over the transmission channel. Also called *transducer.*

travel-stop nut A nut, used in dry-barrel hydrants, that is screwed on the threaded section of the main rod. It bottoms at the base of the packing plate, or revolving nut, and terminates downward travel (opening) of the hydrant valve.

tree system A distribution system layout that centers around a single arterial main, which decreases in size with length. Branches are taken off at right angles, with subbranches from each branch.

trench brace The horizontal member of a shoring system that runs across a trench, attached to the stringers.

tuberculation The growth of nodules (tubercules) on the pipe interior, which reduces the inside diameter and increases the pipe roughness.

tubercules Knobs of rust formed on the interior of cast-iron pipes by the corrosion process.

turbine meter A meter that measures flow rates by measuring the speed at which a turbine spins in water, indicating the velocity at which the water is moving through a conduit of known cross-sectional area.

turbine pump (1) A centrifugal pump in which fixed guide vanes (diffusers) partially convert the velocity energy of the water into pressure head as the water leaves the impeller. (2) A regenerative turbine pump.

unaccounted-for water Water that is pumped into the distribution system but is not delivered to the consumers or otherwise accounted for.

underregistration A condition in which a meter records less water than is actually flowing through the meter.

upper barrel See *upper section.*

upper section   The upper part of the main hydrant assembly, including the outlet nozzles and outlet-nozzle caps. The upper section is usually constructed of gray cast iron. Also known as *upper barrel*, *nozzle section*, or *head*.

upper valve plate   A portion of the main valve assembly of a standard compression hydrant that closes against the seat.

upright   The vertical member of a shoring system, placed against the trench wall.

vacuum   Any absolute pressure that is less than atmospheric, i.e., less than 14.7 psi (101 kPa) at sea level.

vacuum breaker   A mechanical device that allows air into the piping system, thereby preventing backflow that could otherwise be caused by the siphoning action created by a partial vacuum.

valve   A mechanical device installed in a pipeline to control the amount and direction of water flow.

valve-and-hydrant map   A mapped record that pinpoints the location of valves throughout the distribution system. Generally of plat-and-list or intersection type.

valve box   A metal, concrete, or composite box or vault set over a valve stem at ground surface to allow access to the stem so that the valve can be opened and closed. A cover for the box is usually provided at the surface to keep out dirt and debris.

valve key   A metal wrench with a socket to fit a valve operating nut. Inserted into the valve box to operate the valve.

valve stem   The rod used to open or close a valve.

variable frequency   Relating to a type of telemetry signal in which the frequency of the signal varies as the parameter being monitored varies.

vault   An underground structure, normally made of concrete, that houses valves and other appurtenances.

velocity meter   A water meter using a spinning rotor with vanes (like a propeller) and operating on the principle that the speed of the vanes accurately reflects the velocity of the flowing water.

velocity pump   The general class of pumps that use a rapidly turning impeller to impart kinetic energy or velocity to fluids. The pump casing then converts this velocity head, in part, to pressure head. Also known as *kinetic pump*.

Venturi   An hourglass-shaped device based on the hydraulic principle that states that as the velocity of fluid flow increases, the pressure decreases. This device is used in a multitude of ways to measure flow, feed chemicals, and pump water.

Venturi meter   A pressure-differential meter used for measuring flow of water or other fluids through closed conduits or pipes, consisting of a Venturi tube and a flow-registering device. The difference in velocity head between the entrance and the contracted throat of the tube is an indication of the rate of flow.

Venturi tube   See *Venturi*.

vertical turbine pump   A centrifugal pump, commonly of the multistage diffuser type, in which the pump shaft is mounted vertically.

volt   The practical unit of electrical potential (electrical pressure). One volt will send a current of 1 ampere through a resistance of 1 ohm.

voltage   (1) See *volt*. (2) In telemetry, a type of signal in which the electromotive force (measured in volts) varies as the parameter being measured varies.

voltmeter   An instrument for measuring electromotive force (electrical pressure), which is expressed in volts.

volute   The expanding section of a pump casing (in a volute centrifugal pump) that converts velocity head to pressure head.

warm-climate hydrant   A fire hydrant with a two-piece barrel that is the main valve located at ground level.

washout valve   See *blowoff valve*.

water audit   A procedure that combines flow measurements and listening surveys in an attempt to give a reasonably accurate accounting of all water entering and leaving a system.

water-gradient contour map   See *pressure zone map*.

water hammer   The potentially damaging slam, bang, or shudder that occurs in a pipe when a sudden change in water velocity (usually as a result of someone too-rapidly starting a pump or operating a valve) creates a great increase in water pressure.

water meter   A device installed in a pipe under pressure for measuring and registering the quantity of water passing through.

water purveyor   Anyone who sells drinking water to the public. Usually the owner of a public water system.

watt   The practical unit of electric power. In direct current (DC) circuits, watts equals volts times amperes. In an alternating current (AC) circuit, watts equals volts times amperes times power factor.

wattmeter   An instrument for measuring real power in watts.

wear rings   Rings made of brass or bronze placed on the impeller and/or casing of a centrifugal pump to control the amount of water that is allowed to leak from the discharge to the suction side of the pump.

weir   An obstruction to the flow of water, placed in an open channel. Measures flow rate by measuring the depth of the water flowing through a precisely sized and shaped notch in the weir.

well point   A perforated metal tube or screen connected to the bottom of a suction pipe. The device is jetted or driven into the earth, and groundwater is withdrawn through it.

wet-barrel hydrant   A fire hydrant with no main valve. Under normal, nonemergency conditions, the barrel is full and pressurized (as long as the lateral piping to the hydrant is under pressure and the gate valve ahead of the hydrant is open). Each outlet has an independent valve that controls discharge from that outlet. The wet-barrel hydrant is used mainly in areas where temperatures do not drop below freezing. The hydrant has no drain mechanism.

wet tap   A connection made to a main that is full or under pressure. Compare *dry tap*.

wet-top hydrant   A dry-barrel hydrant in which the threaded end of the main rod and the revolving or operating nut are not sealed from water in the barrel when the main valve of the hydrant is open and the hydrant is in use.

wobble meter   See *nutating-disk meter*.

work order   A form used to communicate field information back to the main office. Also used to order materials, organize field crews, and obtain easements.

wound-rotor induction motor   A type of electric motor, similar to a squirrel-cage induction motor but easier to start and capable of variable-speed operation.

yoke   A fitting designed to assist in easy meter installation and to maintain electrical continuity between the incoming and outgoing lines even if the meter is removed for service.

Index